Biophysics

An Introduction

Biophysics

An Introduction

Rodney M. J. Cotterill
Danish Technical University, Denmark

JOHN WILEY & SONS, LTD

Published by   John Wiley & Sons Ltd, The Atrium, Southern Gate, Chichester,
                    West Sussex PO19 8SQ, England

                    Telephone   (+44) 1243 779777

Email (for orders and customer service enquiries): cs-books@wiley.co.uk
Visit our Home Page on www.wileyeurope.com or www.wiley.com

Reprinted April 2003, May 2004

This publication is designed to provide accurate and authoritative information in regard to the subject
matter covered. It is sold on the understanding that the Publisher is not engaged in rendering
professional services. If professional advice or other expert assistance is required, the services of a
competent professional should be sought.

Other Wiley Editorial Offices

John Wiley & Sons Inc., 111 River Street, Hoboken, NJ 07030, USA

Jossey-Bass, 989 Market Street, San Francisco, CA 94103–1741, USA

Wiley-VCH Verlag GmbH, Boschstr. 12, D-69469 Weinheim, Germany

John Wiley & Sons Australia Ltd, 33 Park Road, Milton, Queensland 4064, Australia

John Wiley & Sons (Asia) Pte Ltd, 2 Clementi Loop #02–01, Jin Xing Distripark, Singapore 129809

John Wiley & Sons Canada Ltd, 22 Worcester Road, Etobicoke, Ontario, Canada M9W 1L1

Wiley also publishes its books in a variety of electronic formats. Some of the content that appears in
print may not be available in electronic books.

British Library Cataloguing in Publication Data

A catalogue record for this book is available from the British Library

ISBN 0–471–48538–1

Typeset in 10.5/13pt Times by Kolam Information Services, Pondicherry, India
Printed and bound in Great Britain by Antony Rowe Ltd, Chippenham, Wiltshire
This book is printed on acid-free paper responsibly manufactured from sustainable forestry
in which at least two trees are planted for each one used for paper production.

For my teacher, Herbert C. Daw, and for my biophysics students – past, present and future

Contents

Preface

This book is based on the course in biophysics that I have taught for the past two decades at the Danish Technical University, and it should be suitable for similar courses at other places of higher education.

I originally delivered the lectures in Danish and Henrik Jørgensen, one of my first students, recorded my words in shorthand and then collaborated with my secretary, Carolyn Hallinger, to produce a set of Danish notes. I updated these from time to time, and ultimately translated them into English. There were two subsequent expansions of the text before it acquired the form reproduced here. Meanwhile Ove Broo Sørensen and Bjørn Nielsen provided valuable help with many of the illustrations.

The course now attracts so many students that I have needed the backing of two assistant teachers, in connection with the weekly homework assignments. Henrik Bohr and Bjørn Nielsen have provided this service with great skill and diligence, and it is a pleasure to acknowledge their contribution to the enterprise.

The cause of biophysics at this university has benefited greatly from the support provided by colleagues in other departments, and most notably by Robert Djurtoft, Ole Mouritsen, Knud Særmark and Jens Ulstrup, together with whom I set up what came to be known as the Biophysics Initiative. Professors Mouritsen and Særmark were formerly my departmental colleagues, and I enjoyed close interactions with both of them.

The interest and encouragement of the wider Danish biophysics community has also been invaluable, and I would especially like to mention Salim Abdali, Preben Alstrøm, Olaf Sparre Andersen, Svend Olav Andersen, Christen Bak, Per Bak, Rogert Bauer, Klaus Bechgaard, Kirstine Berg-Sørensen, Myer Bloom, Jakob Bohr, Tomas Bohr, John Clark, Jens Peder Dahl, Tom Duke, Henrik Flyvberg, Christian Frøjær-Jensen, Sonia Grego, John Hjort Ipsen, Karl Jalkanen, Mogens Høgh Jensen, Kent Jørgensen, Carsten Knudsen, Bent Kofoed, Morten Kringelbach, Erik Hviid Larsen, Signe Larsen, Jens Jørgen Led, Per Anker Lindegaard, Jens Ulrik Madsen, Axel Michelsen, Erik Mosekilde, Knud Mørch, Kell Mortensen, Claus Nielsen, Simon Nørrelykke,

Lene Oddershede, Niels Berg Olsen, Steffen Petersen, Dorthe Posselt, Flemming Poulsen, Christian Rischel, Jens Christian Skou, Kim Sneppen, Ove Sten-Knusen, Maria Sperotto, Stig Steenstrup, Thomas Zeuthen and Martin Zuckerman.

Solutions for the exercises can be found on my website:
http://info.fysik.dtu.dk/Brainscience/rodney.html

Rodney Cotterill

1 Introduction

It is probably no exaggeration to say that many regard biophysics as a discipline still waiting to be adequately defined. This conclusion appears to be endorsed by the considerable differences between several of the publications on the subject cited at the end of this chapter. Indeed, in terms of the items they discuss, these barely overlap with each other. But this should be taken as an indication of the sheer multiplicity of things that now belong under the biophysics banner; no single author could reasonably be expected to cover them all. If one considers what these books and articles describe *collectively*, a unified picture does in fact emerge.

Biophysics is simply the application of physics to biology, with a view to furthering the understanding of biological systems. There is a related activity in which methods developed originally for purely physical challenges have been applied to biological (and in some cases medical) issues. Biophysics tends to be studied by those who have a background in physics, and who may thus be bringing useful expertise to the investigation of living things. But there have also been examples of biologists acquiring the requisite knowledge of physics and then using this to solve a specific problem.

Biophysics is not a young subject, but its emphasis has gradually changed over the years. In the first part of the 20th century, biophysicists primarily concerned themselves with things quite closely related to medicine, and many large hospitals had a resident member of this fraternity. The issues of interest were the flow of blood through pumps and the associated tubing (drawing on the work of George Stokes, among others), the monitoring of heart function through the related electrical activity, and later of brain activity with much the same instrumentation (with valuable input from Hans Berger), and also the fracture of bone (with a borrowing of the ideas developed by Alan Griffith, in a quite different context).

Around the same time, the field was gradually acquiring a new type of activity related to processes at the atomic level, this having been provoked by Wilhelm Röntgen's discovery of X-rays. His astonishment at discovering their power of penetrating human tissue, but apparently not bone, soon led to the use of X-rays for diagnostic purposes, of course. Only later did it emerge that there are grave dangers associated with such radiation, and physicists then found their advice being sought in connection with the monitoring of X-ray doses. Through their efforts, recording by film was supplemented by recording with electronic devices. One of the pioneering theoretical efforts in understanding the interaction of radiation with matter was published by Niels Bohr, who had earlier put forward the first successful picture of the atom.

These developments were of obvious importance to medicine, but another use of X-rays, originally confined to the inorganic domain, was later going to have an enormous impact on all of biology, and through this on medicine itself. Max von Laue and his colleagues, Walter Friedrich and Paul Knipping, had discovered the diffraction of X-rays, and William Bragg and Lawrence Bragg were soon applying the phenomenon to the determination of crystal structures. The latter Bragg, William's son, encouraged the extension of the technique to the biological realm, and researchers such as William Astbury, John Bernal, Peter Debye and Max Perutz soon took up the challenge. The early work in the area, before the Second World War, contributed to the determination of the sizes of protein molecules, and within twenty years it was producing pictures of proteins at the atomic level.

Mention of molecular size serves as a reminder that it would be easy to overlook the importance of methods developed for separating different molecular species, and their consequent contribution to biology. These methods would not have emerged had it not been for the prior work on the underlying physics. So the development of techniques such as ultra-centrifugation (invented by Thé Svedberg), electrophoresis (by Arne Tiselius) and partition chromatography (by Archer Martin and Richard Synge) owes much to the earlier efforts of George Stokes, Albert Einstein and Irving Langmuir.

But to return to structure determination by means of X-ray diffraction, this approach reached its zenith around the middle of the 20th century. Max Perutz and John Kendrew set about determining the structures of the oxygen-transporting proteins myoglobin and its larger cousin haemoglobin. Meanwhile, Maurice Wilkins, Rosalind Franklin and Raymond Gosling had turned their attention to deoxyribonucleic acid (DNA). Within a decade, the secrets of these key structures had been exposed, important input having come from the knowledge acquired of the bonding between atoms, thanks to the efforts of physicists.

Another spectacular success for physics was the invention of the electron microscope by Ernst Ruska (following important efforts by Denis Gabor). This played a vital role in the study of the microstructure of muscle, by Hugh Huxley and his colleagues, and of viruses, by Aaron Klug and Robert Horne, and their respective colleagues. And there was still good mileage to be had from X-rays because Allan Cormack and Godfrey Hounsfield applied these to the study of brain tissue, by computer assisted tomography (CAT scanning). This fine lead was subsequently augmented by development of such other brain-probing techniques as positron emission tomography (PET) and functional magnetic resonance imaging (fMRI). And while on the subject of the brain, we have magnetoencephalography (MEG), which owes its existence to something which started as speculation in the purest of physics, namely the work of Brian Josephson on quantum mechanical tunnelling between superconductors.

So the current growth in the application of physics to biology can point to many respectable antecedents. To the names already quoted of physicists who turned their talents to biology we could add Francis Crick (who has latterly shifted his attention from matters molecular to matters mental), Max Delbrück, Walter Gilbert, Salvador Luria and Rosalyn Yalow. These researchers, and many others like them, have brought to biology the quantitative discipline that is the hallmark of physics.

But we should not overlook the unwitting contributions to biology made by physicists of an earlier era – physicists who had probably never even heard of the word biophysics. And in this respect, no advance can quite compete with that which came from the seemingly esoteric study of gaseous discharge. In 1855, Heinrich Geissler devised a vacuum pump based on a column of mercury which functioned as a piston. He and Julius Plücker used this to remove most of the air from a glass tube into which they had sealed two electrical leads, and they used this simple apparatus to study electrical discharges in gases. Their experiments, and related ones performed by Michael Faraday and John Gassiot, probed the influences of electric and magnetic fields on the glow discharge, and it was established that the light was emitted when 'negative rays' struck the glass tube. The discharge tube underwent a succession of design modifications, by William Crookes, Philipp Lenard and Jean Perrin, and this activity culminated with Wilhelm Röntgen's discovery of X-rays in 1895 and Joseph (J. J.) Thomson's discovery of the electron, two years later. These landmarks led, respectively, to the investigations of atomic arrangement mentioned above, and explanations of the forces through which the atoms interact.

These advances were to prove pivotal in the study of biological systems. Moreover, we should not overlook the instruments that owe their existence to those early investigations of the influences of various fields on an electron beam. These led to the cathode ray oscilloscope, with which Edgar Adrian was able to discover the all-or-nothing nature of the nerve impulse. The precision with which that instrument enables one to determine the temporal characteristics of the impulse was vital to Alan Hodgkin and Andrew Huxley's explanation of nerve conduction. The cathode ray oscilloscope presaged the emergence of electron microscopy, which was referred to above.

We should add one more name to the list, because it is nearly always overlooked: John Atanasoff. In the 1930s, confronted with a data analysis problem in his research in solid-state physics, he hit upon the idea of automating his calculations with an electronic machine. This was, indeed, the first electronic digital computer, and the descendants of that device have been indispensable to many of the techniques mentioned above. It would not be eccentric, therefore, to call Atanasoff one of the unsung heroes of biophysics.

Biophysics, then, is an activity that operates within biology, and it contributes to the tackling of some of the major mysteries in that realm. Even though there may be some dispute as to which are the main issues at the current time, few would dispute the claim that protein folding, tissue differentiation, speciation, microscopic recognition and (not the least) consciousness and intelligence are amongst the greatest challenges of our era. It is certainly the case that when we fully understand the physical principles underlying these phenomena, biology as a whole will be very much more advanced than it is today.

Let us briefly consider the nature of these challenges. First, the protein-folding problem has been referred to as the second half of the genetic code. It has long been known that the sequence of bases in the DNA molecule determines the sequence of amino acids in a protein, that is to say the protein's primary structure. It is also well known that the primary structure dictates the final three-dimensional conformation of the protein molecule, but we are unable at the present time to predict that structure, working solely from the primary sequence. The best that one can do is to predict, with a reasonable degree of reliability, certain sub-structural motifs that are frequently observed to be present in the three-dimensional structure. Although this is a notable achievement in its own right, it still falls far short of the desired ability to predict any protein's structure from the primary sequence, and this is an obvious obstacle to full realization of the potential inherent in genetic manipulation. If one were able to overcome that hurdle, this would open up the possibility of tailoring proteins to fulfil specific tasks, for the fact is that what a protein does is determined by its three-dimensional structure.

The tissue-differentiation problem arises from the fact that every cell in a multi-cellular organism contains an identical set of genetic instructions, but for some reason only part of the message is expressed in any one type of cell. In our own bodies, for example, it is this fact that determines that there are different cellular structures in our various parts, and that the same distribution of these bits and pieces is observed in every normal individual. It has long been clear that the differentiation mechanism depends upon the interaction of proteins and nucleic acids, and that it thus hinges on the forces between the constituent atoms. It has also emerged that the differentiation process depends upon the diffusion of certain molecular species in the growing embryo. Biophysics can thus contribute to this topic, through elucidating the microscopic factors that influence the diffusion.

Speciation deals with the questions of why and how a single species occasionally gives rise to two distinct evolutionary branches. It has long been clear that modification of the genetic message lies at the heart of this phenomenon, but the details are still lacking. After all, no two humans have identical sets of genes (unless they happen to be clones or identical twins, triplets, etc.), but we nevertheless all belong to the same species. Here, too, interaction between molecules is of the essence.

Microscopic recognition has to do with the molecular processes that dictate the manner in which different cells mutually interact. Such interactions are important for all the body's cells, but they are particularly important in the case of those that belong to the immune system. The great importance of this system is reflected in the fact that over 1% of a person's body weight is represented by such cells. They must distinguish between those things that belong to the body's tissues and outsiders that might threaten the organism's integrity. So, yet again, one has a mechanism that ultimately depends upon interactions between atoms.

Finally, there is the great mystery of consciousness and the related issue of intelligence. Although there are those who prefer to make a clear distinction between mind and body, there is a growing feeling that it might soon be possible to understand how such ephemeral things as consciousness and the mind arise from the physiological processes that occur in the nervous system. It is by no means clear that adumbration of the physical basis of consciousness would also further our understanding of what underlies intelligence, but it does not appear too optimistic to believe that this could be the case.

It might seem that this list of problems overlooks other pressing issues in the biological domain. One might be tempted to ask why the major scourges of cancer and AIDS have not been included. The fact is, however, that they are implicit in two of the above five categories, because cancer is merely one aspect of the wider issue of tissue differentiation, and AIDS is caused by the human immune deficiency virus (HIV), which undermines the immune system, the latter being categorized under the general heading of microscopic recognition.

The challenging problems identified above have not been listed in an arbitrary order. On the contrary, they show a natural progression from the level of a single molecule, as in the case of the protein-folding issue, to properties of the organism that derive from the behaviour of millions of individual cells. In much the same way, the subject matter in this book follows a logical sequence in which processes at the atomic level are dealt with first, while relevant properties of the nervous system appear toward the end of the book. In between those extremes, the sequence roughly follows that of increasing size. Thus the discussion of molecules leads on logically to properties of organelles, and this in turn is followed by a brief treatment of entire cells. Finally, the issue of neural signalling is discussed, both at the level of the single neuron and ultimately with reference to the functioning of the entire brain. Important items in this latter part of the book are membrane excitability, which underlies that signalling, the changes at the sub-cellular level involved in the laying down of memories, and the process of cognition. On the other hand, no chapters specifically address the three central items in the above list. These are nevertheless mentioned in the relevant places, and representative items in the scientific literature are cited in the Further Reading sections.

There are approximately 10^{12} individual cells in the adult human body. Hopefully, the following chapters will enable the reader to get a good impression of the processes which occur on a number of different size scales, and which lead to the overall functioning of the body. The things described herein should serve to confirm that the quantitative approach has much to recommend it when one is trying to work out how the body's component structures and systems acquire their wonderful functions. Finally, this book aims to endorse what Philip Anderson noted concerning biological phenomena. These are ultimately dependent on Nature's fundamental forces, of course, but the existence of higher levels of organization in living matter implies that there must also be other laws at work. As one makes the transition to each higher level of organization, one must anticipate the emergence of new principles that could not have been predicted on the basis of what was seen at the lower level.

Exercises

1.1 Write an essay on the following question. Will biophysics become one of the major scientific disciplines in the 21st century?

1.2 Max Perutz once referred to 1953 as the *annus mirabilis* of molecular biology. What did he have in mind?

Further reading

Anderson, P. W., (1972). More is different. *Science*, **177**, 393–396.
Cerdonio, M. and Noble, R. W., (1986). *Introductory Biophysics* World Scientific, Singapore.
Cotterill, R. M. J., (2004). *The Material World* Cambridge University Press, Cambridge.
Elsasser, W. M., (1958). *The Physical Foundation of Biology* Clarendon Press, Oxford.
Flyvbjerg, H., *et al.*, eds., (1997). *Physics of Biological Systems: From Molecules to Species* Springer, Berlin.
Glaser, R., (2001). *Biophysics* Springer, Berlin.
Nossal, R. J. and Lecar, H., (1991). *Molecular and Cell Biophysics* Addison-Wesley, Redwood City, CA.
Parak, F. G., (2001). Biological physics hits the high life. *Physics World* **14**(10), 28–29.
Parisi, G., (1993). Statistical physics and biology. *Physics World*, **6**, 42–47.
Parsegian, V. A., (1997). Harness the hubris: useful things physicists could do in biology. *Physics Today*, 23–27, July issue.
Schrödinger, E., (1944). *What is Life?* Cambridge University Press, London.
Setlow, R. B. and Pollard, E. C., (1962). *Molecular Biophysics.* Addeson-Wesley, Reading, MA.
Sybesma, C., (1989). *Biophysics: An Introduction* Kluwer Academic, Dordrecht.

2 Chemical Binding

In this chapter, a *qualitative* account of the electronic structure of atoms will be given (see Appendices A and B), partly because a mathematically precise analysis of groups of atoms is still not possible, but mainly because a qualitative treatment is usually sufficient to provide an understanding of the way in which atoms bind together to form molecules. We thus begin by taking a brief look at an isolated atom.

2.1 Quantum Mechanics

Through his own experimental work on the atomic nucleus, Ernest Rutherford put forward a picture of the atom in which the heavy nucleus is located at the centre, while the electrons, discovered by Joseph (J. J.) Thomson, move in the surrounding space, their characteristic distances from the nucleus being of the order of 0.1 nm (i.e. a tenth of a nanometer, or 1 Ångström unit). The major developments in the theory of atomic structure thereafter were due to Niels Bohr, who realized that only certain energy states would be permitted by the quantum principle postulated by Max Planck in 1900; by Louis de Broglie, who advocated that a dual attitude be adopted toward sub-atomic particles, such that they are regarded as simultaneously having both particle and wave natures (see Appendix A); and by Erwin Schrödinger, whose equation showed how to derive the allowed states of electrons, both regarding their permitted energies and their spatial distribution with respect to the nucleus.

Schrödinger's time-independent equation reads

$$\mathcal{H} \cdot \Psi = \mathcal{E} \cdot \Psi \qquad (2.1)$$

where $\mathcal{H}$ is the Hamiltonian operator, $\mathcal{E}$ is the energy and Ψ is the wave function, the latter being a function of position with respect to the nucleus. This equation appears to be remarkably simple, but one must bear in mind that the Hamiltonian itself will usually be a composite of several terms, while the wave function will include components describable only in terms of complex numbers. The solution of the Schrödinger equation for the very important case of the hydrogen atom is given in Appendix B.

Max Born hit upon the correct interpretation of the distribution yielded by the Schrödinger equation when he suggested that $\Psi\Psi^*$ (where Ψ^* is the complex conjugate of Ψ) gives the probability that an electron will be located at that position. (For our purposes here, $\Psi\Psi^*$ can be regarded as simply being the square of the amplitude of the wave function.) Just as the vibrations of a (one-dimensional) guitar string and the (two-dimensional) skin of a drum can be characterized by a set of numbers which refer to the positions and multiplicity of the nodal points (i.e. positions where the amplitude is zero), so it is with the electron probability distribution around a nucleus. Although we need not go into the details here, different quantum states of an electron in the vicinity of an atomic nucleus are characterized by different spherical and non-spherical probability distributions. Although other factors also come into play, as we will see later, it is the shapes of these distributions that determine the shapes of the molecules formed when two or more atoms form a reasonably permanent mutual liaison.

In the case of the one-dimensional guitar string, the situation can be characterized by a single number, which is related to the number of nodal points located along the string. In the case of the vibrating skin of a drum, two different numbers are required in order to fully characterize the situation: one of these refers to nodal points whereas the other refers to nodal lines (which may be curved). In the three-dimensional space around an atomic nucleus, therefore, it is not surprising that three different numbers are required for a full description of the spatial arrangement of the probability distribution for each electron, there now being nodal points, nodal lines and nodal surfaces. It turns out that a further number is required, because the electron possesses what is known as spin, which is very roughly analogous to the spin of a planet, as it describes its orbit around the sun. Just as the spin of such a planet may be either in a left-handed or a right-handed direction, so too the spin of an electron has one of just two possibilities. The spin quantum number of an electron is usually designated by the letter s, and the other quantum numbers by the letters n, l, and m (see Appendix B). Figure 2.1 shows the spatial distribution of the squared probability amplitude for a number of different situations which can apply to an electron in orbit around an atomic nucleus. These are indeed usually referred to as orbitals.

The lowest energy (ground) state for a hydrogen atom, the $1s$ state, is characterized by a spherically-symmetric wave intensity with a single spherical nodal surface at infinity. The lowest energy excited state is the $2s$, and this has an additional spherical nodal surface centred on the nucleus. The $2p$ states, which have slightly higher energy, have nodal surfaces which pass through the nucleus. There are three of them, corresponding to the three possible values of the magnetic quantum number m. States with lobes extending along one of the

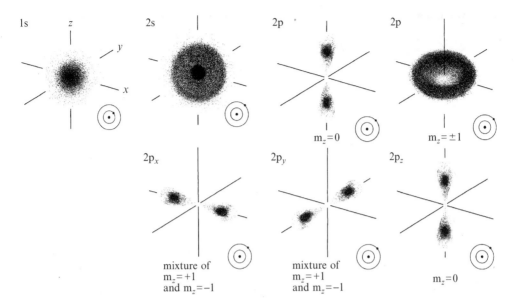

Figure 2.1 Spatial distribution of the squared probability amplitude for a number of different situations which can apply to an electron in orbit around an atomic nucleus

Cartesian axes can be obtained by mixing the three $2p$ states in the correct proportions. The insets in Figure 2.1 and later diagrams indicate schematically the equivalent Bohr orbits.

2.2 Pauli Exclusion Principle

Although we do not need to go into all the details here, there are certain rules which are useful when considering situations in which there is more than one electron present in an atom. For a start, no two electrons can be associated with the same atomic nucleus and have precisely the same values for all four of the quantum numbers. This is known as the Pauli Exclusion Principle (after Wolfgang Pauli), and it is a particularly potent factor when two similar atoms lie sufficiently close to one another. Regarding the actual shapes of the orbitals, the s types all have spherical symmetry, whereas the p types show elongation along an axis (see Figure 2.1). (The s used to designate one type of orbital should not be confused with the symbol for the spin quantum number; see Appendix B.) When all the p orbitals are fully occupied by their permitted complement of electrons, however, these collectively also display spherical symmetry. This is particularly noticeable in the noble gas atoms, which indeed possess only such full shells. Another important property of electron orbitals is

that a linear combination of different possible wave functions is also a possible
solution of the time-independent Schrödinger equation.

Suppose that Ψ_1 and Ψ_2 are two such possible wave functions. We will then
have that

$$\mathcal{H} \cdot \Psi_1 = \mathcal{E} \cdot \Psi_1$$

and

$$\mathcal{H} \cdot \Psi_2 = \mathcal{E} \cdot \Psi_2$$

If we now multiply the first of these equations by the coefficient C_1 and the
second by C_2, and add the two results, we obtain

$$\mathcal{H} \cdot (C_1 \cdot \Psi_1 + C_2 \cdot \Psi_2) = \mathcal{E} \cdot (C_1 \cdot \Psi_1 + C_2 \cdot \Psi_2) \qquad (2.2)$$

This is an equally valid version of the time-independent Schrödinger equation.

We can now proceed to discuss what happens when two atoms approach
each other. It is clear that they must exert a force upon each other, and a
moment's reflection reveals that these forces may be either attractive or repulsive. This conclusion comes from the dual facts that matter does not spontaneously explode or implode. In other words, one meets with resistance if an
attempt is made to squeeze a piece of condensed matter (i.e. a solid or a liquid)
into a smaller volume. Likewise, resistance is encountered if one tries to stretch
a piece of material beyond its quiescent dimensions. This indicates that the
interatomic potential is repulsive at sufficiently short range and attractive at
sufficiently long range, and the implication is thus that there must be an
intermediate distance at which there is neither repulsion nor attraction. This
will correspond to an interatomic separation for which the forces are precisely
balanced, and it is this characteristic distance that essentially determines the
density of a piece of material.

2.3 Ionization Energy, Electron Affinity and Chemical Binding

It is interesting to note that these considerations were well appreciated even
before it had been unequivocally demonstrated that atoms actually exist. It is
not surprising, therefore, that the arguments do not take into account possible
redistribution of the subatomic particles when two atoms approach one another
sufficiently closely. In general, there will be a rearrangement of the electrons
between the two atoms, the notable exception being the case where both atoms
are of the noble gas type. Two quantities are of importance when considering
what might happen in the two-atom case, these being the ionization energy

and the electron affinity. The ionization energy, E_I, is the minimum energy required to remove an electron from an otherwise neutral atom. The situation is described by

$$A \rightarrow A^+ + e^- \qquad (2.3)$$

A denotes the neutral atom, A^+ a positively charged ion, and e^- the electron. The electron affinity, E_A, is the energy gained when a neutral atom acquires an additional electron. This other situation is described by

$$A + e^- \rightarrow A^- \qquad (2.4)$$

A^- denotes the negatively charged atom that results from this process. It is very important to note that these two equations are *not* merely mutual opposites, because the product of the first reaction is a positive ion and a negative electron, whereas the participants in the second reaction are a *neutral* atom and a negative electron.

A good example is seen in the compound lithium fluoride, in which a lithium atom readily donates one of its electrons to a fluorine atom, the latter thereby acquiring an electron structure which resembles that of a noble gas atom, with all of its electron orbitals having the maximum number of permitted electrons (see Table 2.1). The separation of the electron from the lithium atom requires an amount of energy equal to the ionization energy, E_I. When the fluorine

Table 2.1   The electronic characteristics of the elements

Atomic Number	Element	Orbital electronic configuration	E_I(aJ)	E_A(aJ)
1	H	$1s$	2.178	0.120
2	He	$1s^2$	3.938	
3	Li	[He]$2s$	0.863	0.087
4	Be	[He]$2s^2$	1.493	−0.096
5	B	[He]$2s^2 2p$	1.329	0.032
6	C	[He]$2s^2 2p^2$	1.804	0.200
7	N	[He]$2s^2 2p^3$	2.329	−0.016
8	O	[He]$2s^2 2p^4$	2.181	0.235
9	F	[He]$2s^2 2p^5$	2.791	0.553
10	Ne	[He]$2s^2 2p^6$	3.454	
11	Na	[Ne]$3s$	0.823	0.119
12	Mg	[Ne]$3s^2$	1.225	−0.048
13	Al	[Ne]$3s^2 3p$	0.959	0.096
14	Si	[Ne]$3s^2 3p^2$	1.305	0.261
15	P	[Ne]$3s^2 3p^3$	1.762	0.112

continues overleaf

Table 2.1 (*continued*)

Atomic Number	Element	Orbital electronic configuration	$E_I(aJ)$	$E_A(aJ)$
16	S	$[Ne]3s^23p^4$	1.659	0.332
17	Cl	$[Ne]3s^23p^5$	2.084	0.578
18	Ar	$[Ne]3s^23p^6$	2.524	
19	K	$[Ar]4s$	0.695	
20	Ca	$[Ar]4s^2$	0.979	
21	Sc	$[Ar]4s^23d$	1.051	
22	Ti	$[Ar]4s^23d^2$	1.094	
23	V	$[Ar]4s^23d^3$	1.080	
24	Cr	$[Ar]4s3d^5$	1.083	
25	Mn	$[Ar]4s^23d^5$	1.191	
26	Fe	$[Ar]4s^23d^6$	1.266	
27	Co	$[Ar]4s^23d^7$	1.259	
28	Ni	$[Ar]4s^23d^8$	1.223	
29	Cu	$[Ar]4s3d^{10}$	1.237	
30	Zn	$[Ar]4s^23d^{10}$	1.504	−0.144
31	Ga	$[Ar]4s^23d^{10}4p$	0.961	0.029
32	Ge	$[Ar]4s^23d^{10}4p^2$	1.262	0.192
33	As	$[Ar]4s^23d^{10}4p^3$	1.572	0.096
34	Se	$[Ar]4s^23d^{10}4p^4$	1.562	0.272
35	Br	$[Ar]4s^23d^{10}4p^5$	1.897	0.538
36	Kr	$[Ar]4s^23d^{10}4p^6$	2.242	
37	Rb	$[Kr]5s$	0.669	
38	Sr	$[Kr]5s^2$	0.912	
39	Y	$[Kr]5s^24d$	1.041	
40	Zr	$[Kr]5s^24d^2$	1.113	
41	Nb	$[Kr]5s4d^4$	1.085	
42	Mo	$[Kr]5s4d^5$	1.137	
43	Tc	$[Kr]5s^24d^5$	1.166	
44	Ru	$[Kr]5s4d^7$	1.180	
45	Rh	$[Kr]5s4d^8$	1.195	
46	Pd	$[Kr]4d^{10}$	1.334	
47	Ag	$[Kr]5s4d^{10}$	1.213	
48	Cd	$[Kr]5s^24d^{10}$	1.440	−0.096
49	In	$[Kr]5s^24d^{10}5p$	0.927	0.032
50	Sn	$[Kr]5s^24d^{10}5p^2$	1.176	
51	Sb	$[Kr]5s^24d^{10}5p^3$	1.384	
52	Te	$[Kr]5s^24d^{10}5p^4$	1.443	0.352
53	I	$[Kr]5s^24d^{10}5p^5$	1.675	0.490
54	Xe	$[Kr]5s^24d^{10}5p^6$	1.943	
55	Cs	$[Xe]6s$	0.624	
56	Ba	$[Xe]6s^2$	0.835	
57	La	$[Xe]6s^25d$	0.899	
58	Ce	$[Xe]6s^24f5d$	1.107	
59	Pr	$[Xe]6s^24f^3$	0.923	
60	Nd	$[Xe]6s^24f^4$	1.011	

Table 2.1   (*continued*)

Atomic Number	Element	Orbital electronic configuration	E_I(aJ)	E_A(aJ)
61	Pm	$[Xe]6s^24f^5$		
62	Sm	$[Xe]6s^24f^6$	0.897	
63	Eu	$[Xe]6s^24f^7$	0.908	
64	Gd	$[Xe]6s^24f^75d$	0.987	
65	Tb	$[Xe]6s^24f^9$	1.080	
66	Dy	$[Xe]6s^24f^{10}$	1.093	
67	Ho	$[Xe]6s^24f^{11}$		
68	Er	$[Xe]6s^24f^{12}$	0.974	
69	Tm	$[Xe]6s^24f^{13}$	0.931	
70	Yb	$[Xe]6s^24f^{14}$	0.993	
71	Lu	$[Xe]6s^24f^{14}5d$	0.801	
72	Hf	$[Xe]6s^24f^{14}5d^2$		
73	Ta	$[Xe]6s^24f^{14}5d^3$	1.262	
74	W	$[Xe]6s^24f^{14}5d^4$	1.278	
75	Re	$[Xe]6s^24f^{14}5d^5$	1.261	
76	Os	$[Xe]6s^24f^{14}5d^6$	1.394	
77	Ir	$[Xe]6s^24f^{14}5d^7$	1.442	
78	Pt	$[Xe]6s4f^{14}5d^9$	1.442	
79	Au	$[Xe]6s4f^{14}5d^{10}$	1.477	
80	Hg	$[Xe]6s^24f^{14}5d^{10}$	1.671	
81	Tl	$[Xe]6s^24f^{14}5d^{10}6p$	0.978	
82	Pb	$[Xe]6s^24f^{14}5d^{10}6p^2$	1.188	
83	Bi	$[Xe]6s^24f^{14}5d^{10}6p^3$	1.167	
84	Po	$[Xe]6s^24f^{14}5d^{10}6p^4$	1.350	
85	At	$[Xe]6s^24f^{14}5d^{10}6p^5$		
86	Rn	$[Xe]6s^24f^{14}5d^{10}6p^6$	1.722	
87	Fr	$[Rn]7s$		
88	Ra	$[Rn]7s^2$	0.851	
89	Ac	$[Rn]7s^26d$		
90	Th	$[Rn]7s^26d^2$	1.121	
91	Pa	$[Rn]7s^25f^26d$		
92	U	$[Rn]7s^25f^36d$	0.984	
93	Np	$[Rn]7s^25f^46d$		
94	Pu	$[Rn]7s^25f^6$	0.823	
95	Am	$[Rn]7s^25f^7$	0.968	
96	Cm	$[Rn]7s^25f^76d$		
97	Bk	$[Rn]7s^25f^9$		
98	Cf	$[Rn]7s^25f^{10}$		
99	Es	$[Rn]7s^25f^{11}$		
100	Fm	$[Rn]7s^25f^{12}$		
101	Md	$[Rn]7s^25f^{13}$		
102	No	$[Rn]7s^25f^{14}$		
103	Lr	$[Rn]7s^25f^{14}6d$		

atom subsequently acquires that same electron, and thereby develops an electron structure resembling that of neon, the overall system gains an amount of energy equal to the electron affinity, E_A. Because E_I is usually larger than E_A, this electron transfer process might seem to require a net input of energy (see Table 2.2). However, we must remember that the situation does not involve two atoms that are well separated. On the contrary, they remain in the vicinity of each other, and so there are other energy contributions to be taken into account. It transpires that the Coulomb interaction (named for Charles de Coulomb) fully compensates for the above net energy input, so there will indeed be chemical binding between the two atoms involved. The situation is illustrated in Figure 2.2, and one notes that the binding is purely electrostatic in nature; there is no directionality in the binding. This interaction is, of course, a classical example of ionic bonding.

In ionic bonding, electropositive and electronegative atoms combine through the electric attraction between ions that are produced by electron transfer. Good examples are provided by the alkali halides such as LiF. Each neutral (electropositive) lithium atom loses its $2s$ electron and thereby becomes a positive ion, somewhat resembling a positively-charged helium atom. Each neutral (electronegative) fluorine atom gains an extra $2p$ electron and becomes a negative ion, resembling a negatively-charged neon atom.

It is no coincidence that it is the lithium atom in the above reaction that loses an electron. Of the two elements involved, it is lithium that has a lone electron outside shells that are fully occupied. The fluorine atom, on the other hand, has an outermost shell that lacks only one electron. This difference is clearly revealed in the first ionization energies of the two atoms (that is to say the energy which must be expended in removing one electron from the neutral atom). As can be seen in Table 2.1, the first ionization energy of lithium is 0.863 aJ, whereas the first ionization energy of fluorine is 2.791 aJ. Indeed, to find ionization

Table 2.2 Comparison of electronegativity with the sum of ionization energy and electron affinity. All values are for 25°C

	E_I	E_A	Sum	e_N
F	2.8	0.6	3.4	4.0
Cl	2.1	0.6	2.7	3.0
Br	1.9	0.5	2.4	2.8
I	1.7	0.5	2.2	2.5
H	2.2	0.1	2.3	2.1
Li	0.9	0.1	1.0	1.0
Na	0.8	0.1	0.9	0.9
K	0.7	0	0.7	0.8
Rb	0.7	0	0.7	0.8
Cs	0.6	0	0.6	0.7

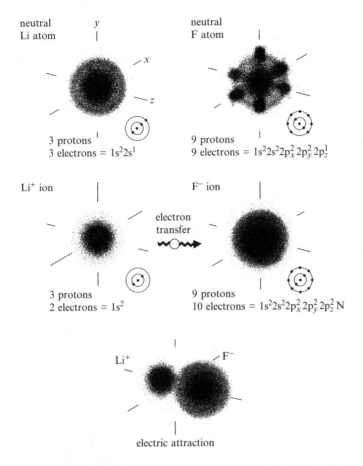

neutral Li atom

y

neutral F atom

x

z

3 protons
3 electrons = $1s^2 2s^1$

9 protons
9 electrons = $1s^2 2s^2 2p_x^2 2p_y^2 2p_z^1$

Li$^+$ ion

F$^-$ ion

electron transfer

3 protons
2 electrons = $1s^2$

9 protons
10 electrons = $1s^2 2s^2 2p_x^2 2p_y^2 2p_z^2$ N

Li$^+$ F$^-$

electric attraction

Figure 2.2   An illustration of ionic bonding in the case of LiF

energies larger than that exhibited by the fluorine atom, one would have to go to the noble gases themselves, with their fully occupied electron orbitals.

2.4   Electronegativity and Strong Bonds

Although the propensity that a given atomic species displays for losing or gaining electrons is determined by the dual factors of ionization potential and electron affinity, an adequate qualitative measure of the same thing is provided by a single parameter known as the electronegativity, e_N. Atoms with large electronegativities tend to capture electrons, whereas the opposite is the case for atoms with small electronegativities (these being said to be electropositive). A reliable scale of electronegativity has been derived by Linus Pauling and it is, as indicated above, a dual measure of ionization energy and electron affinity. Typical values are given in Table 2.2.

If the difference in the electronegativities of two atoms is quite small, there will be no clear tendency for one to lose an electron while the other gains this subatomic particle. There is thus no basis for ionic binding in such a situation. Instead, one has either covalent bonding or metallic bonding, the first of these occurring if the two atoms are both electronegative, and the latter arising when they are both electropositive. An example of covalent bonding is seen if the two atoms involved are both fluorine. The nine electrons in an atom of this element are arranged in such a way that there are two in the $1s$ orbital, two in the $2s$ orbital, two in each of the $2p_x$ and $2p_y$ orbitals, and finally a single electron in the $2p_z$ orbital. It is only the latter orbital, therefore, which lacks an electron and, as we saw in the case of ionic bonding, it can fulfil this need by acquiring an electron from another atom. However, in the case of the covalent bond, it does this not by completely removing an electron from that other atom, but rather by entering into a mutual sharing of atoms, in which the unfilled orbitals of both atoms are filled by the other's lone $2p_z$ electron.

One sees from Figure 2.3, which depicts the situation in a hydrogen fluoride molecule, that the $2p_z$ orbitals alone are involved in the binding. Although the interatomic bond is highly directional, as mentioned earlier, it possesses rotational symmetry and there is very little resistance to rotation of one of the atoms with respect to the other, about the z-axis. This form of covalent bonding is known as a σ-bond (sigma bond) and numerous examples are

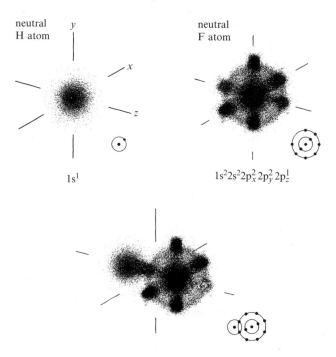

Figure 2.3   A simple example of covalent bonding which occurs in the HF molecule

encountered in molecules having biological relevance. In the case of the HF molecule the incomplete orbitals, the $1s$ of hydrogen and the $2p$ of fluorine, overlap and the two electrons are shared by a sigma-bond.

If we turn to an atom which lies one place lower in the periodic table, namely oxygen, it will lack a total of two electrons from its p orbitals rather than just the one seen in the case of fluorine. In much the same way as before, two oxygen atoms can form a molecule, and the $2p_z$ orbitals will again overlap, producing a sigma bond. However, this leaves the $2p_y$ orbitals still incompletely filled, and they can be imagined as bending toward each other so as to produce further covalent bonding, as seen in Figure 2.4. The latter involves distortion

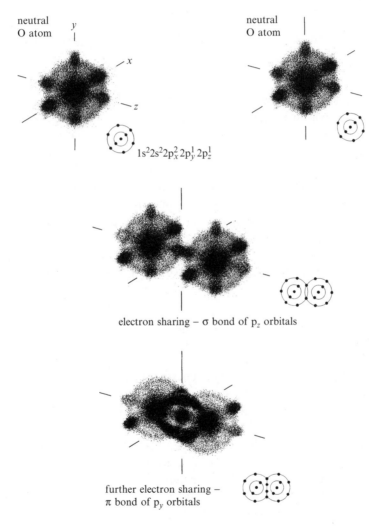

Figure 2.4 The covalent bonding between two oxygen atoms involves both a sigma-bond and a pi-bond

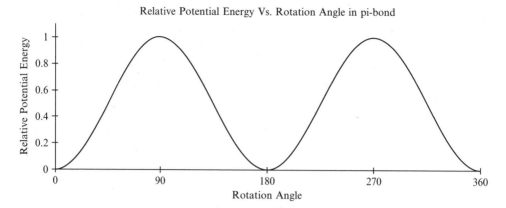

Figure 2.5   An energy versus rotation angle plot of the pi-bond caused by the resistance to rotation

and overlap of the $2p_y$ orbitals and confers axial rigidity on the molecule. Unlike the case of the sigma bond, with its single overlap along the z-direction, this new bond will involve two lobes, neither of which lies on that axis. There will thus now be a two-fold symmetry to the situation, and one can imagine any attempt to rotate one of the atoms with respect to the other encountering resistance, due to the fact that the $2p_y$ orbitals would thereby be stretched. This new type of bond is known as a π-bond (pi bond), and the resistance to rotation is reflected in an energy versus rotation angle plot which is roughly sinusoidal, as seen in Figure 2.5.

Four σ-bonds are present in the methane molecule, CH_4, each one attaching a hydrogen atom to the central carbon atom. The tetrahedral shape of this molecule might seem surprising, given what has been said so far about the mutual orientations of the various p orbitals (see Figures 2.1 and 2.3), lying at right angles to one another. The explanation lies in two factors, namely electron promotion and wave function hybridization. The first of these is viable when the increase of energy involved in promoting an electron to a higher quantum state is less than the subsequent lowering of energy when the wave functions can then be hybridized. Any combination of wave states in an atom is itself a possible wave state and the result is known as a hybrid orbital. The number of the latter must equal the number of original orbitals. Thus an s orbital and a p orbital produce two different sp hybrids (Figure 2.6). The combination of the different hybrids produces a configuration whose geometry determines the spatial arrangement of interatomic bonds. The hybridization principle is captured by Equation (2.2). In the case of the methane molecule, CH_4, one of the $2s$ electrons is promoted to an otherwise-empty $2p$ state, and the resulting three $2p$ orbitals and the one remaining $2s$ orbital hybridize to produce four sp^3 hybrids (see Figure 2.7). It is the latter which produce the four sigma-bonds

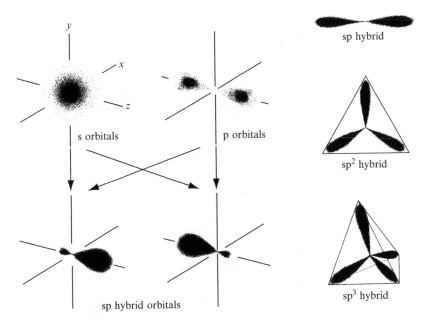

Figure 2.6   Wave function hybridization

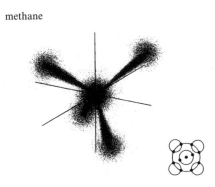

Figure 2.7   Hybridization in the methane molecule, CH_4

with the $1s$ orbitals of the hydrogen atoms, giving the methane molecule its tetrahedral symmetry. A somewhat related example is seen in the water molecule, H_2O, (see Figure 2.8). The dominant factor determining the shape of the water molecule is the hybridization of the $2s$ and $2p$ orbitals of the oxygen atom. The four hybrid orbitals have lobes which project out from the oxygen nucleus in the direction of the four corners of a tetrahedron. Two of these orbitals provide bonds for the two hydrogen atoms, while the remaining two become 'lone-pair' orbitals. The latter are negatively charged and attract the hydrogen atoms on neighbouring water molecules.

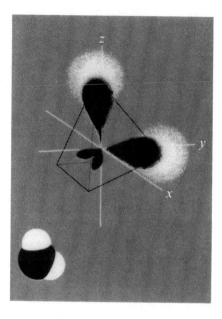

Figure 2.8   Hybridization in the water molecule, H_2O

Although the energy barrier represented by a π-bond is not insuperable, given sufficient thermal or mechanical activation, this type of bond can be regarded as giving a molecule considerable stiffness at its location. In effect, this decreases the number of parts of the molecule that are able to move freely, and in some cases it considerably reduces the alternative conformations that must be investigated by those who wish to determine molecular structure from first principles. We will return to such computations later, when considering the very important protein-folding problem.

In those cases where all atoms present are electropositive, the tendency will be for a metal to be produced. If the number of atoms is sufficiently large, the result will be an assembly of positive ions, held together by a sort of glue of relatively free electrons. The word *relatively* is important here, however, because in practice the electrons are able to move only with those energies consistent with the scattering and interference effects that arise because of their wave nature. This produces the well-known energy bands that are important in distinguishing metals from semiconductors and insulators. Such atomic assemblies are of relatively minor importance in the biological sphere, however, so we will not discuss them further. All the types of bond discussed so far, that is to say the ionic, covalent and metallic types, are collectively referred to as strong bonds. It should be noted that these distinctions are not rigid, and that many assemblies of atoms display binding that is intermediate between two of the above-mentioned categories.

2.5   Secondary Bonds

Because there exist forms of condensed matter comprising assemblies of molecules that retain their individual identity, there must be types of intermolecular force that we have not yet considered. These are known as secondary bonds and they are quite weak compared with those that we have considered until now. In those cases where the molecules have no permanent dipole moment (that is to say, there is no net separation of opposite electrical charges, analogous to the separation of north and south poles in a permanent magnet), the most important type of secondary bond is that which is named after Johannes van der Waals. This type of bond arises because although the atoms involved have no permanent dipole moments, there are small instantaneous dipole moments that arise from the motions of the individual electrons. At any instant, an electron's distribution with respect to the nucleus need not necessarily be as uniform as is its averaged value, and this means that there will be instantaneous dipole moments. These can influence those of the surrounding atoms, and the net outcome can be shown to be a weakly positive attraction (see Chapter 3). Unlike the covalent bond, this van der Waals bond is not directional, and it cannot be saturated; the number of other atoms which can be attracted to a given atom, via the van der Waals force, is determined solely by geometrical considerations. This is why the crystalline forms of the noble gases, which enter into only this type of bond, are usually close-packed.

   The other type of secondary bond that we need to consider here is the hydrogen bond, and its importance in biological structures could hardly be exaggerated. It frequently arises through the interaction of hydrogen and oxygen atoms, as in the configuration NH $\bullet\bullet\bullet$ OC. The electronegativity of the nitrogen atom is so large compared with that of the hydrogen that the hydrogen atom's lone electron tends to spend most of its time between the two atomic nuclei. This means that the (proton) nucleus of the hydrogen atom is not so well screened by the electron as it would be in the isolated neutral version of that atom. The net result is that the hydrogen atom develops a positive pole. Just the reverse happens in the case of the oxygen atom, the electronegativity of which is greater than that of the carbon atom, so it, in turn, functions as a negative pole. The hydrogen bond is not very strong, but it is quite directional, in that the nucleus of the hydrogen atom is exposed only within a limited solid angle. Moreover, the small size of the hydrogen atom limits the number of other atoms that can approach it sufficiently closely. The hydrogen bond is thus also quite saturated.

   Following this brief and qualitative discussion of the various types of interatomic bond, we should now turn to the quantitative aspects of interatomic interactions. These forces dictate the spatial arrangement of the atoms in a molecule. If we wish to make calculations on the relative stabilities of various candidate conformations, therefore, we will need a quantitative description of these forces. It is thus to this aspect of the issue that we will turn in the next chapter.

Exercises

2.1 The atoms of the first five elements of the periodic table contain one (hydrogen), two (helium), three (lithium), four (beryllium) and five (boron) electrons, respectively. Using dots to symbolize the *nuclei* of one atom of each of these five elements, draw the approximate distributions of electron probability density corresponding to the *lowest energy states* of each of these elements.

2.2 Describe the difference between ionization energy and electron affinity.

2.3 Draw the electronic configurations of a σ-bond and a π-bond, respectively, between two appropriate atoms.

2.4 Draw the approximate distributions of electron probability density corresponding to the *lowest energy state* of the nitrogen molecule, N_2. Indicate *all* the electrons in the molecule, and add labels to indicate the types of bond involved.

2.5 The 1976 Nobel Prize in chemistry was awarded to William Lipscomb for his work on boranes, which are composed of atoms of boron and hydrogen. From your knowledge of atomic and molecular orbitals, what would you expect to be the spatial configuration of the nuclei and electron probability densities in one (or more) of these boranes, given that they are rather unstable and chemically highly reactive?

Further reading

Atkins, P. W., (1990). *Physical Chemistry*. Oxford University Press, Oxford.
Moore, W. J., (1972). *Physical Chemistry*. Longman, London
Pauling, L., (1960). *The Nature of the Chemical Bond*. Cornell University Press, Ithaca, NY.
Pauling, L., (1970). *General Chemistry*. Freeman, San Francisco.
Serway, R. A., (1992). *Physics for Scientists and Engineers*. Saunders, Philadelphia.

3 Energies, Forces and Bonds

The discussion in the previous chapter provided the basis for a qualitative understanding of the factors that influence the interactions between atoms. And it showed how consideration of the shapes of electron probability orbitals can lead to an understanding of the three-dimensional shapes of molecules. If one needs quantitative information about these interactions, however, it is necessary to have information about the actual forces through which the various atoms (and indeed molecules) interact. And such information will include facts about both the analytical forms of particular interatomic interactions and the relevant parameters. This will be our primary concern in the present chapter, for we will be considering several well-known interatomic potential functions, as well as the magnitudes of the parameters by which they are characterized. This done, we will also take a brief look at the characteristics of several types of interatomic bonds.

3.1 Interatomic Potentials for Strong Bonds

The arguments given in the preceding chapter regarding the repulsive, attractive and equilibrium aspects of the interatomic potential were actually put forward by Ludwig Seeber in 1824. His conjectures showed remarkable insight, given that they were being made about 50 years before there was any real evidence for the existence of atoms.

The potential energy of one atom with respect to another is often denoted by V, but this raises the possibility of confusion with the symbol generally employed for electrical voltage, which we will need to refer to in several later chapters. In this book, therefore, the interatomic potential will be written $\mathscr{E}$. The potential energy is, as we have been saying, a function of the separation distance, r, between the two atoms, and in cases where this varies only with distance (and not also with angle) we may write $\mathscr{E}(r)$. In such cases, the interactions are said to be purely central forces, and the force between the two atoms, $\mathscr{F}(r)$, will also depend only upon r. The two parameters are related by the well-known equation

$$\mathcal{F}(r) = -\frac{\partial \mathcal{E}(r)}{\partial r} \tag{3.1}$$

where $\partial/\partial r$ represents the partial first derivative with respect to r. The importance of this relationship to all interactions between atoms could hardly be exaggerated, and it has been used in countless computer simulations of the static and dynamic aspects of biological molecules (see Chapter 6). The relationship is shown graphically in Figure 3.1, in which it is important to note that the zero-force distance corresponds to the minimum of the potential energy, whereas the maximum attractive force distance corresponds to the point of inflexon in the attractive portion of the potential energy curve. The force between two atoms must be repulsive at sufficiently small separation distances, because matter does not spontaneously implode, and it must be attractive at

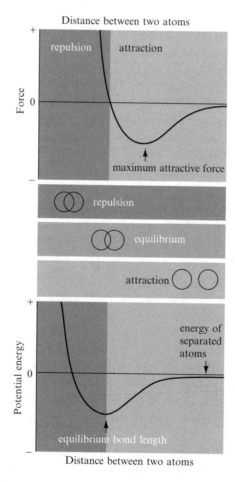

Figure 3.1 Graphical representation of the relationship between the force between two atoms and the distance between them

sufficiently large separations, because matter does not spontaneously fall apart. There must therefore be an intermediate separation distance at which equilibrium prevails.

Given the variety of types of bonding between atoms, which were described in the previous chapter, it is not surprising that various functional forms of interatomic potential are required to describe them. One of the earliest attempts to calculate a *component* of the interatomic interaction was made by Max Born and Joseph Mayer in 1932. Their goal was to derive the functionality in the case of repulsion between two atoms whose electrons were exclusively in completely filled orbitals. Using the techniques of quantum mechanics, they were able to show that this repulsion is purely exponential in nature, and it naturally arises from the Exclusion Principle postulated by Wolfgang Pauli, as discussed in the previous chapter.

This closed-shell repulsion potential has the form

$$\mathscr{E}(r) = A \exp\left(\frac{2r_s - r}{\rho}\right) \tag{3.2}$$

in which r_s and ρ are distance parameters which determine how quickly the repulsion falls away with increasing distance, and A is a coefficient which determines the strength of the interaction at a given distance.

The parameter ρ is approximately equal to the atomic radius, while A depends on which elements are involved in the interaction (see Table 3.1). This component of the interatomic potential is usually referred to as the Born–Mayer potential. For interactions between different sizes of closed shells one can use the average values for A and r_s. The parameter ρ is equal to 0.0345 nm for all sizes of closed shell. It is worth emphasizing that because there will usually be some fully occupied electron orbitals in both interacting atoms, except in the cases of the lightest elements, a Born–Mayer component will usually be present in the interatomic interaction irrespective of what other terms may arise because of the specific type of bonding. The form of the Born–Mayer potential is illustrated in Figure 3.2.

Table 3.1 The energy and distance parameters for the Born–Mayer closed-shell repulsion (*Zeitschrift für Physik*, **75**, 1 (1932))

Closed shell	A (aJ)	r_s (nm)
He	0.200	0.0475
Ne	0.125	0.0875
Ar	0.125	0.1185
Kr	0.125	0.1320
Xe	0.125	0.1455

ENERGIES, FORCES AND BONDS

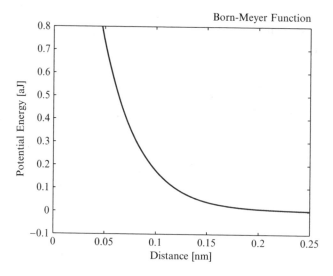

Figure 3.2 The Born–Mayer function is analytically exact as a description of the repulsive interaction between two atoms, each of which exclusively consists of closed electron shells

When two charged atoms, that is to say two ions, interact, a Coulomb contribution to the interatomic potential will always be present. This has the form

$$\mathscr{E}(r) = \frac{1}{4\pi\epsilon_0} \frac{\tau_1 \tau_2 q_e^2}{r} \tag{3.3}$$

in which q_e is the electron charge (1.602×10^{-19} Coulombs), τ_1 and τ_2 are the charge states of the two ions ($1, 2, 3, \ldots$), and ϵ_0 is the permittivity of free space. The value of the latter is 8.85×10^{-12} Farads per meter. It should be noted that, as for the Born–Mayer potential, the ionic interaction decreases with increasing distance. In this case, however, the decrease is merely linear rather than exponential. It must be emphasized that the sign of each charge (plus or minus) is included in each τ, so that the overall potential will be negative if the charges have opposite signs. In such a case there will be attraction between the two ions. The approximate form of the Coulomb interaction potential is shown in Figure 3.3.

Unlike the interactions described in the preceding two paragraphs, the covalent interaction is never repulsive. And whereas the preceding functional forms dealt with in this section can be shown to be exact solutions of the corresponding situations, the functional form that best describes the covalent interaction potential should really be regarded as being empirical. It is

$$\mathscr{E}(r) = \frac{-1}{4\pi\epsilon_0} \frac{q_e^2}{r} \exp\left(\frac{-r}{\zeta}\right) \tag{3.4}$$

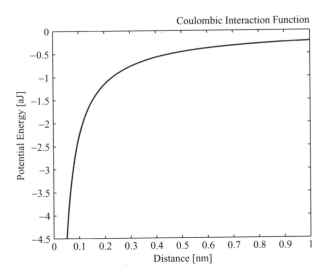

Figure 3.3   The Coulombic interaction potential

in which the distance parameter, ζ, in the exponential has been given that symbol to distinguish it from the corresponding parameter in the Born–Mayer potential. As before, q_e stands for the charge on an electron. We see that this function is essentially a Coulomb interaction modulated by an exponentially decreasing factor, and the negative sign reflects the fact that the potential is always attractive. It is noteworthy that the τ's which appeared in the Coulomb interaction are missing in the present case, because the covalent bond always refers to a single orbital interaction. The form of the covalent interaction function is shown in Figure 3.4 (solid line), and it is interesting to compare it with the corresponding form for the ionic interaction (dotted line), when this involves single charges on each ion. As can be seen from the two plots, the covalent interaction dies away more rapidly than its ionic counterpart.

Although metallic bonding will not interest us further in this book, it is instructive to note how complicated the situation is in such materials. Apart from the closed-orbital terms, which are adequately represented by the Born–Mayer function, there is also the Coulombic repulsion between the (positive) ions. Finally, there are terms which arise from the above-mentioned electron interference and scattering effects. In calculations on the properties of metals, frequent use is made of the Morse potential function (named after its originator, Philip Morse)

$$\mathscr{E}(r) = \mathscr{E}_{Morse}\{\exp[-2\alpha(r - r_0)] - 2\exp[-\alpha(r - r_0)]\} \tag{3.5}$$

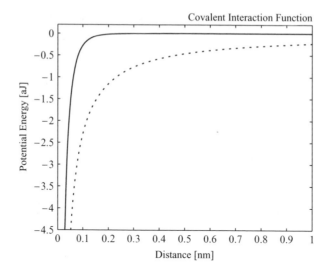

Figure 3.4 The covalent interaction function, with the Coulombic interaction indicated by the dotted line

in which the parameter α is a distance factor not unlike those which appear in the Born–Mayer and covalent functions, although in this case it is an *inverse-distance*. There is a second distance parameter, namely r_0, which does not have a counterpart in those other potential functions. It appears because the Morse potential has both repulsive and attractive terms, and the r_0 is the distance at which the potential function has its minimum value. Finally, $\mathscr{E}_{\text{Morse}}$ is a coefficient that determines the depth of the potential well. It is a useful exercise to differentiate the Morse potential once with respect to r, so as to convince oneself that the potential really does have its minimum value at r_0. That is to say

$$\left. \frac{\partial \mathscr{E}(r)}{\partial r} \right|_{r=r_0} = 0 \tag{3.6}$$

It is also interesting to note, in passing, that the Morse potential has the useful feature that the exponent of the first term is precisely double that of the second term. This makes for economy of computation, which is an important consideration in large-scale calculations on metallic crystals. The form of the Morse potential is indicated in Figure 3.5 which shows a minimum of $-\mathscr{E}_{\text{Morse}}$ at the equilibrium separation.

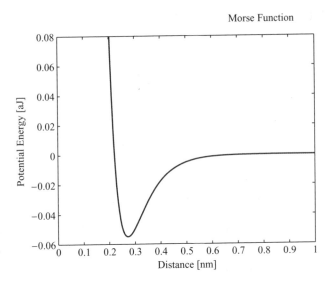

Figure 3.5   The Morse function

3.2   Interatomic Potentials for Weak Bonds

Having covered the various forms of strong interaction, we turn now to the weaker bonds, and begin with the attraction that can occur between filled orbitals. That there should be repulsion when filled orbitals approach sufficiently closely is not surprising, because the Pauli Exclusion Principle will ultimately prohibit overlap of the orbitals belonging to the two atoms. As we saw earlier, this provided the basis for the work of Born and Mayer. That there should also be an attraction, as the distance increases beyond the equilibrium separation, is more surprising. As was first demonstrated by Johannes van der Waals, the interaction arises from the instantaneous dipoles that nevertheless exist even in completely filled orbitals (see Chapter 2). Van der Waals was able to show that this interaction varies as the sixth power of the distance, so we have

$$\mathscr{E}(r) = -\beta/r^6 \tag{3.7}$$

in which β is a coefficient.

The two factors, repulsion and attraction, both appear in the Buckingham potential function (named for **R.A. Buckingham**). It has the form

$$\mathscr{E}(r) = A \exp(-r/\rho) - \beta/r^6 \tag{3.8}$$

We see that this is nothing other than a composite of Born–Mayer and van der Waals terms. Although the Buckingham function is precise with respect to both the repulsive and the attractive terms, it is not particularly suited to computer simulations, because it comprises a mixture of exponential and power expressions. In this respect a potential developed by John Lennard-Jones is much easier to deal with. This function has the form

$$\mathscr{E}(r) = \varepsilon\{(r_0/r)^{12} - 2(r_0/r)^6\} \tag{3.9}$$

There are just two parameters, namely the distance parameter r_0 and an energy parameter ε. As was the case with the Morse potential, we see that the repulsive term is simply the square of the attractive term, though in this case one is dealing with powers rather than exponentials. This simplifying factor has made the Lennard-Jones potential function extremely popular in the computer simulation of all manner of atomic and molecular systems, and it has become the standard choice when the so-called non-bonding interactions must be included in a calculation. The reader should be warned that the Lennard-Jones function is sometimes written in the alternative form:

$$\mathscr{E}(r) = \varepsilon\{(r_0/r)^{12} - (r_0/r)^6\} \tag{3.10}$$

This lacks the coefficient 2 in the second (attractive) term. The advantage of the former version is that it admits of very simple interpretations for the two parameters, namely that the energy minimum occurs at a distance r_0, and that the depth of the potential well is simply $-\varepsilon$ (see Figure 3.6). The reader may find it instructive to check these facts by, as before, taking the first derivative of the potential with respect to r and equating this to zero. That is to say

$$\left.\frac{\partial \mathscr{E}(r)}{\partial r}\right|_{r=r_0} = 0 \tag{3.11}$$

The reader should also make careful note of the fact that the interatomic potential, by convention, always goes to zero as r tends to infinity. That is to say

$$\mathscr{E}(r)|_{r\to\infty} = 0 \tag{3.12}$$

Tables 3.2 and 3.3 list the values of r_0 and ε, respectively, for *non-bonded* pairings of atoms of various elements.

Finally, amongst the purely central forces, there is the hydrogen bond (see Chapter 2). This has been the subject of a great amount of discussion in the literature, and it has not proved possible to derive a functional form which

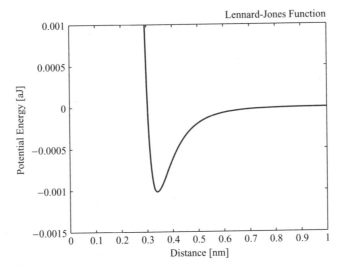

Figure 3.6   The Lennard-Jones function

Table 3.2   The Lennard-Jones distance parameter for the non-bonded interactions between pairings of atoms commonly encountered in the biological domain

r_0 (nm)	C	N	O	H	S
C	0.34	0.33	0.33	0.28	0.35
N		0.31	0.31	0.28	0.33
O			0.30	0.27	0.33
H				0.24	0.30
S					0.30

Table 3.3   The Lennard-Jones energy parameter for the non-bonded interactions between pairings of atoms commonly encountered in the biological domain

$\varepsilon 10^{-3}$ aJ	C	N	O	H	S
C	1.02	1.58	1.41	0.80	2.18
N		2.50	2.22	1.22	3.40
O			2.00	1.17	3.05
H				0.78	1.72
S					2.10

enjoys anything approaching universal acceptance. Hydrogen-bond potential functions have nevertheless been put forward by a number of researchers (notably Frank Stillinger), and one that has been fairly popular is the following

$$\mathscr{E}(r) = \mathscr{E}_{LJ}(r) + \mathscr{E}_{ionic}(r) + \mathscr{E}_s(r) \tag{3.13}$$

in which $\mathscr{E}_{LJ}(r)$ is the Lennard-Jones potential (with parameters chosen to correspond to the non-bonding situation), $\mathscr{E}_{ionic}(r)$ is the Coulombic inter-action given by Equation (3.3), and $\mathscr{E}_s(r)$ is known as a switching function. The effect of the latter is to make the overall $\mathscr{E}(r)$ go smoothly to zero for distances greater than a critical value, which is usually taken to be 0.285 nm.

3.3   Non-central Forces

Let us now move on by taking a brief look at the complexities that arise when one takes non-central components of the interatomic force into account. Faced with such a task, one finds that progress can be made only if one adopts an empirical approach, using approximations where necessary. A study along these lines has produced an expression for the total energy of a molecule that is now widely used in computer simulations of biological molecules. It has the following form

$$\begin{aligned}
\mathscr{E}_{total} = &\sum_b \left[k_2(b-b_0)^2 + k_3(b-b_0)^3 + k_4(b-b_0)^4 \right] \\
&+ \sum_\theta \left[k_2(\theta-\theta_0)^2 + k_3(\theta-\theta_0)^3 + k_4(\theta-\theta_0)^4 \right] \\
&+ \sum_\phi \left[k_1(1-\cos\phi) + k_2(1-\cos 2\phi) + k_3(1-\cos 3\phi) \right. \\
&+ \sum_\chi k_2\chi^2 + \sum_{b,b'} k(b-b_0)(b'-b_0') \\
&+ \sum_{b,\theta} k(b-b_0)(\theta-\theta_0) + \sum_{b,\phi} (b-b_0)[k_1\cos\phi + k_2\cos 2\phi + k_3\cos 3\phi] \\
&+ \sum_{\theta,\phi} (\theta-\theta_0)[k_1\cos\phi + k_2\cos 2\phi + k_3\cos 3\phi] \\
&+ \sum_{\theta,\theta'} k(\theta-\theta_0)(\theta'-\theta_0') + \sum_{\theta,\theta',\phi} k(\theta-\theta_0)(\theta'-\theta_0')\cos\phi \\
&+ \sum_{i,j} \frac{q_i q_j}{r_{ij}} + \sum_{i,j} \varepsilon_{ij} \left[2\left(\frac{r_{ij}^0}{r_{ij}}\right)^9 - 3\left(\frac{r_{ij}^0}{r_{ij}}\right)^6 \right]
\end{aligned} \tag{3.14}$$

Such a potential can be regarded as deriving from two types of interaction: the valence terms and the non-bonded terms, the latter referring to those compon-ents that do not stem from covalent bonds. The valence terms themselves also come from different categories of interaction, namely what are called the diagonal components and the off-diagonal cross-coupling components. In the valence terms, there are the internal coordinates of bond length, b, angle, θ, torsion angle, ϕ, and out-of-plane angle, χ. The cross-coupling terms include the various possible combinations of two and three internal co-ordinates, and they are found to be important if good agreement with experimental vibration spectra is at a premium. This is generally the case, because large biological molecules are not static at body temperature; their dynamic properties are of great significance for metabolic function. The non-bonded terms include a Lennard-Jones type function that has a ninth rather than a 12th power in its repulsive part; the repulsion is thus somewhat softer than the case discussed earlier in this chapter.

3.4 Bond Energies

Before putting interatomic potentials to use, it will be helpful to get an idea of the energies of the bonds commonly encountered in molecules of biological relevance. This will give us a better view of the types of reactions and processes that can occur in the biological domain. We will be considering many cases in which molecules are subjected to one influence or another, and it is important that we develop a feeling for what might happen under a given set of circumstances. We might wish to know whether a given molecule is likely to be able to withstand certain conditions. Why is it, for example, that sitting in the sun can cause a change in the colour of the skin, whereas the same is not the case when we sit in front of a fire, even though the latter may be just as effective at giving us instant warmth?

A few calculations relevant to our own bodies will serve to illustrate the usefulness of being able to make quantitative estimates. It can readily be shown that the total energy of all the bonds in a typical adult body is about 250 times the energy we consume in a single day. Let us make the following approximations. First, we assume that the bonds which keep our bodies together are essentially of the C—C type. This is not true, of course, because there are many C–N and C—O bonds, but we will later discover that these two types of bond have rather similar energies anyway. Now there are roughly 5×10^{22} atoms per gram of tissue, and we can assume that there would be about three bonds per atom (i.e. that the degree of bonding is slightly larger than would be the case if all bonds occurred only in linear chains, with their two bonds per atom). This indicates that an adult human would comprise about 4×10^{27} bonds. A typical covalent bond energy is approximately 0.6 aJ (i.e. 0.6×10^{-18} J), so the total energy of all the bonds in the body is about 2.4×10^9 J or 0.6×10^9 calories, which is indeed about 250 times the normal daily intake of 2400 kilocalories, in the well-fed portion of the world's population at least.

Table 3.4 shows the characteristics of a number of interatomic bonds of biological relevance. It is particularly those that involve a single bond that will

Table 3.4 The interatomic bond energies and interatomic equilibrium separations for a number of pairings and bond types encountered in biological molecules

Bond	Energy (aJ)	Distance (nm)
H–H	0.72	0.074
H–C	0.69	0.109
H–N	0.65	0.100
H–O	0.77	0.096
C–C	0.58	0.154
C=C	1.02	0.135
C≡C	1.39	0.121
C–N	0.49	0.147
C=N	1.02	0.129
C≡N	1.48	0.115
C–O	0.58	0.143
C=O	1.23	0.122
N–N	0.26	0.148
N=N	0.69	0.124
N≡N	1.57	0.100
N–O	0.35	0.144
N=O	0.80	0.120
O–O	0.35	0.149
O=O	0.82	0.128

be of importance, although it is instructive to compare these with the situations in which two, or even three, bonds are involved. It must be emphasized that all the bonds shown in Table 3.4 are of the covalent type. This means that the equilibrium interatomic separation, shown in the final column of the table, would arise from a balance between Born–Mayer repulsion and covalent attraction.

We are now in a position to calculate the total energy represented by the bonds in a simple molecule, assuming that its structure is already known. All we have to do is to work out how many bonds of each type there are and then look up the energies of these bonds in Table 3.4. It is true that we will then be ignoring the non-bonding terms, but this is a reasonable approximation for small molecules, in which remote parts of the molecule are not able to approach one another, as it adopts its three-dimensional conformation. Such close approaches would give extra contributions to the energy. A good example of a molecule in which the non-bonding terms are relatively small is glucose, shown in Figure 3.7. Just below that figure, a table is given which shows the small bookkeeping exercise that underlies the calculation of the total energy of all the bonds in the molecule, namely 15.64 aJ.

Glucose is the molecule that is broken down in the mitochondria (singular, mitochondrion) of every cell, during the respiration process. The final product

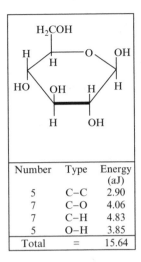

Number	Type	Energy (aJ)
5	C–C	2.90
7	C–O	4.06
7	C–H	4.83
5	O–H	3.85
Total	=	15.64

Figure 3.7 The energy stored in all the bonds present in a glucose molecule

is water and carbon dioxide, together with the liberated energy. Using an approach similar to the one we have just employed in connection with the total binding energy of the glucose molecule, we can derive the following equation for the respiration process

$$C_6H_{12}O_6 \quad + \quad 6O_2 \quad \rightarrow \quad 6CO_2 \quad + \quad 6H_2O \quad + \quad 3.44\,aJ \qquad (3.15)$$

$$-15.64\,aJ \qquad -4.92\,aJ \qquad -14.76\,aJ \qquad -9.24\,aJ$$

in which the total bond energy in each component is shown underneath.

In reality, the mitochondria produce only about 2 aJ of useful energy from each glucose molecule that is broken down, so the efficiency of this process is just over 50%. It ought to be noted, however, that very few industrial processes approach this level. It should also be noted that this simple calculation has ignored non-covalent interactions. A more exhaustive calculation would have taken both primary and secondary interactions into account. In the next chapter, we are going to discover that calculations involving only bond energies are, in any event, inadequate; one really needs to work with what is known as the free energy, which includes, in addition to the energy term, one that accounts for the entropy changes occurring during the reaction. These principles will be illustrated when we consider the physics of protein structure. The reason the weak, non-bonding interactions are important stems primarily from their large number. In principle, each atom in a molecule has this type of weak interaction with all those atoms it is not covalently bonded to, and that number will invariably be much larger than the number of covalent bonds. This fact is illustrated in the following example.

The well-known polymer polyisoprene occurs with two different arrangements of its side-chains. These are known as the *cis*- and *trans*-isomers. They are commonly known as natural rubber and gutta-percha, respectively. The noteworthy thing is that whereas natural rubber is very flexible, gutta-percha is stiff and hard. (Interestingly, both these isomers were used in the golf balls of earlier decades, the natural rubber providing the elastic core and the gutta-percha being used for the stiffer outer cover.)

Because these two polymers have the same backbone structure, and also the same side groups, the difference in their mechanical behaviour must derive solely from the non-bonded interactions. In the *cis* configuration (see Figure 3.8), the methyl group (i.e. the CH_3 group) interacts with the hydrogen atom on the same side of the double bond's axis, this interaction preventing the methyl group and the hydrogen atom from rotating away from each other. As a result, the chain develops a kink, the methyl and the hydrogen repelling each other until equilibrium prevails. The kink in the backbone can be straightened out by the application of sufficient tensile stress, but as soon as this stress is released the repulsive forces between the two side groups in question cause the kink to develop again in the same place. In the *trans* configuration (see Figure 3.9), the methyl group and the hydrogen are located on opposite sides of the double bond's axis, so the backbone does not develop kinks. It is for this reason

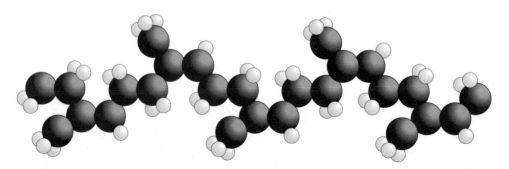

Figure 3.8   The naturally kinked molecular configuration of *cis*-polyisoprene gives it a high degree of elasticity

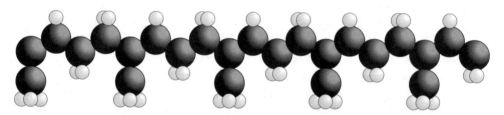

Figure 3.9   Molecules of *trans*-polyisoprene do not have the kinked structure of their *cis* counterparts and they are far more rigid

that gutta-percha does not show the same elastic effects. It is true, of course, that gutta-percha is not ideally stiff; with the application of sufficient tensile stress, it can be extended, but this requires elongation of the *covalent* bonds, and that requires a far greater force. The same is true of the *cis* form, which can be further extended after the normal elastic extension has been reached, but again the further extension requires application of much greater force.

As we have just seen, the *cis* and *trans* forms of polyisoprene differ with respect to rotation about a double bond. The resistance to such rotation is known as steric hindrance, and it has actually been measured for some simple polymers. For polyethylene, for example, the variation of energy as a function of the angle of rotation about the double bond was found to be as shown in Figure 3.10, in which the *trans* form is defined as having zero rotation angle, while the *cis* form corresponds to an angle of 180°. The inset to the figure shows the situations viewed parallel to the backbone, and it is clear that the *cis* form will give rise to larger non-bonded energy contributions, because of the close proximity between the various hydrogen atoms. It is interesting to note, both from the experimental curve and also from the inset, that the 120° case also corresponds to a local minimum of non-bonded interaction energy called the gauche form.

As an exercise in the application of potential energy functions, one can attempt to justify the general form of Figure 3.10 by directly computing the various non-bonded energy contributions. As can be seen in Figure 3.10 (inset)

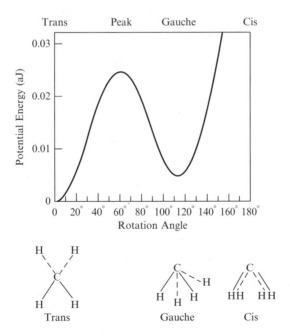

Figure 3.10 The experimental rotational energy function for twisting about one of the C–C bonds in polyethylene

there are four hydrogen atoms to be taken into account, these occurring in two pairs. Within either pair, the mutual positions of the hydrogen atoms do not vary, so one needs to take into consideration only those energy contributions that arise from the inter-pair interactions. One needs to know the precise geometry of the situation, in order to first calculate the relevant distances and then compute the corresponding energies. Each carbon atom can be imagined as being positioned at the centre of a tetrahedron, as far as the mutual angles of the corresponding bonds are concerned. On the other hand, the various atoms with which it shares covalent bonds will not in general lie at the corners of a rhombohedron, because the distances involved are not all the same. This is made clear in Table 3.4, in which we see that the C—H bonds are 0.109 nm long, whereas the C—C bond has a length of 0.154 nm. One can assume that the non-bonding interactions are adequately described by a Lennard-Jones potential function of the type

$$\mathscr{E}(r) = \varepsilon\{(r_0/r)^{12} - 2(r_0/r)^6\} \tag{3.16}$$

with $\varepsilon = 3 \times 10^{-4}$ aJ.

It is important to note that these energy and distance values differ considerably from those listed for the H—H bond in Table 3.4. The reason for the difference lies in the fact that those table entries are for covalent bonds, whereas we are here dealing with the much weaker *non-bonded* interaction. Although the geometry is not particularly easy, one can work out the distances of the various non-bonded interactions, and the corresponding energies. The results of this book-keeping exercise are shown in Figure 3.11 for the *trans*,

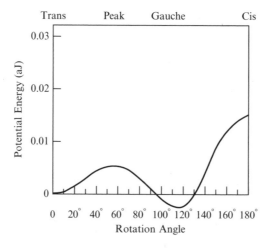

Figure 3.11   The calculated rotational energy function for twisting about one of the C–C bonds in polyethylene using Tables 3.2 and 3.3

peak, *gauche* and *cis* configurations, and the resulting energy values permit one to draw an approximate curve. Although the values of the different points are considerably lower than the corresponding points on Figure 3.10, the curve does at least have the right general shape, with the *cis* form having the highest energy, and the next highest value corresponding to what has been labelled *peak*. The reason all the calculated energies are lower than the experimental values lies in the fact that we have been neglecting the influence of the configuration changes on the underlying molecular orbitals, and perturbation of these would lead to increases of energy. However, as mentioned earlier, this small computational exercise has limited value in any case, because one should really be working with the free energy rather than merely the potential energy. It is thus appropriate that we suspend further discussion until we have considered those aspects of thermodynamics and statistical mechanics that are relevant to molecules and situations of biological interest.

3.5   Spring Constants

Until now in this chapter, we have been considering the forces that act at the atomic and molecular level. In many biophysical problems, however, we need to know something about the forces that act on a more macroscopic level. These forces are ultimately due to those atomic and molecular level forces, of course, but we need to be able to approximate the interactions that occur at the grosser level, without having to think about their microscopic roots.

These were the issues that occupied materials scientists of earlier centuries, in fact, and they made commendable progress despite the primitive nature of science in those days. One of the early giants in this type of endeavour was Robert Hooke, who established that there is a linear relationship between the stress applied to a piece of material and the resulting strain. This wisdom is embodied in what came to be known as Hooke's Law, and he showed that the simple proportionality holds as long as the degree of strain is small. As will be discussed more fully in Chapter 5, it breaks down when the material is strained beyond what is known as the elastic limit. The science of those early days tended to be written up in Latin, and Hooke's result is immortalized in the phrase *ut tensio sic vis* (the elongation is as the force).

When dealing with such multi-atom structures as the components of muscle fibres, for example, it is customary to invoke Hookian behaviour, and to assume that the macroscopic restoring force is linearly related to the strain, by an expression of the type

$$\mathcal{F}_{restoring} = -\kappa \Delta x \qquad (3.17)$$

the negative sign arising because the force acts in the direction counter to the extension Δx. The coefficient κ is known as a spring constant, and we will encounter examples in Chapters 5, 6 and 10.

Systems obeying Equation (3.17) are said to be harmonic. If (following one of Isaac Newton's laws) one replaces the force on the left-hand side with its equivalent, namely mass times acceleration, and then solves the resulting differential equation, one obtains a solution that describes simple harmonic motion. The classic example is, of course, the harmonic motion of a simple pendulum. We should note also that Equation (3.17) can be used in conjunction with Equation (3.1) to produce the (macroscopic) potential energy function that underlies Equation (3.17). We then find by integration that the potential energy varies as the square of the displacement.

When the amplitude of a pendulum is large, the harmonic approximation breaks down, and analysis of the motion then requires the use of elliptic integrals. A fascinating finding of such analysis is that under certain conditions the simple periodic solution is replaced by one describing a solitary wave. The latter is now referred to as a soliton, in fact, and there have been suggestions that such a mode of motion could be important to the dynamics of an enzyme.

Exercises

3.1 Many artificial joints now include metallic parts. The interactions between the atoms in these components are well described by the Morse function

$$\mathscr{E}(r) = \mathscr{E}_{Morse}\{\exp[-2a(r - r_0)] - 2\exp[-a(r - r_0)]\}$$

where typical values of the three parameters are $\mathscr{E}_{Morse} = 0.056 \times 10^{-18}$ J, $a = 1.4 \times 10^{10} \text{m}^{-1}$ and $r_0 = 2.8 \times 10^{-10}$ m. What is the maximum tensile force that a pair of such interacting atoms can withstand, without being torn apart?

3.2 Two atoms in a biological molecule are found to interact in a manner that can be described by the Lennard-Jones potential

$$\mathscr{E}(r) = \varepsilon\{(r_0/r)^{12} - 2(r_0/r)^6\}$$

in which the parameters ε and r_0 are constants. It is found that $\varepsilon = 0.49 \times 10^{-18}$ J, and that the minimum force required to totally separate the atoms is 8.96×10^{-9} Jm^{-1}. At what separation distance would the two atoms be in mutual equilibrium?

3.3 During a simulation of a multi-atom molecule, an analytical function was needed to represent the interactions between the various pairs of atoms. There was the requirement that the chosen function should have the correct value of the minimum energy, $\mathscr{E}_{min}$, and that this should lie at the correct atomic separation distance, $r_{\mathscr{E}_{min}}$. Moreover, the atomic separation distance corresponding to the maximum attractive force also had to be the specific value $r_{\mathscr{F}_{max}}$.

The initial choice was the Lennard-Jones function, and this was reasonably successful because a proper fit to all three of the above-named numbers was possible. But the *shape* of the Lennard-Jones function was not quite suitable, so it was decided to try the Morse function instead. The hope was that the larger number of parameters in the Morse function (three, compared with the Lennard-Jones function's two) would lead to a more suitable shape. The question then arose, however, as to whether it would *still* be possible to fit the function to those above-mentioned three numbers.

After analysing the situations corresponding to the two different potential functions, do you believe that the second fit would have been possible, and if so, under what conditions?

3.4   The recently-developed technique known as 'optical tweezers' permits measurement of the forces applied to single biological molecules. Calculate the maximum tensile force that a polypeptide chain can withstand, assuming that the covalent bonds which hold the backbone together can be described by the Lennard-Jones potential function:

$$\mathscr{E}(r) = \varepsilon\{(r_0/r)^{12} - 2(r_0/r)^6\}$$

in which the values of ε and r_0 for carbon–carbon bonds are respectively 0.56×10^{-18} J and 0.152×10^{-9} m, while the corresponding values for carbon–nitrogen bonds are respectively 0.51×10^{-18} J and 0.149×10^{-9} m.

Further Reading

Atkins, P. W., (1990). *Physical Chemistry*. Oxford University Press, Oxford.

Baker, J. J. W. and Allen, G. E., (1974). *Matter, Energy and Life*. Addison-Wesley, Reading, MA.

Barua, A. K., (1959). Force parameters for some nonpolar molecules on the exponential 6–8 model. *Journal of Chemical Physics* **31**, 957–960.

Berendsen, H. J. C. *et al.*, (1981). Interaction models for water in relation to protein hydration. In B. Pullman (Ed.) *Intermolecular Forces*. Reidel, Dordrecht, pp 331–342.

Born, M. and Mayer, J. E., (1932). Zur Gittertheorie der Ionenkristalle. *Zeitschrift für Physik* **75**, 1–18.

Buckingham, R. A., (1938). The classical equation of state of gaseous helium, neon and argon. *Proceedings of the Royal Society (London)* **A168**, 264–283.

Hirschfelder, J. O., Curtiss, C. F. and Bird, R. B., (1954). *Molecular Theory of Gases and Liquids*. John Wiley and Sons, New York.

Janzen, A. and Leech, J. W., (1976). Elastic constants of the inert gas solids. *Canadian Journal of Physics* **54**, 227–238.

Keener, J. and Sneyd, J., (1998). *Mathematical Physiology*. Springer-Verlag, Berlin.

Lennard-Jones, J. E., (1924). The determination of molecular fields from the equation of state of a gas. *Proceedings of the Royal Society (London)* **A106**, 463–477.

Lennard-Jones, J. E., (1931). Cohesion. *Proceedings of the Physical Society (London)*, **A43**, 461–482.

Margenau, H. and Kestner, N. R., (1969). *Theory of Intermolecular Forces*. Pergamon, Oxford.

Mason, E. A. and Rice, W. E., (1954). The intermolecular potentials for some simple nonpolar molecules. *Journal of Chemical Physics* **22**, 843–851.

Morse, P. M., (1929). Diatomic molecules according to the wave mechanics II Vibrational levels. *Physical Review* **34**, 57–64.

Moore, W. J., (1972). *Physical Chemistry*. Longman, Englewood Cliffs, NJ.

Murray, J. D., (1989). *Mathematical Biology*. Springer-Verlag, Berlin.

No, K. T. *et al.*, (1995). Determination of non-bonded potential parameters for peptides. *Journal of Physical Chemistry* **99**, 13019–13028.

Pauling, L., (1960). *The Nature of the Chemical Bond* Cornell University Press, Ithaca, NY.

Pauling, L., (1970). *General Chemistry* Freeman, San Francisco.

Price, S. L., (1996). Applications of realistic electrostatic modelling to molecules in complexes, solids and proteins. *Journal of the Chemical Society, Faraday Transactions* **92**(17), 2997–3008.

Salviucci, P., ed., (1967). *Molecular Forces* North-Holland, Amsterdam.

Sun, H., (1998). COMPASS: an ab initio force-field optimized for condensed-phase applications – overview with details on alkane and benzene compounds. *Journal of Physical Chemistry B* **102**, 7338–7364.

4 Rates of Reaction

The discussion in the two preceding chapters was strictly applicable to a state that is not actually attainable, namely the absolute zero of temperature, for it ignored the thermal motions that all atoms and molecules must have when they are at a finite temperature. And when considering the motions of atoms and molecules in biological matter, it is never permissible to regard the individual entities as moving completely freely, as they are imagined to do in an ideal gas. Cooperative effects are much in evidence, and this invariably makes the situation complicated. Understanding the collective properties of interacting groups of atoms and molecules was a challenge that occupied theoretical physicists and chemists in the late 19th century, and despite much progress, the situation remains less than perfect even to this day. At the turn of the millennium, there was still no universally accepted theory of the liquid state. Indeed, Philip Anderson expressed the opinion that understanding the glass–liquid transition had become the most pressing issue in all of condensed-matter science.

In the present chapter, we will see that the energies invoked in the previous two chapters must be replaced by free energies. And we will then go on to consider the factors that dictate the most probable forms of molecules, under given thermal conditions.

4.1 Free Energy

When one is concerned with changes in the state of a system, a full description of the situation requires knowledge of what is known as the free energy. There are actually two different forms of free energy, one first proposed by Helmholtz and the other by Gibbs.

The Helmholtz free energy, F (named for Hermann von Helmholtz), is a function of the internal energy, $\mathscr{E}_{int}$, the absolute temperature, T, and the entropy, S (see Appendix C). The relationship is

$$F = \mathscr{E}_{int} - TS \tag{4.1}$$

The Gibbs free energy, G (named for Josiah Willard Gibbs), includes an additional term that depends upon both the pressure and the volume. This more inclusive form of the free energy is described by the following equation

$$G = \mathscr{E}_{int} + PV - TS \qquad (4.2)$$

We will discuss the nature of the internal energy shortly. The general rule is that a system can spontaneously alter its status if the change results in a lowering of the free energy. And as a corollary of this, a system will be in equilibrium if its free energy is at a minimum. Possible changes in a system can thus be anticipated if one has a complete knowledge of all the components of the relevant free energy, as a function of the independent variables. Given the functional forms of these components, one can differentiate with respect to the relevant variables, to see whether equilibrium prevails or whether spontaneous changes in the system can be expected. If the changes are to occur under constant temperature and volume, it is usually adequate to work in terms of the Helmholtz free energy. But the reactions that occur in the body usually take place at constant pressure, so it is more common to work through the Gibbs free energy.

An important feature of both forms of free energy is the final term, and its negative multiplier. When the temperature is very low, this term will be insignificant of course, but with increasing temperature it makes its influence felt, the degree of this depending upon the entropy. Unlike the other terms in these free energy expressions, entropy is still an incompletely-understood concept. That fact too will be discussed later in this section (as well as in Appendix C). Suffice it to say that the final term in either of the free energy expressions will have a greater influence if the entropy is greater.

Returning to the internal energy, the first thing to be noted is that this would consist entirely of kinetic energy if there were no interactions between the particles in a system. As we saw in the preceding chapter, when considering interatomic potentials, these interactions certainly do exist, and many can be described by reasonably simple functions. Those interatomic potentials were all defined in such a way that the interaction energy became zero at very large distances. This is mathematically reasonable, of course, but we should not confuse very large separations with equilibrium conditions. On the contrary, two atoms will attract one another even if the distance between them is quite large, although the force of attraction will admittedly be very small indeed. For our purposes here, the minimum in the interaction potential is more relevant. This *does* represent a position of equilibrium. When calculating the internal energy, the potential energy component is indeed referred to this equilibrium situation. In other words, *thermodynamic potential energies are calculated relative to the minimum in the potential well.* Internal energy is defined by the following equation

$$\mathscr{E}_{int} = \mathscr{E}_{pot} + \mathscr{E}_{kin} \tag{4.3}$$

in which $\mathscr{E}_{pot}$ is the potential energy, and $\mathscr{E}_{kin}$ is the kinetic energy.

4.2   Internal Energy

To see how the various components in the internal energy arise, let us begin by considering a system of just two atoms. When they are at their equilibrium separation, and stationary, there is neither potential nor kinetic energy, so the internal energy is simply zero. If we now disturb the system by increasing the interatomic separation, by a few per cent say, the system will acquire potential energy. If the atoms are then released, they are free to move and they will approach each other with increasing velocity, ultimately passing the equilibrium separation and then overshooting this, so that the separation is now less than the equilibrium distance. As the equilibrium point is passed, the potential energy will be momentarily zero, but the kinetic energy will be finite (and at its maximum value for the given starting conditions). The initial potential energy of the displacement will thus have been fully converted into kinetic energy. Ultimately, when the overshoot is such that the potential energy has again reached the value it had when the atoms were released, the kinetic energy will once more have become zero. The direction of motion will then be reversed, and the interatomic separation will again start to increase. As it passes the equilibrium separation, there will again be zero potential energy but finite (and maximum) kinetic energy. If the system is not further disturbed, these oscillations will continue indefinitely, the mutually staggered situations of $\mathscr{E}_{pot} = 0$ and $\mathscr{E}_{kin} = 0$ regularly alternating. In between these extremes, both potential energy and kinetic energy will of course be finite.

As can readily be imagined, the situation becomes far more complicated if the number of atoms is increased, and it is then best handled statistically. In an arbitrarily disturbed system, it will be highly unlikely that all atoms simultaneously have zero kinetic energy, and in any event the zero potential energy situation cannot prevail unless the number of participating atoms is very small. This is not to say that equilibrium cannot be attained, however, because a balance between repulsive and attractive forces can still be achieved even though some of the interatomic spacings differ from the equilibrium two-body value.

As we have just seen, the kinetic energy arises from the motions of the individual atoms, and the formal definition of this parameter for a system comprising N atoms is simply

$$\mathscr{E}_{kin} = \sum_{i=1}^{N} \frac{m}{2} v_i^2 \tag{4.4}$$

where v_i is the speed (not velocity) of the ith atom. As is shown in Appendix C, and assuming that what are known as Maxwell–Boltzmann statistics (named for James Clerk Maxwell and Ludwig Boltzmann) apply, there is a relatively simple relationship between these atomic velocities and the temperature. If the system is a three-dimensional one, which is to say that the positions of all the atoms require three different variables to describe them completely, the relationship between the velocities and the temperature is as given by the following equation

$$\mathscr{E}_{\text{kin}} = \left(\frac{3}{2}\right) Nk_B T \tag{4.5}$$

in which k_B is the Boltzmann constant. The latter has the value 1.38054×10^{-23} JK^{-1}. Because the above equation does not take quantum effects into account, it is said to define temperature in the classical limit. The factor $3/2$ in the final expression in this equation deserves some comment. Because the system is three-dimensional, we see that each degree of freedom has $k_B T/2$ of energy associated with it. When a situation of equipartition prevails, there will be equality between the potential energy and the kinetic energy. However, this can be the case only if the particles interact through a strictly harmonic potential (that is to say that the inter-particle interaction varies as the square of the separation distance).

4.3 Thermodynamics and Statistical Mechanics

In our initial example of just two atoms, the kinetic energy and the potential energy showed wild (and counter-phase) fluctuations between their maximum and zero values. As mentioned earlier, with increasing numbers of atoms in the system, these fluctuations would become less and less pronounced, because the individual motions would not be likely to be all in phase with one another. The system is then best described statistically, and for a sufficiently large number of particles, the temperature, for example, will not fluctuate to any appreciable degree. One of the conceptual difficulties encountered by those who study biological molecules is related to the limited number of atoms in such an entity. It might then appear to be the case that fluctuations in potential energy and kinetic energy really do occur. This ignores the fact that a biological molecule is not normally an isolated system, unless it happens to be floating freely in a vacuum (as in interstellar space, for example). Even then, energy can be received from and emitted to the surroundings, in the form of radiation. In general, the molecule will be in thermal contact with its surroundings, which might consist of other molecules, or water, or indeed a combination of both.

The product PV defines the external energy, and the sum of the internal and external energies is known as the enthalpy, H, which is thus defined by the equation

$$H = \mathscr{E}_{\text{int}} + \mathscr{E}_{\text{ext}} = \mathscr{E}_{\text{pot}} + \mathscr{E}_{\text{kin}} + PV \qquad (4.6)$$

One thus sees that the Gibbs free energy is obtained simply by replacing the internal energy in the Helmholz free energy by the enthalpy.

Now let us turn to the more difficult issue of entropy. This can be approached in two different ways, one being that due to Boltzmann, who was the first to appreciate its existence. Right from the outset let us correct a very widespread misconception regarding this elusive parameter. *It is a mistake simply to equate entropy to the degree of spatial disorder in a system.* On the contrary, as has become appreciated in recent years, the ordered and disordered states of a system, at the same density and temperature, can actually have essentially the same entropy (see Appendix C). The popular misconception regarding entropy arises from a failure to distinguish between the mere configuration of a system and its distribution of possible states, the latter including the states of motion. When Boltzmann carried out his pioneering studies on the subject, it was the latter attribute that he focused on.

One could say that concentrating solely on the instantaneous configuration of a system ignores the way in which this influences the motions of the individual atoms. The point is that the individual atoms are not simply free to move as they please. On the contrary, their motions will be constrained by the instantaneous atomic configuration of the entire assembly. In other words, entropy is determined by the fundamental cooperativity in the system, and probing this can be a very complicated matter indeed. That this is true is attested to by the fact that we still do not have a completely satisfactory description of the liquid state (see Appendix C). There are theories that treat liquids as if they were disordered solids, and there are other theories which look upon a liquid as a condensed gas. Neither of these approaches is really satisfactory, and there is now a growing appreciation of the fact that the liquid state is fundamentally different from either of those other two extremes (except above the critical point, of course).

Boltzmann's statistical treatment of entropy defined this by the relationship

$$S = k_B \ln \Omega \qquad (4.7)$$

in which Ω is related to the probability of a state, as measured by the number of ways in which it can be realized. It is important to appreciate the fact that the word state involves both the positions and the velocities of all the atoms in the system; as noted above, one cannot draw conclusions regarding the entropy merely from consideration of the instantaneous atomic arrangement.

The states referred to in the above definition of entropy occur in what is know as phase space, rather than simple configuration space, the difference being that phase space includes dimensions for velocity as well as position. The nature of entropy can be appreciated by considering various extreme cases, one of which was our two-atom system discussed earlier. It will be recalled that the regular oscillations of this system caused the positions and velocities to systematically sweep through a continuous set of values. However, there was a relationship between position and speed that was always rigidly adhered to; a given position always implied one given speed. In the other extreme of a many-atom gas, a given set of positions for all the atoms could conceivably be associated with many different instantaneous speeds of these particles, so the system would explore a much larger region of phase space. If one compares the situation in a crystal and in a glass, at the same density and the same low temperature, the individual atoms perform oscillations in local potential-energy wells that are approximate multi-dimensional parabolas, that is to say hyper-parabolas (not to be confused with hyperbolas, of course).

The entropies of those two states may thus be quite similar. However, as the temperature rises, there are co-operative effects in the crystal that force the individual atoms to move coherently with respect to one another, these motions being collectively described by the waves known as phonons. The disorder in the glassy state prevents cooperativity occurring to the same extent, and there is more randomness in the individual motions. With increasing temperature it is this, rather than mere disorder in the atomic positions, which gives the glass a higher entropy than the crystal, at the same density and temperature.

The existence of entropy became clear in the nineteenth century, even before it was confirmed that individual atoms actually exist. The need to invoke such an entity arose from considerations of the efficiencies of machines, and the pioneers of thermodynamics (particularly Lord Kelvin – born William Thomson – Sadi Carnot and Rudolf Clausius) realised that the departure from ideal efficiency is intimately linked to the existence of such a randomising factor. Indeed, that branch of science worked with a different definition of entropy, namely that

$$dS = dQ/T \qquad\qquad (4.8)$$

where dS is the increment of entropy and dQ is the amount of heat energy *absorbed* by a system, in what is known as a reversible fashion. The point to be made here is that if the heat absorbed by the system merely produces greater incoherence between the atomic motions, this energy input will not be recoverable as useful work, and the entropy of the system will have increased.

This brings us to what is known as the First Law of Thermodynamics. It states that if a system *absorbs* heat energy dQ, and if it has an amount of work dW

performed *on it*, the total *increase* in the internal energy will simply be the sum of these two components. That is to say

$$d\mathscr{E}_{\text{int}} = dQ + dW = dQ - PdV/T \qquad (4.9)$$

where the last term in the second form of the equation applies if the work done on the system is carried out at constant pressure. It is important to notice the negative sign in the second form of this equation. We see a familiar example of this fact every time we use a bicycle pump. Both the pump and the tyre become warmer as more and more air is pumped into the tyre.

Having made a brief contact with thermodynamics, let us return to the statistical nature of systems, which is the domain of that branch of science known as statistical mechanics. We have already seen that we must think in terms of phase space rather than simple configuration space, so the little diagrams shown in Figure 4.1 are not quite as simple as they might at first appear to be. They show a container that is divided into two parts by a barrier that has a small hole in it. The left-hand diagram shows the rather unlikely situation in which all the points representing the position and velocities of the individual atoms, in an N-atom assembly, are in the same half of the container. The right-hand picture, by contrast, shows a more probable situation in which half of the points are in one part of the container and the other half are in the other part. This is a combinatorial problem of the type that asks how many different ways n items can be chosen from amongst a total of N items. The difference is that in the present case these 'items' are actually states in phase space. The principles are nevertheless the same, since we may interpret N as being the number of elementary regions of phase space which can be occupied at any one time by the states in the system. The number of ways, Ω, in which the

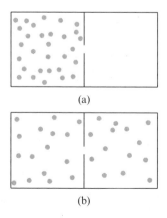

(a)

(b)

Figure 4.1   These two different situations correspond to different degrees of filling the available phase space

N items can be distributed in two groups comprising, respectively, n_1 and n_2 states is given by the simple combinatorial expression

$$\Omega = \frac{N!}{n_1! \times n_2!} \tag{4.10}$$

where

$$N! = N(N-1)(N-2) \cdots \times 2 \times 1 \tag{4.11}$$

Irrespective of the initial state of the system, the passage of a sufficiently long time will produce an equilibrium condition in which, apart from small fluctuations, there will be equally many molecules in the two halves of phase space, as shown on the right-hand side of Figure 4.1.

It is a relatively simple matter to demonstrate that the right-hand side of the above equation is greatest if n_1 and n_2 are equal, whereas it continually decreases as the difference between n_1 and n_2 increases. Because the most probable situation will indeed correspond to approximate equality between n_1 and n_2, we see that the tendency is for the entropy of the system, that is to say $k_B \ln \Omega$, to increase. In other words, there will always be a tendency for the states of a system to occupy as much of phase space as they are permitted.

Armed with this physical background, we can proceed to develop some useful relationships between the various parameters. For a start, let us imagine that a system of the above type is described by a set of states $1, 2, 3, \ldots, i, \ldots$ which have corresponding energies $\varepsilon_1, \varepsilon_2, \varepsilon_3, \ldots, \varepsilon_i, \ldots$ Furthermore, let the number of elements in state 1 be n_1, in state 2 be n_2, and so on. The total number of elements will simply be

$$N = \sum_i n_i \tag{4.12}$$

Moreover, generalising what was stated earlier, the number of different ways in which the N indistinguishable elements can be distributed is given by

$$\Omega = \frac{N!}{n_1! \times n_2! \times n_3! \times \cdots \times n_i! \times \cdots} \tag{4.13}$$

We now wish to enquire as to which values of the different n_i will give the maximum entropy. Remembering that both 0! and 1! equal unity, the maximum value of entropy will be obtained when all the n_i equal unity or zero. The entropy will then simply be $k_B \ln N!$

Another relationship that follows automatically concerns the total energy, $\mathscr{E}_{\text{total}}$. This is simply given by

$$\mathscr{E}_{total} = \sum_i n_i \varepsilon_i \tag{4.14}$$

Moreover, as is shown in Appendix C, we obtain the interesting result that

$$n_i = N \times \exp\left(-\varepsilon_i/k_B T\right)/\sum_i \exp\left(-\varepsilon_i/k_B T\right) \tag{4.15}$$

and thus that the total energy is given by

$$\mathscr{E}_{total} = N \times \sum_i \varepsilon_i \exp\left(-\varepsilon_i/k_B T\right)/\sum_i \exp\left(-\varepsilon_i/k_B T\right) \tag{4.16}$$

The denominator in the above two equations is a frequently occurring quantity in statistical mechanics, and it has been given the name partition function, which is usually denoted by the letter Z. That is

$$Z = \sum_i \exp\left(-\varepsilon_i/k_B T\right) \tag{4.17}$$

A formula that can be used to calculate the number of elements in the various energy levels is known as the Boltzmann distribution. It gives the relationship between those different elements (that is to say, states). Let us imagine that n_i and n_j are the number of elements with energies ε_i and ε_j respectively. Using the above analysis, the desired proportionality is simply given by

$$\frac{n_i}{n_j} = \exp\left(-(\varepsilon_i - \varepsilon_j)/k_B T\right) \tag{4.18}$$

The Boltzmann distribution naturally applies only to systems that are in thermal equilibrium.

It is useful to rewrite a number of the expressions that appeared earlier, using the partition function. We thus have

$$S = Nk_B \ln Z + Nk_B T(\delta \ln Z/\delta T)V = Nk_B \ln Z + \mathscr{E}/T \tag{4.19}$$

because

$$\mathscr{E} = \frac{Nk_B T^2}{Z} \cdot \frac{\delta Z}{\delta T} \tag{4.20}$$

we get

$$F = -Nk_B T \ln Z \tag{4.21}$$

One of the best-known applications of statistical mechanics is the derivation of an expression which gives the probability of a particle being able to surmount an energy barrier of a given height. Indeed, it would be no exaggeration to call this the classic example in this branch of science. Let us imagine that the particle is in an energy well, and let its current energy be $\mathscr{E}_1$. Furthermore, let the top of the energy barrier correspond to the energy $\mathscr{E}_2$. The height of the energy barrier will therefore be $\Delta\mathscr{E} = \mathscr{E}_2 - \mathscr{E}_1$. We assume that the energy varies continuously up the energy barrier, so that we can replace the discrete summation normally found in the partition function with a continuous integral. The probability that the particle has an energy in the interval $[\mathscr{E}, \mathscr{E} + d\mathscr{E}]$ is, using an expression derived earlier,

$$f(\mathscr{E})d\mathscr{E} = \frac{\exp(-\mathscr{E}/k_BT)d\mathscr{E}}{\int_0^\infty \exp(-\mathscr{E}/k_BT)d\mathscr{E}} \tag{4.22}$$

Likewise the probability that the particle has an energy lying between $\mathscr{E}_1$ and infinitely high energy is

$$f(\mathscr{E} > \mathscr{E}_1) = \int_0^\infty f(\mathscr{E})d\mathscr{E} \frac{\int_{E_1}^\infty \exp(-\mathscr{E}/k_BT)d\mathscr{E}}{\int_0^\infty \exp(-\mathscr{E}/k_BT)d\mathscr{E}} = \exp(-\mathscr{E}_1/k_BT) \tag{4.23}$$

and similarly

$$f(\mathscr{E} > \mathscr{E}_2) = \exp(-\mathscr{E}_2/k_BT) \tag{4.24}$$

We now note that of all the particles which have energy greater than $\mathscr{E}_1$ a fraction of them have an energy greater than the larger energy $\mathscr{E}_2$. This fraction is simply given by

$$\frac{\exp(-\mathscr{E}_2/k_BT)}{\exp(-\mathscr{E}_1/k_BT)} = \exp(-\Delta\mathscr{E}/k_BT) \tag{4.25}$$

This is the desired result because it simply gives the probability that a particle will be able to surmount the energy barrier. The right-hand side of this equation is commonly referred to as the Boltzmann factor, and the reader would be well advised to commit it to memory. Its chief consequences are also worth remembering, namely that the probability of surmounting the energy barrier increases as the height of the barrier decreases, and also as the temperature increases.

As an example of the application of the entropy concept, we can use a familiar little demonstration with a rubber band. The reader will find that the demonstration is easier if a flat band is employed. As was noted earlier, in our discussion of interatomic bonds, normal rubber is the name given to the polymer *cis*-polyisoprene, in which the individual chains are kinked and inter-woven in the relaxed form of the substance. There are numerous different ways in which the disorder can be achieved, and they have associated with them a relatively large volume of available phase space. This configuration of the polymer chains is thus characterized by high entropy. If the rubber band is stretched out to its full extent, all the polymer chains will lie roughly parallel to one another and their kinks will have been removed. The access of the constitu-ent atoms to phase space will thereby be greatly reduced, and so the entropy will fall.

If the stretching process is performed rapidly, so that no heat can flow out of the system, the free energy will rise, and this will manifest itself through increased heat content. The result will be a rising temperature, as can easily be detected if the rubber band is pressed against one's moistened lips. If the temperature of the band is now equilibrated with the lips, while maintained in its fully-stretched condition, a subsequent release of the tension will cause the opposite sequence to occur, and the band will suddenly feel cooler. (If the reader performs this part of the experiment, care should be taken to avoid pinching the lips as the band is released.)

In the second half of this little experiment, we have demonstrated that the entropy rises again, and this causes a lowering of the free energy. A common example of the use of the Boltzmann factor is encountered in the analysis of particle diffusion in condensed matter. As we saw when considering the motions of a pair of atoms, temperature manifests itself through atomic motion, and if the atoms are constrained to move in a local potential energy well, these motions will take the form of reasonably regular vibrations.

It will invariably be the case that there are escape routes from these energy wells, the routes involving energy barriers of the type that were considered above. A common problem thus arises as to what the escape rate will be, given that the atomic vibration frequency is known. Since the probability of a particle surmounting the energy barrier is simply given by the Boltzmann factor, we can easily write an expression for the number of successful attempts to surmount the barrier in a time t, given a vibration frequency of ν_{atom}. The number of successful jumps, n_{jumps}, is simply given by

$$n_{\mathrm{jumps}} = \nu_{\mathrm{atom}} t \exp\left(-\Delta\mathscr{E}/k_B T\right) \qquad (4.26)$$

where $\Delta\mathscr{E}$ is the height of the energy barrier. This type of expression is frequently encountered in analyses of atomic motions in crystals, and although the energy barriers in biological molecules are not as well defined, the analysis could certainly be extended to those cases as well. It should be noted, however, that there are forms of condensed matter in which the energy-barrier concept might not be particular useful. In a liquid, for example, although some analyses have been couched in terms of similar energy barriers, there is evidence that this is *not* what happens (see Appendix C). We touched on the issue earlier, when noting that a liquid probably does not resemble a highly defective solid, as some have conjectured.

We know the order of magnitude of several characteristic energies, so it should be relatively straightforward to imagine how molecules are influenced by various conditions. We earlier saw that the energy per degree of freedom is given by $\mathscr{E} = k_B T/2$. Because each particle will have six such degrees of freedom, in a three-dimensional system, namely three for kinetic energy and three for potential energy, the total energy per particle will be $3k_B T$. Now the temperature in the human body is about 310 K, so each atom will have energy of the order of 10^{-2} aJ at its disposal.

Table 4.1 shows the latent heats of melting and evaporation for a number of different substances, of various bonding classes. The transition energies naturally reflect the differences in the underlying interatomic interactions. Looking at the entries in this table, and given the energy per atom which we have just calculated, it is not surprising that some organic substances will melt when held in the palm of one's hand. Conversely, one would not expect a hydrogen bond to break at body temperature, even though the margin of safety is not all that high. As can be seen from Figure 4.2, that margin is only about a factor of three at body temperature. This is a serious consideration, given that the two strands in each of our DNA molecules are held together by the hydrogen bonds that bind the pyrimidine and purine portions in each base pair (see Chapter 7). Different types of process in many-atom systems are naturally associated with different characteristic energies. In general, the breaking of interatomic bonds requires more energy than the mere redistribution of these. The stronger types of bond obviously require more energy expenditure if they are to be ruptured.

Table 4.1   The transition energies naturally reflect the differences in the underlying interatomic interactions

	Element or compound	Latent heat of melting (aJ)	Latent heat of evaporation (aJ)
Argon	(Ar)	0.002	0.011
Chlorine	(Cl)	0.011	0.034
Germanium	(Ge)	0.053	0.557
Gold	(Au)	0.021	0.541
Iron	(Fe)	0.026	0.585
Mercury	(Hg)	0.004	0.099
Sodium	(Na)	0.004	0.148
Uranium	(U)	0.026	0.704
Water	(H_2O)	0.010	0.079
Salt	(NaCl)	0.050	1.275
Benzene	(C_6H_6)	0.016	0.058

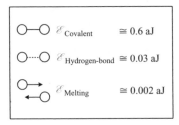

Figure 4.2 Comparative energy levels

If the bond-breaking is to be accomplished thermally, the stronger bonds will need warming to higher temperatures for a given rate of bond-breaking.

In fact, the situation is more complicated because many atoms can act co-operatively to break a bond that would be beyond the disruptive capability of a single moving atom. This is seen in enzymes, in which the energy of many atoms is concentrated at a particular place. A fuller discussion of this important process must be left until Chapter 7, however, when we discuss protein structure. One might think that covalent bonds, which are much stronger than hydrogen bonds, would not be vulnerable to breaking in this manner, but the work of Alexander Fleming on bacteria suggested that this can indeed happen. One day, when he was suffering from a cold, a few drops of nasal mucus fell from his nose onto a culture of bacteria. Ever inquisitive, he decided not to throw the culture away, and a few days later he was surprised to find that the bacteria had been killed in the very area where the mucus had fallen. These bacteria were protected by an outer shell, which comprised many covalent bonds. Something in the mucus had apparently been capable of breaking these bonds, and the enzyme in question is now known as lysozyme (the structure of which was ultimately determined by David Phillips and his colleagues). Later on, we will encounter enzymes that are similarly capable of breaking the covalent bonds in the backbones of DNA molecules.

4.4 Reaction Kinetics

The rates at which chemical reactions proceed are governed by the principles discussed above. However, in order to establish contact with what can be observed experimentally, we need to discuss a number of formal definitions. The reaction

$$A + B \rightarrow C \tag{4.27}$$

is said to be mono-molecular, while the reverse reaction

$$C \rightarrow A + B \tag{4.28}$$

is said to be bi-molecular. The concentration of a given reactant has tradition-ally been indicated by the relevant symbol contained within square brackets, so the rate of consumption of species A in Equation (4.27) is given by

$$v_A = -\frac{d[A]}{dt} \tag{4.29}$$

while the rate of formation of species C is given by

$$v_C = \frac{d[C]}{dt} \tag{4.30}$$

It is also useful to define the reaction fluxes, J, and this introduces the concept of the rate constant, customarily denoted by the letter k. If we write

$$A + B \xrightarrow{k_1} C \tag{4.31}$$

and

$$C \xrightarrow{k_2} A + B \tag{4.32}$$

we then have

$$J_C = k_2[C] \tag{4.33}$$

and

$$J_A = J_B = k_1[A][B] \tag{4.34}$$

It is worth noting that the rate constant is the inverse of a time constant, τ. The order of a reaction, n, is defined by

$$J = k[C]^n \tag{4.35}$$

So we see that Equation (4.33) describes a first-order reaction whereas the reaction in Equation (4.34) is second-order.

Let us now make contact with our thermodynamic foundations, and enquire as to the influence of temperature on the reaction rate. A convenient starting point is to define the chemical potential, which for a given component (sub-stance) i is given by

$$\mu_i = \mu_i^0 + RT \ln a_i \tag{4.36}$$

where R is the gas constant ($= 8.314 \ \mathrm{J\,K^{-1}\,mol^{-1}}$) and μ_i^0 is the chemical potential of component i in the standard state. The chemical activity a_i is related to the concentration through the corresponding coefficient of activity, γ_i, by

$$a_i = \gamma_i[i] \tag{4.37}$$

These preliminaries completed, we can now consider the general reaction

$$\nu_A A + \nu_B B \rightleftharpoons \nu_C C + \nu_D D \tag{4.38}$$

in which the ν_i are the stoichiometric coefficients. If the changes take place under constant pressure and constant temperature, and the changes are thus limited to variations in the quantities of the reactants, the change in the Gibbs free energy will be given by

$$\Delta G = \nu_C \mu_C + \nu_D \mu_D - \nu_A \mu_A - \nu_B \mu_B \tag{4.39}$$

Using Equation (4.36), this can be rewritten as

$$\Delta G = \Delta G^0 + RT(\nu_C \ln a_C + \nu_D \ln a_D - \nu_A \ln a_A - \nu_B \ln a_B) \tag{4.40}$$

in which the changes have been referred to the molar standard reaction free energy ΔG^0, and where

$$\Delta G^0 = \nu_C \mu_C^0 + \nu_D \mu_D^0 - \nu_A \mu_A^0 - \nu_B \mu_B^0 \tag{4.41}$$

Gathering the logarithmic terms in Equation (4.40), we obtain van't Hoff's equation (first derived by Jacobus van't Hoff), namely

$$\Delta G = \Delta G^0 + RT \ln \left(\frac{a_C^{\nu_C} a_D^{\nu_D}}{a_A^{\nu_A} a_B^{\nu_B}} \right) \tag{4.42}$$

If thermodynamic equilibrium prevails, the change in the Gibbs free energy will be zero, and Equation (4.42) can then be rewritten with the help of Equation (4.37) (putting all the coefficients of activity to unity, which is justifiable if the concentrations are small). This produces an expression for the important quantity known as the equilibrium constant, K_{eq}, namely

$$K_{eq} = \frac{[C]_{eq}^{\nu_C}[D]_{eq}^{\nu_D}}{[A]_{eq}^{\nu_A}[B]_{eq}^{\nu_B}} = e^{-\frac{\Delta G^0}{RT}} = A e^{-\frac{\Delta \mathscr{E}_{int}^0}{RT}} \tag{4.43}$$

The final step in this multiple equation moves the entropy term in the Gibbs free energy – see Equation (4.2) – into a pre-exponential frequency factor. This permits us to obtain a relationship between the equilibrium constant and the relevant activation energy for the reaction. The final term in Equation (4.43) is indeed reminiscent of Equation (4.25). Finally, in view of Equation (4.38), we note that

$$K_{eq} = \frac{\text{(Rate constant for forward reaction)}}{\text{(Rate constant for reverse reaction)}} \tag{4.44}$$

We have arrived at our goal because, as Svante Arrhenius first pointed out, the rate constants tend to vary exponentially with temperature. Indeed, in view of the earlier developments in this chapter, we are not surprised to find a Boltzmann factor appearing in Equation (4.43). (It is true that the exponent in Equation (4.43) involves the gas constant, rather than Boltzmann's constant, but these are simply related by $R = N_A k_B$, where N_A is Avogadro's number, which equals $6.022 \times 10^{23} \, \text{mol}^{-1}$.)

Let us put these ideas into practice by considering the modification of a substrate S to produce a product P, in the mediating presence of an enzyme E. (We recall that a catalyst – of which an enzyme is an example – helps to accelerate a chemical reaction but is not itself modified by that reaction.) We have

$$E + S \underset{k_1'}{\overset{k_1}{\rightleftharpoons}} ES \overset{k_2}{\longrightarrow} E + P. \tag{4.45}$$

We have again assumed that the concentrations are so low that the reverse of the second reaction can be ignored. From the above analysis, it is clear that we may write

$$[ES] = \frac{k_1[E][S]}{k_1' + k_2} = \frac{[E][S]}{K_M} \tag{4.46}$$

K_M is known as the Michaelis constant, in honour of Leonor Michaelis, who carried out this classic analysis with Maud Menten. Reactions compatible with Equation (4.46) are said to display Michaelis–Menten kinetics.

4.5   Water, Acids, Bases and Aqueous Reactions

Water is the most common substance on earth, and it is one of only two liquids which occur naturally in appreciable quantities, the other being petroleum. About 60% of the weight of the human body is water, it being present in the interior of every cell. It also accounts for the bulk of such specialized media as blood and mucous. It is an almost universal solvent, and when it provides the medium for acid–base reactions, it is not just a bystander but an active participant.

The structure of the H_2O molecule gives water its special properties. The O—H bond length is $= 0.095718$ nm and the H—O—H angle is $104.52°$. The former number is simply explained because it is essentially the sum of the Pauling radii for hydrogen (0.030 nm) and oxygen (0.066 nm). But the angle might seem more problematic. It arises from something we encountered in Chapter 2, namely hybridization. The electronic configuration of oxygen is $1s^2 2s^2 2p^4$, with the full complement of two electrons in the $2p_x$ orbital, but only one electron in each of the $2p_y$ and $2p_z$ orbitals. As first suggested by Linus Pauling, these latter become hybridized with the two $2s$ electrons to produce an electron probability distribution having four lobes which point roughly toward the corners of a slightly distorted tetrahedron, the oxygen nucleus lying at its centre. (The corresponding angle in a perfect tetrahedron is $109.5°$.) Two of these lobes combine with the respective $1s$ orbitals of the two hydrogen atoms, to produce the O—H bonds, the hydrogen atoms becoming positive electrical poles because of the consequent incomplete screening of their proton nuclei. The remaining two lobes do not quite reach to the remaining two corners of the distorted tetrahedron (see Figure 2.8). They are negatively charged, as would be expected from the overall neutrality of the molecule. The net charge on each of the four poles is approximately 20% of that of an electron. It is the existence of these poles which gives the water molecule its electrical dipole moment, the magnitude of which can be determined by a method devised by Peter Debye. The dielectric constant of water, at $20°C$, is approximately 80 times that of the vacuum value.

The near-tetrahedral structure of the water molecule provides the basis for the most common crystal structure of ice, first determined by William Bragg in 1922. John Bernal and Ralph Fowler noted the similarity between many of the physical properties of liquid water and ice, and concluded that the H_2O molecules in the latter are essentially intact. The continuing lack of a satisfactory picture of water's liquid state is attributable to the fact that liquids in general are still imperfectly understood (see Appendix C). But the Bernal–Fowler picture of a liquid state in which hydrogen bonds are perpetually being formed and broken has continued to enjoy support. Our knowledge of the relevant energies endorses this concept. The hydrogen bond has an energy of about 0.03 aJ whereas each rotational and translational degree of freedom of a water molecule can easily be shown to be associated with an energy, at body temperature, about a third of that value. The Boltzmann factor for the rupture of hydrogen bonds will thus be favourably high.

If this was the complete story, water would be less interesting than it is. The vital additional factor is the spontaneous break up of water molecules into hydrogen ions (H^+) and hydroxyl ions (OH^-). The rate at which this dissociation occurs has been found to be 2.5×10^{-5} per s, which means that each molecule in liquid water breaks up on average once in about 11 hours. At any instant, there are approximately 2.5×10^{16} hydrogen ions in 1 l of neutral water, and of course the same number of hydroxyl ions. The number of non-dissociated water molecules in 1 l is approximately 3×10^{25}, so the hydrogen ions are well spaced; about 800 water molecules could lie along a line running from one ion to its nearest neighbour. The same is true of hydroxyl ions, so the average spacing between adjacent hydrogen and hydroxyl ions is about 400 water molecules.

As a further complication, there is good evidence that both types of ion are strongly associated with neutral water molecules, the hydrogen ions forming hydronium (H_3O^+); but the ions are known also to be rather mobile. A hydrogen ion, for example, stays associated with a given water molecule for only about 1 ps before moving on. This means that a given water molecule will catch a hydrogen ion once every 0.5 ms or so, and hold on to it for about 1 ps. The mobility of the hydrogen and hydroxyl ions is considerably higher than, for example, Na^+ and K^+ ions because the latter attract some of the water molecules in their vicinity, and are thus forced to drag this little retinue along with them as they move. This occurs because of the relatively high surface-charge density on these ions, and the resulting hydration shell comprises between five and 10 water molecules. (Sodium has the larger hydration shell because it is the smaller ion of the two, and thus possesses the higher surface-charge density.)

Measurement of the very short lifetimes of ions in water is an impressive achievement. The breaking up of any molecule involves the overcoming of interatomic forces, of course, so there is an energy barrier to be surmounted. In the processes we have just been considering, the source of the required energy is the thermal energy of the atoms themselves, and the rate will thus be increased by a rise in temperature. However, because the establishment of equilibrium in a reaction involves movement of the participants, it cannot be established immediately. Inevitably, a relaxation time is involved. (One sees an analogous lag, on a much longer time scale, in the relatively slow accommodation of the sea's temperature to the changing seasons.) In order to probe such relaxation times, and thus obtain a measure of reaction rates, it is necessary to change the temperature within a period that is short compared with the relaxation times. In flash photolysis, developed by George Porter and Ronald Norrish, and independently by Manfred Eigen, the rapid changes were provoked by discharging an electrical condenser through the solution being investigated.

Let us now bring in the formalism used in earlier sections of this chapter. The overall reaction of interest is

$$H_2O \rightleftharpoons H_3O^+ + OH^- \qquad (4.47)$$

so the equilibrium constant will be given by

$$K_{eq} = \frac{[H_3O^+][OH^-]}{[H_2O]} \qquad (4.48)$$

Given the small concentrations mentioned above, it is clear that the concentration of neutral water is changed only imperceptibly by changes in the concentrations of the various ions. So Equation (4.48) can be replaced by what is known as the ion product for water, namely

$$K_{H_2O} = [H_3O^+][OH^-] \qquad (4.49)$$

The value of this parameter, which is a measure of the rate at which hydrogen and hydroxyl ions collide with one another, is 1.0×10^{-14} at 25°C. Under neutral conditions, the concentration of either ion is thus 10^{-7} moles, but in an aqueous solution there can be other sources of hydrogen and hydroxyl ions. However, the presence of these other ions will not affect the rate at which water molecules dissociate. So if the concentration $[H_3O^+]$ rises to 10^{-2}, for example, the concentration $[OH^-]$ must fall to 10^{-12}.

A convenient shorthand way of denoting hydrogen-ion concentration was suggested by Søren Sørensen in 1909, and it uses the now familiar symbol pH. It is simply the negative power of ten in the hydrogen-ion concentration. So we have, formally,

$$pH = -\log_{10}[H_3O^+] \qquad (4.50)$$

Thus a pH of 1 corresponds to a concentration of 10^{-1}, while a concentration of 10^{-14} gives a pH value of 14. The ammonia used in smelling salts has a pH of about 12, while orange juice and vinegar have approximate values of 4.5 and 3 respectively. The acid in a fully charged car battery has a pH of roughly 1, a value that is closely approached by the fluid in the human stomach.

The examples just given refer to what are commonly known as acids and alkalis, of course, so let us turn to the formal definitions associated with those concepts. According to a suggestion by Svante Arrhenius in 1887, an acid is a chemical compound that dissociates in solution to produce hydrogen ions, while an alkali, formally referred to as a base, is a compound that yields hydroxyl ions. As we saw above, the pH of neutral water is 7, and the presence of an acid increases the concentration of hydrogen ions and thus lowers the pH value. As dictated by Equation (4.49), it will simultaneously decrease the concentration of hydroxyl ions. Conversely, a base increases the pH of an aqueous solution.

In 1922, Nicolaus Brønsted and Thomas Lowry independently broadened the definitions of acid and base by suggesting that the former is any compound that can lose a proton while a base is any compound that can acquire a proton. These days, one would call them proton donors and proton acceptors, respectively. Finally, in the following year, Gilbert Lewis defined an acid as a substance that can accept a share in a pair of electrons held by a base, so as to form a covalent bond. (We may note the equivalence, in electrical terms, of losing a positive charge and acquiring a share in a negative charge.) This is reminiscent of one meaning of the word oxidation: the process of removal of an electron from a compound. The other meaning of that term is the obvious one of adding an oxygen atom so as to produce an oxide. So an oxidizing agent is one that tends to accept electrons, and the substance from which it acquires them will thereby become an acid. Indeed, the name oxygen means literally acid generating.

The importance of acids and bases in the biological domain could hardly be exaggerated. The rearrangement of molecular fragments in chemical reactions occurring in an aqueous solution is determined by the relative affinities they have for each other and for the water molecule. The greater the latter affinity, the stronger will be the tendency for the compound to dissociate and interact with water, and the stronger will be the resulting acid or base. Acids are sour and they turn the organic dye litmus from blue to red. A sufficiently strong acid will react with a metal and release hydrogen. Bases have a bitter taste and they turn litmus from red to blue. They feel soapy because of their interaction with the oily substances in the skin. Finally, we note that an acid neutralizes a base to produce a salt and water.

Let us put these ideas on a formal basis. We denote an acid by the symbol HA, to emphasize its propensity for losing a hydrogen atom. Ionization of the acid is described by

$$HA \rightleftharpoons H^+ + A^- \tag{4.51}$$

and the equilibrium constant K is given by

$$K = \frac{[H^+][A^-]}{[HA]} \tag{4.52}$$

Rearranging this, noting that $[H_3O^+]$ and $[H^+]$ are equivalent, and finally taking base-10 logarithms throughout, we arrive at the expression

$$\log_{10} \frac{1}{[H_3O^+]} = \log_{10} \frac{1}{K} + \log_{10} \frac{[A^-]}{[HA]} \tag{4.53}$$

The left-hand term is, of course, the pH, and by analogy the first term on the right-hand side is abbreviated to pK. So we have

$$pH = pK + \log_{10} \frac{[A^-]}{[HA]} \qquad (4.54)$$

This is known as the Henderson–Hasselbalch equation (in honour of Lawrence Henderson and Karl Hasselbalch). It provides a measure of the pH for a given ratio of acid to base in a solution. And from Equation (4.54) we see that the pK of an acid is the value of pH at which the acid is dissociated to a degree of 50%. Plotting the value of pH as a function of the ratio of acid to base produces what is known as a titration curve; it has a characteristic sigmoidal shape, centered on the point pH $=$ pK, where there is an inflexion point. The substance formed by ionization of an acid is referred to as the acid's conjugate base. The corollary is that protonation of a base produces that base's conjugate acid.

An example of a conjugate acid–base pair, common in the biological realm, is acetic acid and the acetate ion. Such a pair has the valuable property that it resists changes of pH in a solution, which is to say that it can act as a buffer. The pK value for acetic acid is 4.8, and it can readily be shown that a two-fold change in the acid–base ratio would produce a mere 0.5 change in the pH. One of the most important classes of buffer in biological systems is in fact the proteins. In the blood, for example, the buffering work-horses are haemoglobin and serum albumin.

In the previous chapter we briefly discussed the transient excursions made by vibrating atoms when they move in each other's force fields, the movements being provoked by the absorption of thermal energy. When the distance of separation between two atoms is greater than the equilibrium distance, the two atoms will move toward each other if they are free to do so. Although such vibrations absorb thermal energy, therefore, the energy is not permanently stored. Oxidation–reduction reactions, also known as redox reactions, provide a system with a means of longer-term energy storage, and they are thus of paramount importance in the energy acquisition and transportation processes in biology. In photosynthesis, for example, the absorption of the energy from light photons produces a separation of electrical charges. However, if these charges were able to rejoin each other immediately, no benefit would accrue. It is the job of various molecules to ensure that the instantaneously separated charges are prevented from recombining, by permitting attachment of those charges and by moving to regions relatively remote from the site of generation.

A track can be kept of the relevant change in free energy, ΔG^0, by noting that

$$\Delta G^0 = -nFV^0 \qquad (4.55)$$

In this expression, n is the number of transferred electrons, F is the Faraday constant (named in honour of Michael Faraday) and V^0 is the standard redox potential. The latter is a measure of the ability of a redox pair of molecules to pass electrons between them. Whence the alternative term: electron transfer potential. Just as free energy is a relative concept, requiring one to specify the standard value (see Equation (4.41) above, for example) so does the redox potential. The standard in this case is the hydrogen electrode, which is set to 0V for zero pH. (The volt was, of course, named in honour of Alessandro Volta.) Substances with negative redox potentials have lower affinities for electrons than does H_2, and they are reducing agents. Substances with positive redox potentials have higher affinities for electrons than does H_2, and they are oxidizing agents. We will return to redox reactions when considering biological energy in Chapter 9.

Let us close this long water-oriented section with a brief look at another factor of great biological significance, namely hydrophobicity. It is another topic that has been the victim of a certain amount of misunderstanding. Substances of biological importance can be broadly divided into two classes according to whether they do or do not readily mix with water. The latter are said to be hydrophobic while the former are hydrophilic. There has been an understandable tendency to think in anthropomorphic terms, and to say that hydrophobic substances 'do not like water'. This is rather misleading. It might, for example, lead one to believe that a water molecule would find it difficult to penetrate that paradigm of a hydrophobic structure – the interior of a biological membrane. This is not the case, however, because a water molecule can pass through such a membrane with relative ease.

The misunderstanding arises because of a lack of appreciation of how the hydrophobic property actually arises. It is primarily an effect related to the entropy of water. As we saw at the start of this section, the water molecule has two positive poles and two negative poles. In the liquid state, therefore, one molecule can easily be oriented so as to match the charge region presented by one of its neighbouring molecules with an appropriately opposite charge of its own. The molecules are in perpetual movement, of course, so they are constantly having to rotate in order to optimize this charge-neutralizing effect. If a neutral molecule of another species, or merely a fragment of such a molecule, intrudes into a region of liquid water, the water molecules' ability to play their instantaneous charge-neutralizing role will be diminished because the number of effective arrangements will be decreased. This will manifest itself in a decrease in the entropy of the water, and consequently in a rise in free energy. The upshot is an effective repulsion. So it is not so much a case of the hydrophobic substance not 'liking' water as water not 'liking' the hydrophobic substance. We will return to such considerations when discussing biological membranes in Chapter 8. After discussing numerous effects that are related to thermal energy, in one way or another, let us close this chapter by considering cases in which the provoking agency lies closer to the quantum regime.

4.6   Radiation Energy

The molecules of our bodies are constantly being subjected to radiation. This comes in a variety of wavelengths, some of which correspond to radiation that is quite harmless, for example the radiation we call cosmic rays. Radiation at X–ray wavelengths is useful as a diagnostic tool, as long as it is employed in moderation. Similarly, we need a certain amount of ultraviolet radiation, which amongst other things produces vitamin D in the appropriate layer of the skin, but here also too much radiation is harmful. The radiation which is most directly useful to the human being is that which has a wavelength falling in the visible region, because this provides the basis for our sense of vision.

The reason cosmic radiation is not dangerous lies in the fact that there are no energy transitions in the molecules of our bodies which correspond to the energy range of the typical cosmic ray. Our bodies could thus be said to be transparent to that radiation. The same is true of radiation in the radio bands, which simply passes through our bodies, and also through the walls of our houses. Infra-red radiation has energy which lies in a band that corresponds to the thermal motions of our constituent molecules, so it is not surprising that such radiation can lead to heating effects in the body.

Because we have already considered thermal phenomena, we need not discuss further the effects of infra-red radiation. We have not yet considered changes in the electronic states of atoms and molecules, however, so it would be useful to delve deeper into what occurs when radiation of an appropriate wavelength impinges on an atom. In order to make any progress, we will first have to consider some of the fundamental facts that have been revealed concerning the behaviour of such particles. A good starting-point is the work of Louis de Broglie, who is cited in Appendix A. It was he who hit upon the remarkable idea that such a particle is simultaneously a particle and a wave. These two attributes are linked in de Broglie's fundamental equation

$$\lambda = \frac{h}{mv} = \frac{h}{p} \tag{4.56}$$

where λ is the wavelength and h is Planck's constant ($= 6.63 \times 10^{-34}$ Js), m is the particle's mass and v its speed, and finally p is the particle's momentum. Planck's constant itself was defined through his beautifully simple equation which expressed the quantization of energy. That equation is

$$\mathscr{E} = h\nu \tag{4.57}$$

where ν is the wave frequency (which can be loosely equated with the frequency of vibration of the equivalent particle), while $\mathscr{E}$ is the energy.

The third and final fundamental relation which we will need in this analysis embodies a principle first discovered by Werner Heisenberg. It expresses the fact that there is a fundamental limitation on the accuracy with which one can simultaneously determine the position and momentum of a particle. That relationship is

$$\Delta r \times \Delta p \geq \frac{h}{2\pi} \tag{4.58}$$

where r is the position, in respect to a suitable origin of co-ordinates. We should note various things about this third relationship. For a start, it involves an inequality rather than an equivalence, this expressing the fact that we are now dealing with a limitation on the size of two parameters, rather than an identity. We also note that this expression, just like the preceding two, involves Planck's constant; this indicates that here too we are dealing with something which emerges from the quantization that makes itself so strongly felt at the atomic level. Finally, we note that the expression merely says something about the product of the positional and momentum uncertainties, rather than their individual magnitudes. Thus one can increase the precision with which position is determined, but this will automatically decrease the precision with which one can measure the momentum. Conversely, one can make one's measurements of momentum more precise, but only at the expense of tolerating greater uncertainty with respect to the particle's position.

We are now going to discover that the above three principles, taken in conjunction, permit us to make an estimate of the approximate size of the hydrogen atom (and of its first ionization energy). This, the simplest of all atomic species, consists of a single electron orbiting around the positive nucleus, which is a single proton. Because these two particles bear electrical charges (both having the magnitude of the electronic charge q_e), their interaction potential will be of the Coulomb type, which we encountered in Chapter 3. We can thus immediately write the potential energy of the electron, with respect to the nucleus, as follows

$$\mathscr{E}(r) = \frac{1}{4\pi\epsilon_0} \frac{q_e^2}{r} \tag{4.59}$$

where $q_e = 1.602 \times 10^{-19}$ Coulombs. Our goal is to discover which value of the distance, r, gives a stable orbit. The total energy of the electron is, of course, the above potential energy plus the electron's kinetic energy, the latter being $mv^2/2$.

Rewriting this latter term so as to bring in the symbol p, we can write the total energy in the following form

$$\mathscr{E} = \frac{1}{4\pi\epsilon_0} \frac{q_e^2}{r} + \frac{p^2}{2m} \tag{4.60}$$

where m is the electron's mass ($m = 9.11 \times 10^{-31}$ kg).

Because we are attempting to find the minimum possible radius for the hydrogen atom, we can use the equality sign in Heisenberg's uncertainty principle (Equation 4.58) and thus rewrite the above equation in the form

$$\mathscr{E} = \frac{1}{4\pi\epsilon_0}\frac{q_e^2}{r} + \frac{h^2}{8\pi^2 mr^2} \tag{4.61}$$

which must now be minimized with respect to r. When this has been done, and when $\delta\mathscr{E}/\delta r$ has been equated to zero, we obtain

$$r = \frac{h^2 \epsilon_0}{\pi m q_e^2} = 0.053 \text{ nm} \tag{4.62}$$

which approximately corresponds to

$$\mathscr{E} \cong -2 \text{ aJ} \tag{4.63}$$

It is interesting to note that this energy is of the same order as (but even larger than) that of the typical covalent bond. We can, furthermore, use this result to find the frequency (and the wavelength) of a photon which is just able to eject an electron from a hydrogen atom. This is given by

$$h\nu \geq 2 \times 10^{-18} \text{ J} \tag{4.64}$$

and thus

$$\nu \geq 3 \times 10^{15} \text{s}^{-1} \tag{4.65}$$

This frequency is in that part of the ultraviolet spectrum that is quite close to the visible region (i.e. it is in the near ultraviolet), which makes one realize how thankful we have to be for the presence of the atmosphere's ozone layer.

Exercises

4.1   A typical hydrogen bond has been measured to have an energy of approximately 0.03×10^{-18} J. Find the largest permissible wavelength of incident electromagnetic radiation, if this is to be used to rupture a hydrogen bond.

4.2   Find the ratio of the rates of thermal rupture of hydrogen bonds at the two temperatures 40°C and 5°C.

4.3   The emergence of HIV/AIDS has caused great concern amongst the commercial
      suppliers of blood, especially because it has been found that the heat treatments
      used by the various firms are not all totally effective in killing the virus. Three of
      the major companies in the field are USA's 'Armour' (here called 'A'), the UK's
      'Blood Products Laboratory' ('B') and the USA's 'Cutter Biological' ('C'). Their
      respective heat treatments are A: 60°C for 30 h, B: 80°C for 72 h, and C: 68°C for
      84 h. Treatment B is the most effective against the virus, and the efficiencies of the
      other two, relative to B, are given by 56:10000 for A:B and 931:10000 for C:B.
      Assuming that each virus is destroyed if just one particularly susceptible intera-
      tomic bond is broken, what is the energy of this vulnerable bond?

4.4   A full quantum mechanical treatment of the hydrogen atom produces an electron
      probability density distribution which has its maximum at a radius which is in
      impressively good agreement with the lowest-energy Bohr orbit for that element,
      namely 0.053 nm. Moreover, it produces a value for the first ionization energy for
      hydrogen which is in good agreement with the experimental result, namely 2 aJ.
      Use these data to calculate the minimum wavelength of electromagnetic radiation
      which would produce such ionization of a hydrogen atom.

Further reading

Atkins, P. W., (1990). *Physical Chemistry*. Oxford University Press, Oxford.
Eisenberg, D. and Kauzmann, W., (1969). *The Structure and Properties of Water*. Clarendon
      Press, Oxford.
Flory, P. J., (1969). *Statistical Mechanics of Chain Molecules*. Interscience, New York.
Landau, L. D. and Lifshitz, E. M., (1969). *Statistical Physics*. Pergamon, Oxford.
Lindsay, R. B., (1968). *Introduction to Physical Statistics*. Dover, New York.
Mayer, J. E. and Goeppert-Mayer, M., (1940). *Statistical Mechanics*. John Wiley and Sons,
      New York.
Moore, W. J., (1972). *Physical Chemistry*. Longman, Engelwood Cliffs, NJ.
Pauling, L., (1970). *General Chemistry*. Freeman, San Francisco.
Pippard, A. B., (1964). *Elements of Classical Thermodynamics*. Cambridge University Press,
      Cambridge.
Ter Haar, D., (1956). *Elements of Statistical Mechanics*. Rinehart, New York.
Yvon, J., (1969). *Correlations and Entropy in Classical Statistical Mechanics*. Pergamon,
      Oxford.
Zemansky, M. W., (1951). *Heat and Thermodynamics*. McGraw-Hill, New York.

5 Transport Processes

Movement is one of the defining characteristics of the biological world. It is seen in one form or another in all living things. The movements associated with animals are of various types, and they range from the relatively subtle movements of their various internal organs to the more obvious locomotion of entire organisms. These large-scale examples have their counterparts in the smaller movements seen in single-celled creatures, and indeed in the individual cells of multi-celled animals. Movement is slower in plants, but it is indispensable if the plant is to succeed in its competition with other plants, and win its place in the sun.

At a still more subtle level, movement is of paramount importance at the molecular level, for without it the substances vital to an organism's survival would not find their way to the locations at which they are needed. The more gross aspects of such flow are handled by the vascular systems of both animals and plants, while diffusion mediates the necessary transport at the molecular level.

The fundamental phenomena that underlie these motions ultimately depend upon the collective behaviour of molecules in liquids. It is thus disappointing that there is still no universally accepted theory of the liquid state. This problem is addressed in Appendix C, where the fundamental aspects of molecular motion are described. However, despite this lack of knowledge, one can account for the overall behaviour of liquids through the formal treatment of diffusion and viscosity by appealing to the continuum limit. This is the task of the present chapter. Some of the effects to be discussed here are clearly influenced by the presence of an electric field. However, consideration of that complication will be deferred until Chapter 11, with its discussion of the basis of cellular excitability.

5.1 Diffusion

Although diffusion is clearly caused by the motion of particles through space, for our biological purposes here it can be regarded as the mixing of particles amongst one another. The phenomenon was investigated first by the English botanist Robert Brown, who in 1828 observed pollen particles in a glass of water, with the help of a microscope, and noticed their irregular and apparently erratic movements. These movements are now known as Brownian motion.

In 1905, the year when he also produced his epoch-making papers on special relativity and the photoelectric effect, Albert Einstein published a quantitative explanation of this motion. He showed that it is caused by the impacts, on the pollen particles, of the even smaller (and thus *sub*-microscopic) water molecules. In the Einstein analysis, the water molecules would be more mobile than the pollen particles, because of their smaller size, and this is why it is valid to think in terms of the pollen particles being hit by them, and not the reverse. The analysis is not limited to such situations of size disparity; it is equally applicable when the various types of particle have roughly the same size. Moreover, there is not a restriction to the liquid domain first studied by Brown. The Einstein principles are just as valid for gases and solids (though there are some additional correlation factors in the case of the latter – we need not go into them here).

If we regard the movements of a specific particle, we discover that we can divide its trajectory into small straight paths; the particle appears to jump from point to point, in an essentially ballistic manner. If each jump is independent of the particle's prior history, the movement process is said to be stochastic, and the movements themselves are said to describe a random walk. (The totality of the stochastic movements is known as a Markov chain – in honour of Andrei Markov.) The classic result pertaining to the random walk (for steps of unit length), in the absence of any directional bias, is that the mean distance, $\bar{r}$, travelled in any number of steps, m, equals zero, whereas the root mean squared distance, r_{rms} equals $\sqrt{m}$.

A particularly useful first illustration is provided by diffusion via what is called the vacancy mechanism in a crystal, even though this has limited relevance in the biological domain. The atomic arrangement in the two-dimensional hexagonal close-packed crystal shown in Figure 5.1(a) gives each atom six nearest-neighbours, these being located precisely at the minimum of the interatomic potential (r_0 in Figure 5.1(b)) if all interactions other than those between nearest neighbours can be ignored. Diffusion in this system occurs through the movement of the vacancy (and on the larger scale, by many vacancies). Each diffusion event involves the movement into the vacancy of one of the six atoms that surround it, the vacancy simultaneously moving in the opposite direction.

It is easy to see that the mid-point of each of these elementary movements brings the moving atom closer than r_0 to two other atoms, the latter forming a sort of window which the moving atom has to pass through. And because the distance in question is less than r_0, it is equally clear that the moving particle will be exploring the repulsive region of the interatomic potential (see Figure 5.1(b)). The movement thus involves passage over an energy barrier. In liquids,

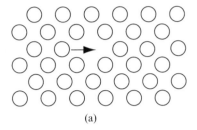

(a)

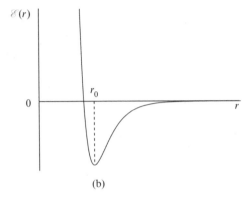

(b)

Figure 5.1 Diffusion via the vacancy mechanism in a crystal

and in biological situations in general, the energy barriers are of a more extended type. Indeed, in many important situations, it is not even certain that there *are* interatomic energy barriers of the type that occur in the above example (see Appendix C). There will, on the other hand, usually be barriers of *free* energy.

The parameter that quantitatively characterizes diffusion is the diffusion coefficient, D, which is defined as the net flow of particles per unit time across an imaginary plane of unit area lying at right angles to the concentration gradient, that gradient also having unit strength. For a stochastic process, it can be shown that the root mean square distance r_{rms} travelled, in time t, by a diffusing particle is given by

$$r_{rms} = (Dt)^{1/2} \tag{5.1}$$

Let us see what size of diffusion distance Equation (5.1) predicts for an easily describable situation. Imagine that we liberate a small amount of radioactive gas somewhere in a large room, and place a Geiger counter 10 m from the point of release. We ask when the counter will be observed to maximally register the radioactivity. Table 5.1 lists the typical values of D encountered in the solid, liquid and gaseous states (and it should be carefully noted that the units of D traditionally use *centimetres* rather than metres). We find that this maximal situation will not have been reached until 4 months have passed! This might appear astonishing, until we realize that the effects of convection have not been taken into account. From Table 5.1, we also find that the distances diffused in the liquid and solid states, during the same time, would be 10 cm and 10^{-1} cm, respectively.

We will note in Chapter 8 that small membrane-bounded vesicles deliver neurotransmitter molecules to the pre-synaptic membranes of nerve axons. If the diffusion of these vesicles from their place of origin in the somatic region of the nerve cell were a three-dimensional random walk, their delivery to the pre-synaptic membrane would be unacceptably slow. In order to speed up the process, the system fabricates protein microtubules, which lie along the direction of the axis of the axon, and which guide the passage of the vesicles (see Chapters 10 and 11). This essentially one-dimensional diffusion thus efficiently marshals the vesicles to the place where they are needed. The situations with and without the microtubules are indicated in Figure 5.2.

We now turn to the equations that quantitatively describe diffusion, and begin by considering the situation in which there is a concentration gradient only in the direction of the long axis of a cylinder having a cross-section of unit area (see Figure 5.3). Let us regard this cylinder as being divided into imaginary slabs of thickness b, this being equal to the length of a single diffusive jump. (In a gas, that length would be equal to the mean free path between collisions – see Appendix C.) The number, n, of particles in a given slab will thus be Cb, where C is the concentration. If f is the frequency with which the particles make their jumps, the flux J of particles across the imaginary plane that divides adjacent slabs, in which the concentrations are C_1 and C_2, will be given by

Table 5.1 Typical values of the diffusion coefficient D

state	$D\ [\mathrm{cm}^2\mathrm{s}^{-1}]$
solid	10^{-9}
liquid	10^{-5}
gas	10^{-1}

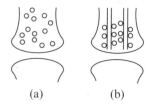

(a)   (b)

Figure 5.2 Transport of vesicles to the pre-synaptic membrane with (b) and without (a) the provision of microtubules

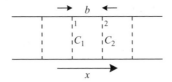

Figure 5.3   Diffusion is caused by non-uniformity in the concentration of the diffusing species

$$J = \frac{1}{2}(n_1 - n_2)f = \frac{1}{2}bf(C_1 - C_2), \quad C_1 < C_2 \tag{5.2}$$

C_1 is less than C_2, and the flux is the number of particles passing the imaginary plane in unit time. The factor $1/2$ is required because it is assumed that there is equal probability of a given particle jumping to the left or to the right, at a given instant. Noting that $(C_1 - C_2) = -(\delta C/\delta x)b$, we find that

$$J = -D(\delta C/\delta x) \tag{5.3}$$

with

$$D = -\frac{1}{2}b^2 f \tag{5.4}$$

Equations (5.3) and (5.4) are collectively known as Fick's First Law (after Adolf Fick).

In general, one would also like to know how concentration varies as a function of time, when things are not in their steady state. To accomplish this, we again imagine two of the above elemental slabs, but this time we let them be separated by a distance l which is large compared with the length of a diffusion jump, yet nevertheless small enough to permit us to take things to the limit, in the manner of the calculus. That is to say l is so small that the associated change in concentration is infinitesimal.

If the concentration at the first slab is C, the concentration at the second slab will be $C + (\delta C/\delta x)l$. The flux of particles leaving the first slab in the direction of the second slab will be

$$J = -D(\delta C/\delta x) \tag{5.5}$$

The flux of particles leaving the second slab in the *same* direction, that is to say *in the direction away from the first slab*, will be

$$J + l\left(\frac{dJ}{dx}\right) = -D\left(\frac{dC}{dx}\right) + l\frac{d}{dx}\left(-D\frac{dC}{dx}\right) \tag{5.6}$$

By defining the directions in this manner, we have made it easy to find the rate of change of the concentration of particles in the portion, of length l, of our imaginary tube of unit cross-section, because those particles that have entered the portion at the one end and have not also left it at the other end will still be in the portion. In other words, the desired net flux into the portion will be given by the difference between the right-hand parts of Equations (5.5) and (5.6), namely

$$-\left(\frac{\delta J}{\delta x}\right)l = \frac{\delta\left(D\frac{\delta C}{\delta x}\right)}{\delta x}l \qquad (5.7)$$

This quantity will, by definition, be identical with $(\delta C/\delta t)l$, so we find that

$$\frac{\delta C}{\delta t} = \frac{\delta\left(D\frac{\delta C}{\delta x}\right)}{\delta x} \qquad (5.8)$$

If, further, it is assumed that D does not vary with concentration (which is usually a safe assumption), this simplifies to

$$\frac{\delta C}{\delta t} = D\frac{\delta^2 C}{\delta x^2} \qquad (5.9)$$

Equations (5.8) and (5.9) express Fick's Second Law. Solution of Equation (5.8), or (5.9) (whichever applies), using the relevant boundary conditions is often mathematically difficult, and entire books have been devoted to the derivation of solutions for particular geometries.

Let us close this section by considering the classic situation shown in Figure 5.4, which is also one of the simplest commonly encountered. The initial situation (i.e. at $t = 0$) is one in which the concentration at all values of x less than zero is C_0, while at all x equal to or greater than zero it is zero. Solution of Fick's Second Law (see Appendix E), in the version (5.9), for these boundary conditions, produces

$$C(x, t) = \frac{1}{2}C_0\left(1 - \frac{2}{\pi^{1/2}}\int_0^{x/2(Dt)^{1/2}} e^{-y^2}dy\right) \qquad (5.10)$$

This gives the concentration as a function of (positional) distance and time. The second term within the parentheses is known as the Gauss error function (after Karl Gauss), which is readily available in tabulated form, the parameter y

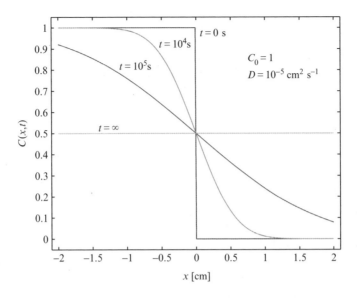

Figure 5.4 Because of diffusion, the initial confinement of the particles of a diffusing species to one region of space gradually leads to a situation in which the species uniformly fills all the available space

being a dummy variable, that is to say one not specifically related to the situation under consideration.

It is interesting to note the appearance of our expression (5.1) in the denominator of the upper limit of the integral in (5.10). For very long times, this denominator will cause that upper limit to tend to zero, and the integral itself will similarly tend to zero. This indicates that the infinite-time condition will be one in which the concentration will be $C_0/2$, uniformly throughout. By looking up the values of the error function in the above-mentioned tabulations, one can easily verify that the intermediate-time solutions, for various (positive and negative) values of x are as indicated qualitatively in Figure 5.4.

As mentioned at the start of this chapter, we will defer until Chapter 11 a proper analysis of the additional complications that arise when some of the diffusing species bear electrical charges, and when an electric field is present. So suffice it to note that an appropriately directed field will oppose the diffusive drift of charged ions. One can think in terms of such a field pushing in the opposite direction to the concentration gradient shown in Figure 5.4. It is also clear that when the applied field is sufficiently strong, the diffusive drift can be reduced to zero, despite the continued existence of the concentration gradient. This is the situation at the bounding membrane of an excitable cell – except that the electrical field is produced by the ions themselves.

Let us now turn to another type of transport, namely that caused by mechanical stress.

5.2 Viscosity

When a solid is subjected to a sufficiently small shear stress, σ, a shear strain γ will be produced that is linearly related to the stress, the constant of proportionality being the shear modulus, μ. We thus have

$$\gamma = \frac{\sigma}{\mu} \qquad (5.11)$$

While in this regime, the solid is said to obey Hooke's Law (first propounded by Robert Hooke). Release of the stress permits the solid to regain its original shape. If the solid is strained beyond the elastic limit, however, the shape is no longer restored exactly when the stress is removed. It is found, instead, to have acquired a permanent deformation. When this happens, the solid is said to have undergone plastic flow and, while the stress is still applied, the rate of plastic flow is found to be proportional to the stress.

The defining characteristic of a liquid is that it is unable to withstand a shear stress. It starts to flow immediately the stress is applied. One could say that a liquid has an elastic limit of zero. Isaac Newton derived a relationship between the shear stress applied to a liquid and the resulting temporal rate of change of strain, $d\gamma/dt$. This is

$$\frac{d\gamma}{dt} = \alpha\sigma = \frac{\sigma}{\eta} \qquad (5.12)$$

These equations imply a reciprocal relationship between the fluidity, α, and the dynamical viscosity, η. It is often useful to normalize the viscosity in terms of the density, to produce what is known as the kinematic viscosity, thus

$$\nu = \frac{\eta}{\rho} \qquad (5.13)$$

The viscosity of water at $20°C$ is $1.0 \times 10^{-3} \, N \, s \, m^{-2}$, and it falls to about a third of this value at the boiling point.

James Clerk Maxwell accounted for the difference in the rate of flow of solids and liquids by thinking in terms of a stress relaxation time constant, τ. If a substance is subjected to instantaneous elastic strain γ_0 and then maintained at that strain level, the stress will instantly rise, to the level σ_0 say. Thereafter, the substance will relax, as the Hookean regime is gradually replaced by the Newtonian regime. It will thus be the case that

$$\frac{d\gamma_{\text{Hookean}}}{dt} = -\frac{d\gamma_{\text{Newtonian}}}{dt} \qquad (5.14)$$

Using the relationships listed above, this implies that

$$\frac{d\sigma/dt}{\sigma} = -\frac{\mu}{\eta} \tag{5.15}$$

and this leads to

$$\sigma = \sigma_0 e^{-t/\tau} \tag{5.16}$$

in which

$$\tau = \frac{\eta}{\mu} \tag{5.17}$$

If, alternatively, the stress level is held constant, we will have

$$\frac{d\gamma_{\text{Newtonian}}}{dt} = \frac{\sigma}{\eta} = \frac{\sigma}{\mu\tau} \tag{5.18}$$

In other words, the relaxation time constant can be regarded as determining the time interval over which the elastic (Hookean) regime is replaced by the viscous (Newtonian) regime. Bulk liquid water is clearly in the latter regime, but water in the immediate vicinity of a protein molecule or a biological membrane, for example, is more confined in its movements. It will not have become a solid, of course, but some have likened such bound water to a softened glass.

The analysis so far has treated substances in the continuum limit, which is adequate for the macroscopic domain. It has made no reference to the molecular level. In that microscopic realm, we cannot ignore the stochastic buffeting from its neighbours to which each molecule is subjected. We have also to take into consideration other factors, such as the viscous drag on a moving molecule and the restoring force that tends to draw it back to its original position when it is displaced. The situation was analysed by Paul Langevin in 1908, and he arrived at the relationship

$$m\frac{d^2x(t)}{dt^2} + \gamma_{\text{Drag}}\frac{dx(t)}{dt} + \kappa x(t) = \mathcal{F}_{\text{Stochastic}}(t) \tag{5.19}$$

In this Langevin equation, m is the particle's mass, γ_{Drag} is the drag coefficient and κ is usually referred to as the spring constant (see Chapters 3, 6 and 10). It is clear that the latter is related to the shear modulus referred to above.

When Einstein successfully accounted for Brownian motion, as discussed in the previous section of this chapter, he provided a link between the microscopic and macroscopic aspects of diffusion. This triumph is embodied in the Einstein relation, which links the diffusion coefficient to the drag coefficient thus

$$D = \frac{k_B T}{\gamma_{\text{Drag}}}$$ (5.20)

The numerator on the right-hand side is, of course, familiar from the extensive discussions of thermal effects in the previous chapter (see also Appendix C).

A full treatment of liquid flow is complicated by the need to take into account such factors as the pressure P and the presence of what are referred to as body forces, an example being the influence of gravity. The complete situation was first analysed by Claude Navier and George Stokes, and the Navier–Stokes equation reads

$$\frac{\partial v}{\partial t} + (v \cdot \nabla)v = -\frac{\nabla P}{\rho} + \nu\nabla^2 v + \mathcal{F}_{\text{body}}$$ (5.21)

The velocity v is of course a vector, and the two vector operators are given by

$$\nabla v_i = \frac{\partial v_i}{\partial x}x_1 + \frac{\partial v_i}{\partial y}y_1 + \frac{\partial v_i}{\partial z}z_1$$ (5.22)

(x_1, y_1 and z_1 are unit vectors) and

$$\nabla^2 v_i = \frac{\partial^2 v_i}{\partial x^2} + \frac{\partial^2 v_i}{\partial y^2} + \frac{\partial^2 v_i}{\partial z^2}$$ (5.23)

The Navier–Stokes equation can be considerably simplified for specific situations. The first term on the left-hand side vanishes, for example, under conditions of constant flow. Moreover, the second term can be shown to be negligible for favourable ratios of object size (that is to say the relevant dimension of a moving body or a static conduit) to velocity of flow. Let us enquire as to these conditions. The term $(v \cdot \nabla)v$ is of order v^2/r, where r is the characteristic length (of the object or conduit). The term $\nu\nabla^2 v$, on the other hand, is of order $\nu v/r^2$. So we have the ratio

$$\frac{(v \cdot \nabla)v}{\nu\nabla^2 v} = \frac{\text{order}(v^2/r)}{\text{order}(\nu v/r^2)} = \text{order}\left(\frac{vr}{\nu}\right)$$ (5.24)

The quotient in the final parentheses was first invoked by Osborne Reynolds, and it has come to be known as the Reynolds number, R

$$R = \frac{vr}{\nu} \tag{5.25}$$

The Reynolds number gives a measure of the relative importance of the inertial and viscous forces. Table 5.2 shows the Reynolds number for various objects over a huge range of size, and common experience tells us that the inertial factor completely dominates in the case of the freighter. (Indeed, many an accident could be avoided if the captains of small boats better appreciated how long it takes a freighter to come to a stop.) At the other extreme, the Reynolds number for a bacterium is some 10^{14} times smaller than that of the freighter, and its motion is dominated by viscosity. When we return to the swimming bacterium in Chapter 10, we will note the consequences of its low Reynolds number. Although we will not discuss that topic here, we should note the importance of the Reynolds number to flow though narrow conduits, such as the capillary blood vessels in the animal vascular system.

When the Reynolds number is small, the drag force is linearly related to the speed. George Stokes derived an expression for the drag force on a moving sphere of radius r, for these conditions. The Stokes Law states that

$$\gamma_{\text{drag}} = 6\pi\eta r \tag{5.26}$$

and, as just noted, this will be valid only if the viscous forces dominate over the inertial forces. If we combine Stokes Law with the Einstein relation, Equation (5.20), we obtain the important Stokes–Einstein formula

$$\eta = \frac{k_B T}{6\pi r D} \tag{5.27}$$

This equation shows what relatively large effect thermal motion will have on a moving particle, and it can be used to determine the degree to which such particles will be suspended when in aqueous solution (that is to say that it is useful in calculations concerning sedimentation). It is also applicable to the issue of centrifugation of course. Let us now turn to a third example of a transport phenomenon.

Table 5.2   The Reynolds numbers for various objects moving through water (density $10^3 \, \text{kg m}^{-3}$ and dynamical viscosity $10^{-3} \, \text{Nsm}^{-2}$)

Object	Length	Speed	Reynolds Number
Freighter	$150\,\text{m}$	$30\,\text{ms}^{-1}$	5×10^9
Human	$2\,\text{m}$	$1\,\text{ms}^{-1}$	2×10^6
Bacterium	$2\,\mu\text{m}$	$25\,\mu\text{ms}^{-1}$	2×10^{-5}

5.3 Thermal Conduction

In common with other types of material, biological substances conduct heat. It is true that constancy of temperature normally prevails at the microscopic level within organisms, but thermal conduction is important at the macroscopic level of the tissues of entire organisms. A vigorously used muscle is found to heat up, for example, as was first studied by Archibald Hill. It is appropriate, here, just to touch on a few of the more salient factors.

The law of heat conduction states that the rate of heat transfer across a given area is proportional to the temperature gradient. The relationship is

$$\frac{dQ}{dt} = -k_{\text{thermal}} A \frac{dT}{dx} \tag{5.28}$$

the generalization to three dimensions being straightforward. Here, A is the area and Q is the heat energy, so the units of the left-hand side are W, that is to say J s^{-1} (those units being named after James Watt and James Joule, of course). k_{thermal} is the coefficient of thermal conductivity.

In the cause of making this section uniform with the previous sections of this chapter, we should ask how these macroscopic parameters are related to the microscopic domain. The most direct way of accomplishing this is to note that the coefficient of thermal conductivity will be given by

$$k_{\text{thermal}} = \frac{\rho C_{V,1} c_{\text{mean}} \ell_{\text{mfp}}}{3} \tag{5.29}$$

In this expression, $C_{v,1}$ is the specific heat of unit mass at constant volume, c_{mean} is the mean speed of the energy-transporting particles and ℓ_{mfp} is the mean free path between their mutual collisions. Although we will not go into the details, it may be noted that the energy per degree of freedom will be the defining ingredient in the specific heat. Finally, to make contact with both of the previous sections of this chapter, we should note the two relationships

$$k_{\text{thermal}} = \alpha \eta C_{V,1} \tag{5.30}$$

The numerical factor α is found to lie between 1 and 3 for gases, and

$$D \simeq \frac{c_{\text{mean}} \ell_{\text{mfp}}}{3} \tag{5.31}$$

The difficulty of extending the argument to the more relevant domain of liquids is made complicated by the present lack of an adequate theory of the liquid state, as is discussed in Appendix C.

Exercises

5.1 After they have influenced the receptors in the post-synaptic membrane, those neurotransmitter molecules which are not broken down by enzymes diffuse back to the pre-synaptic membrane, which lies 2×10^{-8} m away, with a diffusion coefficient which has been determined to be $5 \times 10^{-6} \, cm^2 \, s^{-1}$. How long does it take the molecules to make this short journey?

5.2 The wing cycling frequency of a humming bird, during flight, is approximately 260 Hz (that is to say, around middle C on the musical scale). Assuming that the diffusion coefficient for Ca^{2+} in water (and in the intracellular-fluid) is $10^{-5} \, cm^2 \, s^{-1}$, calculate the radius of such a bird's typical muscle cell if these ions are able to reach the very centre of the cell.

5.3 A honeybee of length 10^{-2} m flies at a speed of $10^{-1} \, m \, s^{-1}$ in air. What is its Reynolds number, given that the density and dynamical viscosity of air are, respectively, $1.3 \, kg \, m^{-3}$ and $1.8 \times 10^{-5} \, N \, s \, m^{-2}$?

5.4 What is the Reynolds number for blood flowing through an artery 10^{-3} m in diameter, assuming that the density and dynamical viscosity of blood do not differ appreciably from the corresponding values of water (see Table 5.2)? Assume that the speed of blood flow is roughly $0.1 \, ms^{-1}$.

Further reading

Carslaw, H. S. and Jaeger, J. C., (1959). *Conduction of Heat in Solids* Clarendon Press, Oxford.

Cottrell, A. H., (1964). *The Mechanical Properties of Matter* Wiley, New York.

Glaser, R., (2001). *Biophysics* Springer, Berlin.

Howard, J., (2001). *Mechanics of Motor Proteins and the Cytoskeleton* Sinauer, Sunderland, MA.

Landau, L. D. and Lifshitz, E. M., (1959). *Fluid Mechanics* Pergamon, Oxford.

Nossal, R. and Lecar, H., (1991). *Molecular and Cell Biophysics* Addison-Wesley, Redwood City, CA.

Serway, R. A., (1992). *Physics for Scientists and Engineers* Saunders, Philadelphia.

6 Some Techniques and Methods

Numerous experimental and theoretical approaches are now available for exploring the structure and function of biological molecules, and the cellular components for which they provide the foundation. There is not room here to provide an exhaustive coverage of these techniques and methods, but we should take a closer look at a few representative examples. They all emerged during the 20th century, but in different decades, by far the oldest being that of structure determination by X-ray diffraction. Let us begin with that technique.

6.1 X-Ray Diffraction and Molecular Structure

An individual biological molecule typically comprises hundreds or even thousands of atoms, and it is naturally important to know how these are arranged with respect to one another, for it is this arrangement that dictates the molecule's physical behaviour under a given set of conditions. The most straightforward technique for determining the relative positions of the various atoms in a molecule is to consolidate a large number of such molecules into a crystal, that is to say an orderly array in which the individual molecules have identical surroundings (except those at the crystal surface, of course). It will then be the case that the equivalent atoms in the various molecules will themselves be arranged in a crystalline arrangement having the same symmetry as the overall pattern of molecules. The positions of the individual atoms in each and every molecule can then be determined by causing the crystal to diffract X-rays, and by making measurements on the resulting diffraction pattern. However, before considering the details, we should familiarize ourselves with some of the underlying crystallographic principles.

The painstaking cataloguing of the shapes of crystals of numerous chemical compounds, by the early crystallographers, revealed that there are just seven fundamentally different symmetry systems, by which three-dimensional space can be divided up into regular units. These crystal systems range from the cubic system, in which all the angles of the elementary unit are right angles and all the side lengths are equal, to the triclinic system, in which all the angles and all the side lengths are mutually different. Intermediate between these extremes are the tetragonal system, the orthorhombic system, the rhombohedral system, the hexagonal system and the monoclinic system. In 1848, Auguste Bravais established that there are just 14 different ways of arranging identical items so as to fill three-dimensional space in an orderly manner. These are now known as the Bravais lattices, and they are a subdivision of the above-mentioned seven crystal systems (see Table 6.1). In two dimensions, the number of different possibilities drops to five, and this imposes a fundamental limitation on the available symmetries in wall-paper, for example.

However, because the pattern of lines (and colours) within the elementary unit of a wall-paper design can be arbitrarily complicated, the resulting design possibilities are unlimited in their scope for variety. Similarly, in a three-dimensional crystal, the arrangements of atoms within the unit cell can be arbitrarily complex (though it is naturally constrained by the underlying interatomic forces), so here too there are essentially unbounded possibilities. The arrangement of atoms within the unit cell is referred to as the basis, so the crystal structure can be concisely described by the relationship:

$$Crystal\ Structure = Lattice + Basis$$

Table 6.1   The seven crystal systems and fourteen space lattices

System	Defining characteristics	Space lattices	Examples
Cubic	Three axes at right angles, all equal in length.	Simple	Cesium chloride
		Body-centered	Sodium
		Face-centered	Copper
Hexagonal	Two equal axes subtend 120° angle, each at right angles to third axis of different length.	Simple	Zinc
Tetragonal	Three axes at right angles, two of equal length.	Simple	Barium titanium oxide
		Body-centered	Indium
Trigonal (rhombohedral)	Three equally inclined axes, not at right angles, all equal in length.	Simple	Calcite
Orthorhombic	Three axes at right angles, all of different lengths.	Simple	Lithium formate monohydrate
		Base-centered	Uranium
		Body-centered	Sodium nitrite
		Face-centered	Sodium sulfate
Monoclinic	Three axes, one pair not at right angles, all of different lengths.	Simple	Lithium sulfate
		Base-centered	Tin fluoride
Triclinic	Three axes, all at different angles, none of which is a right angle, all of different lengths.	Simple	Potassium dichromate

A three-dimensional lattice is defined by its three fundamental translation vectors a, b and c, and these are said to be primitive when the vector position r_n of any point in the crystal is given by

$$r_n = n_1 a + n_2 b + n_3 c \qquad (6.1)$$

n_1, n_2 and n_3 all being integers. In this case, the unit cell is known as a primitive cell and the pattern generated has a lattice point at every corner, and at no other places. For all unit cells, primitive or otherwise, the unit cell is a parallel-epiped whose edges are the fundamental translation vectors. That parallelopiped's volume is clearly $a \cdot (b \times c)$.

X-ray diffraction is carried out in order to determine the basis, the prior determination of the Bravais lattice now being regarded as a routine and straight-forward matter. It is important to bear in mind that X-rays are primarily diffracted from electrons, rather than from atomic nuclei, so the determination of the basis actually involves determination of the distribution of the electrons throughout the primitive cell. It was Lawrence Bragg who first exploited the possibilities for structure determination inherent in X-ray diffraction, in 1913, and it was he who realized that one could think in terms of the X-ray waves being diffracted from parallel atomic planes.

We will derive Bragg's condition for strong diffraction in a more general fashion, but before addressing that task let us consider in a more general manner the distribution of potential scattering centres in a crystal. The planes invoked by Bragg are not confined to those of the above-mentioned parallelopiped of the unit cell. On the contrary, there is an infinite set of planes lying in different directions, the density of points in a plane of given direction varying with the angle by which that plane is skewed with respect to planes defining the unit cell. It is clearly convenient to have a short-hand way of referring to a plane's direction, and the standard prescription was first used by William Miller. It is related to the concept of direction cosines, familiar from three-dimensional coordinate geometry.

A specific set of parallel planes is unambiguously defined by the number of parts into which it divides the three axes lying along the directions of the fundamental translation vectors. The Miller indices, h, k and l, are directly related to the respective numbers of parts, and they are found by converting those numbers to the smallest set of integers, by multiplication or division throughout. When this is done, it is found that the integers can have both positive and negative values, and the negative examples are denoted by a bar above the number, thus $\bar{1}$. A specific direction is indicated by a set of three numbers enclosed within brackets, examples being [112], [10$\bar{1}$] and [315]. It is clear that certain sets of these indices describe equivalent directions; they are referred to collectively through the use of angled brackets. Thus $< 110 >$ refers collectively to [110], [1$\bar{1}$0], [$\bar{1}$10], [101], [10$\bar{1}$] and so on.

It is also convenient to have a short-hand way of referring to the crystallographic planes, and the general and specific references employ braces and parentheses, respectively. Thus {110} refers collectively to the planes (110), (1$\bar{1}$0), ($\bar{1}$10), (101), 0$\bar{1}$1) and so on. In the cubic system, but not generally otherwise, the direction [hkl] is normal (i.e. perpendicular) to the plane (hkl). One great advantage of the Miller-index system is that it readily permits one to determine the spacing between adjacent planes, for we will see that this spacing has a direct bearing on the observed X-ray diffraction pattern. Using the above-mentioned direction cosines, it is easy to show that the inter-planar spacing d_{hkl} is given by

$$d_{hkl} = \left(\frac{h^2}{a^2} + \frac{k^2}{b^2} + \frac{l^2}{c^2} \right)^{-1/2} \tag{6.2}$$

As mentioned earlier, the usefulness of X-ray diffraction to the study of biological molecules lies in its ability to yield information about the arrangement of atoms within the molecule. And because the diffraction is primarily from the constituent electrons, a diffraction pattern is interpreted in terms of the electron density distribution that gives rise to it. Imagine two points i and j within the basis, that is to say within the molecule. The electron density at these two positions will in general be different, and we would like to have information about the relative densities. If the line joining i and j has the direction [hkl], the set of planes (hkl) will lie parallel to a set of planes containing i and equivalent points in other unit cells, and the same will be true for j and its equivalent points in other unit cells.

The question then remains as to how the relative electron densities can be described. This is most conveniently accomplished by imagining the density ρ at a given point as being composed of Fourier components. We recall that the most general form of Fourier series (named for Joseph Fourier) involves a sum of weighted exponentials of positive and negative multiples of an angle ϕ. We have

$$\rho_\phi = \sum_{n=-\infty}^{\infty} K_n e^{in\phi} \tag{6.3}$$

The K_n are the Fourier coefficients, which are in general complex numbers. The efficacy of employing the Fourier representation will emerge shortly, but meanwhile let us turn to the issue of X-ray scattering by electrons.

Joseph (J. J.) Thomson, who discovered the electron, was the first to derive an expression for the amplitude A of an electromagnetic wave (of which an X-ray is an example) scattered from an electron. This is found to be a function of both the distance r and the angle χ between the incident and scattering direction. The expression is

$$A = A_0 \frac{q_e^2}{m_e c^2 r} \sin \chi \qquad (6.4)$$

A_0 is the amplitude of the incident wave, q_e and m_e are the charge and mass of the electron, respectively, and c is the speed of light. An atom will in general contain a number of electrons, so it will give rise to a scattering amplitude that is greater by a factor f than that from a single electron. This factor is known by the equivalent names atomic scattering factor and form factor, and we will use the latter, to avoid confusion with the scattering factor that we will encounter later. If $\rho_e(r)$ is the electron density at the position defined by the vector r, we will have

$$f = \int \rho_e(r) e^{-i\phi} dV \qquad (6.5)$$

The integration must be made over the entire volume of the atom, of course, and ϕ is the phase difference relative to the origin of the position vector, which may conveniently be taken to be the atom's nucleus. (It is useful to bear in mind that the phase difference can be found by multiplying the path difference by the factor $2\pi/\lambda$, λ being the wavelength.)

As a final preliminary, we recall from Appendix A that we may describe an X-ray wave by its time- and position-dependent wave function. We thus have

$$\Psi(r, t) = A e^{-i(\omega t - k \cdot r)} \qquad (6.6)$$

The vector k is known as the wave vector. Returning to Equation (6.5), the phase difference between waves scattered at the origin and by the scattering element located at r will be given by

$$\phi = -(k - k_0) \cdot r = -q \cdot r \qquad (6.7)$$

From Appendix A, we see that q is the momentum transferred to the wave because of the scattering. The form factor (Equation (6.5)) can thus be rewritten as

$$f(q) = \int \rho_e(r) e^{iq \cdot r} dV \qquad (6.8)$$

The usefulness of the Fourier representation now becomes apparent, because comparison of Equations (6.3) and (6.8) shows that we may write the electron density in terms of the inverse Fourier transform of the form factor, thus

$$\rho_e(r) = \frac{1}{(2\pi)^3} \int f(q) e^{iq \cdot r} d^3q \tag{6.9}$$

For scattering in the forward direction, that is to say in the direction of the incident beam ($\phi = 0$), we have

$$f(0) = \int \rho_e(r) dV = Z \tag{6.10}$$

Z being the atomic number.

Let us now return to the task of deriving Bragg's Law for strongly diffracted beams, and to the issue of the structure factor. Let p be the unit vector in the direction of the incident beam and q the unit vector in the direction of the diffracted beam. From Figure 6.1, we see that a wave scattered from the origin O travels the distance $r \cdot (q - p)$ farther than the wave scattered at the point r_n, and this implies that the phase of the latter scattered wave is advanced with respect to the wave scattered at the origin by $2\pi r_n \cdot (q - p)/\lambda$, λ being the wavelength of the X-rays. We will find it convenient to define a new vector S (see Figure 6.1) by the equation

$$\lambda S = q - p \tag{6.11}$$

because the condition for a maximally scattered beam in the direction of q is then that $2\pi r_n \cdot S$ is an integral multiple of 2π for all values of n_1, n_2 and n_3.

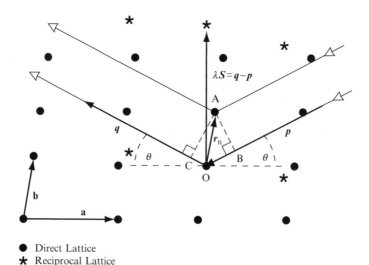

● Direct Lattice
★ Reciprocal Lattice

Figure 6.1 Calculation of the path difference in the direct lattice for rays scattered at the origin (O) and at r_n (A), together with the corresponding reciprocal lattice and the derivation of Bragg's law

This requires that $n_1a \bullet S + n_2b \bullet S + n_3c \bullet S$ be an integer, the latter being positive, zero or negative.

This requirement can be satisfied for all values of the ns only if the individual dot products $a \bullet S$, $b \bullet S$ and $c \bullet S$ are themselves integers. That is to say

$$a \bullet S = h, \quad b \bullet S = k, \quad c \bullet S = l \tag{6.12}$$

Since each of these simultaneous conditions is an equation for a series of parallel planes, the implication is that there will be strongly diffracted beams in the directions of all the intersections of these planes. Moreover, the spacings of those planes is such that the perpendicular distance from any one of them to the origin is inversely proportional to the corresponding edge length of the primitive cell. For example, the perpendicular distance from the origin to the hth plane in the a system is $h/|a|$, or h/a. In other words, these inter-planar intersections themselves lie on a new lattice that is inversely related to the original lattice, and it is referred to as the reciprocal lattice (see Figure 6.1). This new lattice naturally has its own primitive cell, and we may define the edges of that cell by the new vectors a^*, b^* and c^*, the equation analogous to (6.1) being

$$S_{hkl} = ha^* + kb^* + lc^* \tag{6.13}$$

As Paul Ewald first pointed out, the efficacy of the reciprocal lattice stems from its particularly intimate relationship with the observed diffraction pattern. The planes that give rise to the beam having the direction specified by a particular set of the integers hkl will have a spacing d_{hkl}, such that $d_{hkl} = 1/S_{hkl}$, and if the angle between the vector of incidence p and the vector of diffraction q is 2θ, we will have, using Equation (6.12)

$$\lambda S_{hkl} = |q - p| = 2\sin\theta \tag{6.14}$$

Using our expression for the spacing, this can be rewritten to produce Bragg's Law

$$\lambda = 2d_{hkl}\sin\theta \tag{6.15}$$

It is important to note that the directions of the strongly diffracted beams depend only on the dimensions and symmetry of the primitive cell, and not on the basis, that is to say not on the distribution of atoms (and thus electrons) within that cell.

The integers h, k and l are, of course, the Miller indices we encountered earlier, where we saw that they provide a short-hand way of referring to a particular set of parallel planes in the crystal. We also saw that their values are easily found by taking the reciprocals of the intercepts made by such a plane on the axes of the primitive cell. Figures 6.2(a) and 6.2(b) show the various lines and planes in a two-dimensional and a three-dimensional crystal, respectively, corresponding to various Miller indices. If one of these lines or planes lies parallel to a particular axis, the intercept can be imagined as lying at infinity, and the corresponding Miller index is zero.

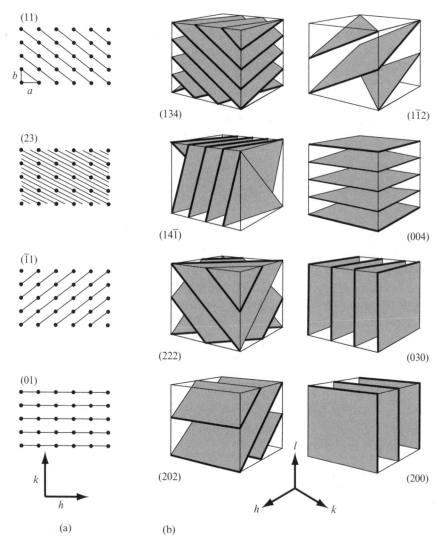

Figure 6.2 Examples of the arrangement of lines (a) and planes (b) corresponding to the Miller indices of various sets of parallel lines in a two-dimensional lattice (*hk* in a) and various sets of parallel planes in a three-dimensional lattice (*hkl* in b). (Figure adapted from an original by Max Perutz. *Scientific American*, November 1964, pp64–76. Copyright 1964 by W. H. Freeman and Company.)

Determination of the directions of the strongly-diffracted beams is a relatively minor matter, of course, and the really important issue centres on the distribution of scattering matter (electrons) within the primitive cell, that is to say on the basis. The density of such matter within a given volume element dV, at the point x, y, z, will be $\rho(x, y, z)$. As we noted earlier, and following the original suggestion by Lawrence Bragg's father, William Henry Bragg, in 1915, we represent that density of scattering matter by means of a Fourier series. Thus for any given set of Miller indices h, k, l, the scattering power may be written as a complex number F_{hkl}, where

$$F_{hkl} = |F_{hkl}|e^{i\alpha(hkl)} = \frac{V}{abc}\int_{-a/2}^{a/2}\int_{-b/2}^{b/2}\int_{-c/2}^{c/2}\rho(x, y, z)e^{2\pi i(hx/a+ky/b+lz/c)}dxdydz \quad (6.16)$$

V is the volume of the primitive cell, the axes of which have lengths a, b and c, those axes being mutually inclined at any angles. $Vdxdydz/abc$ is thus the element of volume between x and $x + dx$, and so on.

We now represent $\rho(x, y, z)$ by a three-fold Fourier series, and substitute this in Equation (6.9), to obtain

$$\rho(x, y, z)$$
$$= \frac{1}{V}\sum_{-\infty}^{\infty}\sum_{-\infty}^{\infty}\sum_{-\infty}^{\infty}|F_{hkl}|\cos\{2\pi hx/a + 2\pi ky/b + 2\pi lz/c - \alpha(hkl)\} \quad (6.17)$$

the summation being made over all positive and negative values of h, k and l. Enlarging on what was noted earlier, the term $F(000)$, that is to say the amplitude of the structure factor for $h = k = l = 0$, is equal to Z, the total number of electrons in the primitive cell, this being required by the need to satisfy

$$\frac{V}{abc}\int_{-a/2}^{a/2}\int_{-b/2}^{b/2}\int_{-c/2}^{c/2}\rho(x, y, z)dxdydz = Z \quad (6.18)$$

The term $\alpha(hkl)$ is the phase constant for the diffracted wave with Miller indices h, k and l. If its value was known for each set of Miller indices it would be a relatively straight-forward matter to reconstruct the pattern of electron density throughout the primitive cell. As Lawrence Bragg noted, the reconstruction would start with a uniform distribution Z/V spread throughout the cell, and to this would be added a series of sinusoidal density variations, oriented according to the appropriate Miller indices, and all mutually overlapping. Figure 6.3 is a schematic depiction of such periodic variations in two dimensions, for various Miller indices and phases. Figure 6.3(b) shows the

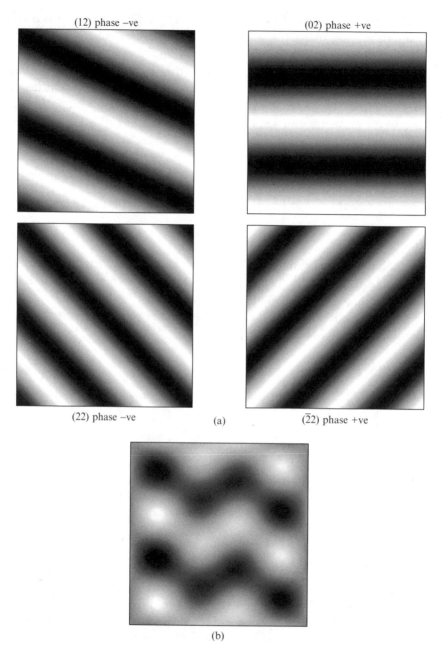

Figure 6.3 The application of Fourier synthesis to the determination of crystal structure is illustrated in this grossly simplified two-dimensional analogue. We assume that there were only four (equally intense) diffraction spots with non-zero intensity, and the sinusoidal alternations show the Fourier components corresponding to their Miller indices and phases. These four plots were superimposed to produce the final picture shown at the bottom. In practice, the various Fourier components would be superimposed with different exposures, because of the varying intensities of the diffraction spots, of which there would be several hundred or more

image of the crystal structure projected on the (100) face. This was obtained by superimposing the band patterns corresponding to the various indices and phases, with the individual exposures corresponding to the intensities of the various diffraction spots.

The final illustration in Figure 6.3 shows what results from the superposition of several individual Fourier components. In practice, a typical structure determination involves the recording of approximately 1000 different diffracted beams, so calculation of the electron density at just one set of coordinates x, y and z, requires a roughly thousand-fold overlap of Fourier components. If this were the whole story, structure determination would be a routine matter that could be fully automated without difficulty. Unfortunately, it is only the amplitude of the structure factor that can be measured directly, not the corresponding phase. This is the celebrated phase problem, and one method of overcoming it is to record and compare the diffraction intensities obtained with and without a heavy atom replacing an identified atom in the molecule under investigation. Because the diffraction amplitude tends to scale with atomic number, such a heavy atom will have a disproportionately large influence on the scattering, and the various phases can be worked out from a before-and-after comparison. (In practice, one may use several different such heavy-atom substitutions, in order to remove residual ambiguities of sign.) It is a vital prerequisite of this approach that the substitution of the normal atom by the alien atom does not cause appreciable structural re-arrangements in the crystal under investigation; whence the technical term isomorphous replacement (the alternative name being the heavy atom method). The first reported use of this approach was by J. West, in 1930, the investigated compound being potassium dihydrogen phosphate.

With the advent of powerful computers, it became possible to circumvent the phase problem by what is known as the direct method. This is a statistical approach, in which the computer homes in on a structure that is consistent with the constraint that the electron density cannot be negative at any point. It is made possible by the fact that the equations relating the observed intensities to the unknown structure outnumber the unknown parameters by at least a factor of 10. Unfortunately, the viability of this method decreases with an increasing number of atoms, and its pioneers, Herbert Hauptman and Jerome Karle, found that it becomes unreliable when the molecule comprises more than a few hundred atoms.

6.2 Nuclear Magnetic Resonance

Although X-ray diffraction is a highly useful technique for studying the structure of biological molecules, it does suffer from the disadvantage that the molecules have to be investigated in an environment that contrasts with the one in which they normally operate; the diffraction method requires crystals whereas the individual molecules normally function in aqueous or membranous surroundings. Moreover, molecules such as proteins are dynamic structures whose function depends to a considerable extent on the flexibility of their outer regions. It is true that the individual atoms in a protein, for example, are not prevented from vibrating just because that protein is part of a crystalline array, but the vibrational excursions will not be the same as those observed when the molecule is floating in solution. The technique of nuclear magnetic resonance permits study of biological molecules in their natural environments, and in this respect it can be said to complement the X-ray method.

If one gently gives a sideways tap to a freely suspended compass needle, it will oscillate about its original quiescent position of alignment in the earth's magnetic field. And the oscillations will gradually decrease in amplitude because of frictional forces. Ultimately, the needle will once again be stationary. It is well known that there are magnetic effects associated with some processes at the atomic level. The motions of electrons in incomplete electronic shells of an atom give rise to such effects, for example, and it is also found that atomic nuclei can have magnetic moments. These too will be aligned when subjected to an external magnetic field, and they can be perturbed by an applied force, just as in the case of the tapped compass needle. Indeed, although atomic nuclei are not subject to the sort of mechanical friction we are familiar with in the macroscopic domain, their oscillations will nevertheless gradually decay if there are retarding forces. Such forces are in fact present because the nuclear moments display coupling to those associated with the electrons, and indeed to other nuclei. The usefulness of the nuclear magnetic resonance technique (which was developed through the collective efforts of Isidor Rabi, Felix Bloch and Edward Purcell) stems from the dependence of this coupling to the surrounding environment, the energy absorption characteristics being exquisitely sensitive to the local magnetic neighbourhood. The local variations in the degree of coupling gives rise to what are known as chemical shifts, and the complexity of the observed spectra increases with increasing molecular size. It is these facts that permit distinction between the different atoms in a molecule, even if they are of the same element, and one may think in terms of 'signatures' that the expert learns to 'read'.

In Chapter 2, we discussed the angular momentum associated with an electron's spin. A nucleus, likewise, has an angular momentum L whose magnitude is $h\sqrt{[I(I+1)]}/2\pi$, where I is the nuclear spin quantum number. This spin is zero for nuclei having even mass numbers (the mass number being the total number of neutrons and protons) and for nuclei of even atomic numbers (the atomic number being equal to the number of protons, of course). The isotopes ^{12}C and ^{16}O are examples of zero-spin nuclei. Otherwise, the spin may be an integer, as in the cases of ^2H and ^{14}N, for which $I = 1$, or a half-integer, as in the cases of ^1H, ^{13}C, ^{15}N and ^{31}P, for which $I = 1/2$. For ^{23}Na and ^{39}K, the spin is 3/2, while it is 5/2 for ^{17}O and ^{25}Mg. The majority of investigations using this technique focus specifically on ^{1}H, with its 99.98% isotopic abundance and its pronounced sensitivity.

The magnetic moment of a nucleus is usually expressed in terms of the nuclear magneton, whose value is $\mu_n = q_e h/4\pi m_p = 5.05 \times 10^{-27} \text{JT}^{-1}$ (the symbol T denoting Tesla, a unit of magnetic field strength), m_p being the mass of the proton. The magnetic moment of a free proton is $g_n \mu_n$, where g_n is known as the nuclear g-*factor* ($g_n = 2.7928$). (The fact that the proton *has* a magnetic moment was discovered by Otto Stern.)

Now when a particle with angular momentum also carries an electric charge, a magnetic dipole moment is produced, with a magnitude equal to the angular momentum multiplied by a factor γ, known as the gyromagnetic ratio. Thus

$$\mu = \gamma(h/2\pi)\sqrt{I(I+1)} \tag{6.19}$$

Our macroscopic compass needle was able to adopt any angle with respect to the earth's magenetic field, but only discrete values are permitted in the quantum domain. As a consequence of this quantization (see Appendix A), the magnetic moment m_z in the z direction (along which a magnetic field could be directed) is constrained to adopt the values

$$m_z = m_I \gamma h/2\pi \tag{6.20}$$

with $m_I = I, I - 1, I - 2, \ldots, 2 - I, 1 - I, -I$. The proton has a spin of 1/2, so there are just two possible values of m_I, namely $+1/2$ and $-1/2$. This, in turn, means that there are only two possible orientations of the magnetic moment with respect to an applied field, and they lie at 180° with respect to each other. (This is consistent with there being, in general, $2I + 1$ permitted orientations.)

The energy $\mathscr{E}$ of any dipole moment, macroscopic or atomic-scale, with respect to a magnetic field H is given by $\mathscr{E} = -\boldsymbol{\mu} \cdot \boldsymbol{H} = -\mu H \cos\theta$, θ being the angle between the magnetic moment and the field. For the proton, therefore, we have the two possible energies

$$\mathscr{E} = \pm \gamma h H/4\pi \tag{6.21}$$

the negative sign corresponding to the higher energy level, in contrast to what is observed for electrical dipoles. When a magnetic dipole is acted upon by a magnetic field, it experiences a torque of magnitude $\boldsymbol{\mu} \times \boldsymbol{H}$, and this is equal to the temporal rate of change of the angular momentum, $d\boldsymbol{L}/dt$. Because, as we saw earlier, $\boldsymbol{\mu} = \gamma \boldsymbol{L}$, we have

$$d\boldsymbol{\mu}/dt = \gamma(\boldsymbol{\mu} \times \boldsymbol{H}) \tag{6.22}$$

The torque will be perpendicular to both μ and H, so the upshot will be precession of the dipole about the magnetic field, in much the same way that a spinning top precesses in the field of gravity. And because the field and moving charge are mutually perpendicular, the system expends no energy. In other words, the precession will continue without decay, unless there is something analogous to the friction acting on the above-mentioned compass needle. The precession frequency is known as the Larmor frequency (named after Joseph Larmor), and this is given by

$$\omega = \gamma H \tag{6.23}$$

For a magnetic field strength of 1 T, the Larmor frequency is 42.577 MHz.

Now let us turn to the situation when numerous dipoles are present, and denote the collective dipole's magnetic moment with the capitalized M. If all the spins were perfectly aligned, and if they all had $m_I = 1/2$, the total energy would be $N\gamma hH/2\pi$, N being the number of dipoles. For such a situation to prevail, the dipoles would have to be isolated from their surroundings. In reality, they will be in equilibrium with that environment, and thus subject to the usual thermodynamic principles. This tells us that the equilibrium magnetic moment will be given by

$$M_0 = (N\gamma h/2\pi)e^{-\Delta\mathscr{E}/k_B T} \tag{6.24}$$

in which the exponential is the familiar Boltzmann term, and where $\Delta\mathscr{E} = \gamma hH/2\pi$, this being the energy difference between the two possible states of each dipole. T is the absolute temperature.

If the system has been disturbed, and the current z-component of the collective magnetic moment is M_z, this will relax back toward the above-noted equilibrium value at a rate given by

$$\frac{d}{dt}M_z(t) = -\frac{M_z - M_0}{\tau_1} \tag{6.25}$$

where τ_1 is the longitudinal (or spin-lattice) relaxation time constant. The solution of Equation (6.25) is

$$M_z = M_0(1 - e^{t/\tau_1}) \tag{6.26}$$

In practice, the molecular system under investigation is subjected simultaneously to a fixed field, H_0, and a sinusoidally varying field whose peak value is H_1, and there is then a second relaxation time constant, τ_2, which is known by the term transverse (or spin-spin). This latter involves relaxation of the spin components in the plane perpendicular to H_0, and these do not depend upon interactions with the spins of the neighbouring lattice.

When the magnitude of H_0 is such that γH_0 matches the Larmor frequency, transitions between the quantized magnetization states can take place, and these can be detected by radio-frequency coils. And given that H_0 will normally include influences from the surroundings, there is an attendant possibility for detection of the earlier-mentioned chemical shifts. In a typical set-up, H_1 takes the form of a brief pulse directed normally to H_0, in the plane of the radio-frequency field, and one monitors the subsequent relaxation. The shape of the resulting resonance peak has the form of a Lorentzian (named for Hendrik Lorentz)

$$S(\omega) = \frac{A\tau_2}{1 + \tau_2^2(\omega - \omega_0)^2} \tag{6.27}$$

while the temporal variation of the resulting signal follows

$$f(t) = \left(\frac{\pi}{2}\right)^{1/2} A e^{-t/\tau_2} e^{i\omega_0 t} \tag{6.28}$$

in both cases, A being a constant. Equation (6.28) is the Fourier transform of Equation (6.27); whence the alternative name for pulsed nuclear magnetic resonance, Fourier transform NMR. One should note that the shape of the Lorentzian is independent of the field strength, whereas the positions of the individual peaks will become more spaced out as the field increases. This means that the resolution will increase for increasing field strength, and it is for this reason that the technique now tends to use very powerful magnets (from which heart patients fitted with pace-makers should stand well clear).

Richard Ernst pioneered what has become known as two-dimensional Fourier transform NMR, in which the specimen is subjected to two brief pulses, rather than one, relaxation data then being collected as a function of the two delays. This produces a plot of the type shown in Figure 6.4. The ω_1 and ω_2 axes correspond to the Fourier transforms of the pulse durations t_1 and t_2, and the prominent diagonal corresponds to resonances in the one-dimensional spectrum shown along the top of the figure. The off-diagonal features of the two-dimensional plot yield information concerning certain conformational aspects of the molecule. Finally, mention should be made of the nuclear Overhauser effect (NOE), named after Albert Overhauser, which arises from magnetic dipole–dipole interactions. This effect is observed by irradiating the sample at a frequency corresponding to the Larmor frequency of just one of the nuclei present. If the incident field strength is sufficiently large, and the coupling to the non-irradiated nuclei sufficiently strong, the magnetization of these latter will also be disturbed from its equilibrium value, and its subsequent

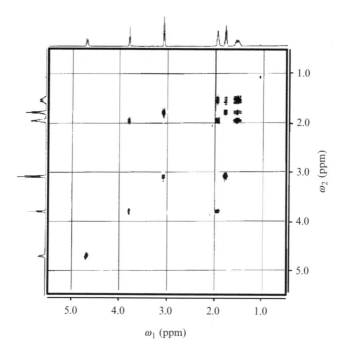

Figure 6.4 A 500 MHz two-dimensional NMR correlation (COSY) spectrum of Lysine in D_2O. The ω_1 and ω_2 axes correspond to the Fourier transforms of the two pulse durations, and the prominent diagonal corresponds to resonances in the one-dimensional spectrum (shown along two orthogonal sides). The diagonal peak at 4.77 ppm is the residual HDO signal. The off-diagonal features of the two-dimensional plot yield information concerning certain conformational aspects of the molecule, and it is not surprising that plots for much larger molecules are rather complicated. (Data reproduced with the kind permission of Jens Jørgen Led.)

return to equilibrium can be monitored. This effect is limited by the rapid spatial attenuation of the dipole–dipole coupling, which falls away with the inverse sixth power of the distance.

6.3 Scanning Tunnelling Microscopy

As was discussed in Appendix A, a quantum mechanical wave has a finite probability of tunnelling through an energy barrier even though this obstacle has a height that exceeds that of the wave's energy. In practice, however, the barrier width must be of near-atomic dimensions if the tunnelling current is to be measurable. In the early experimental demonstrations of quantum mechanical tunnelling, by Ivar Giaever around 1960, the barrier was typically a very thin layer of (insulating) oxide, on a metal surface, the sandwich being completed by a thin metallic film positioned on top of the oxide by vapour deposition. In the scanning tunnelling microscope (STM), invented by Gerd Binnig and Heinrich Rohrer, the barrier is simply empty space. This does not mean that the microscope can function only in high vacuum. On the contrary, because the two electrical conductors are separated by a distance comparable to an atomic diameter, there is simply not enough room for unwanted atoms to enter the space between those conductors. So whereas electron micro- scopes require the specimen to be in a vaccum, this is not the case with the scanning tunnelling microscope. It is even the case that biological surfaces can be studied while they are still in their natural element, that is to say surrounded by water.

The principle of operation is remarkably simple. A very sharp electrically- conducting tip is brought into such close proximity with the specimen that a tunnelling current is produced. The strength of this current is extremely sensi- tive to the distance of the tip to the specimen, so if it is arranged for the sharp probing tip to be moved in such a way that the tunnelling current is constant, its motions will in effect map out the contours of the specimen surface at atomic resolution. This obviously requires that the tip can be moved in a controlled fashion with atomic-level precision, and that can be done by exploit- ing the piezoelectric effect, in which the physical dimensions of a suitable crystal are changed by varying the voltage applied across its opposing faces. This type of adjustment is used to control the position of the tip in the two dimensions lying in the plane of the specimen surface, and also the dimension normal to that surface (see Figure 6.5).

The energy required to extract an electron from a surface whose potential energy is $-\mathscr{E}_{pot}$ with respect to the vacuum would be that same number if the electron had zero kinetic energy. In practice, however, the electron will have a finite kinetic energy, $\mathscr{E}_{kin}$, so the extraction energy will be $(\mathscr{E}_{pot} - \mathscr{E}_{kin})$ this being known as the work function. A typical work function for a metallic surface is about 4 V, that is to say that the energy has a value of 4 multiplied by the charge on the electron. Using the analysis given in Appendix A, we see that tunnelling operates at a characteristic length scale δ, where

$$\delta = \frac{h/2\pi}{\sqrt{2m_e\left(\mathscr{E}_{pot} - \mathscr{E}_{kin}\right)}} \qquad (6.29)$$

and where m_e is the mass of the electron ($= 9.1095 \times 10^{-31}$ kg). This indicates that δ is indeed approximately 10^{-9} m, that is to say a distance comparable to an atomic diameter. We can also use the analysis given in Appendix A to approximate the tunnelling transmission coefficient χ by the expression

$$\chi \approx e^{-2\ell/\delta} \qquad (6.30)$$

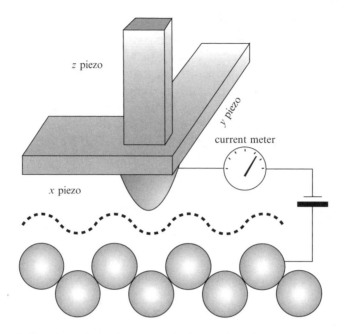

z piezo

y piezo

current meter

x piezo

Figure 6.5 Highly schematic (and not-to-scale) illustration of the manner in which the tip of a scanning tunnelling microscope probes the atomic-scale irregularities of the surface of a specimen, through its piezo-mediated ability to manoeuvre in three dimensions. The associated electronics (not shown) permit the tunnelling current to be maintained at constant strength as the surface is scanned raster-fashion. (Figure adapted from an original by P.K. Hansma, V.B. Ellings, O. Marti and C.E. Bracker, *Science* **242**: 209–216, 1988. Copyright 1988 by the AAAS.)

where *l* is the width of the barrier. The tunnelling current is given by

$$j = \frac{q_e^2 V \chi}{2\pi h l \delta} \qquad (6.31)$$

where q_e is the charge on the electron and V is the voltage applied between the specimen and the probing tip. The exquisite sensitivity of the device can be appreciated by asking what decrease in the tunnelling current would be observed if the specimen-tip distance was increased by just 1% of an atomic diameter, that is to say by about 10^{-11} m. Using Equations (6.30) and (6.31), we can easily establish that this decrease would be about 2%, a change that can easily be detected. We thus see that it is not difficult to arrange for the tunnelling current to be maintained at a constant value, by varying the specimen-tip distance as the tip is made to scan the specimen surface, raster fashion.

With this degree of control, it is not difficult to reconstruct an image of the specimen surface at atomic resolution, and the technique can draw on the sophisticated graphics now made available by modern electronic computing.

Indeed striking pictures of all manner of surfaces produced by this technique have latterly been augmented by equally impressive movies. Let us turn now to a related instrument.

6.4   Atomic Force Microscopy

The atomic force microscope (AFM) was invented in 1985 by one of the inventors of the scanning tunnelling microscope, Gerd Binnig, together with his two colleagues Ch. Gerber and C. F. Quate. It differs from the STM in that its atomic-scale tip is cantilever-spring mounted. Imaging by this instrument requires that the sample be rigidly clamped, and that it is robust against the forces that are applied to the tip. One of the great advantages of the instrument is that it does not require the specimen to be electrically conducting; this contrasts with the case of the STM.

The cantilever is usually made of silicon or silicon nitride, and its length typically lies in the range $1-5 \times 10^{-4}$ m, while the width is about $0.5-5 \times 10^{-6}$ m. Its deflexion is magnified optically, by reflecting a laser beam from its upper surface. The probing tip is located at the outer extremity of the cantilever's lower surface. The reflected laser beam is used to control the distance between the tip and the specimen surface, by feeding the signals produced by the detected light into a computer (see Figure 6.6). As in the STM, the specimen is moved raster-fashion with respect to the tip, with a suitable piezoelectric scanner, the typical scanning range being as large as 10^{-4} m. A good-quality scanner can reproducibly give displacements as small as 10^{-10} m.

The AFM permits measurement of the forces between individual molecules. One of the molecules participating in the interaction under investigation is first attached to the tip, while the other is present on the to-be-probed surface. The tip-surface separation distance is decreased until the two molecules form a bond, and the tip is then gently raised away from the surface. Just before the bond breaks, the magnitude of the cantilever deflection, Δx, gives a direct indication of the strength of the bond, for we have

$$\mathcal{F}_{bond} = \kappa_{cantilever} \Delta x \qquad (6.32)$$

The cantilever's spring constant, $\kappa_{cantilever}$, can be as low as 10^{-3} N m^{-1}. A typical tip radius is 3×10^{-8} m.

The AFM holds particular promise for the study of biological membranes, because it can resolve the individual head regions of the lipids. It has also been used to study membrane-bounded proteins, the channels and receptors in the nervous system being of obvious interest. The instrument has even been used to study DNA molecules (see the example reproduced as Figure 6.7) and DNA-protein complexes.

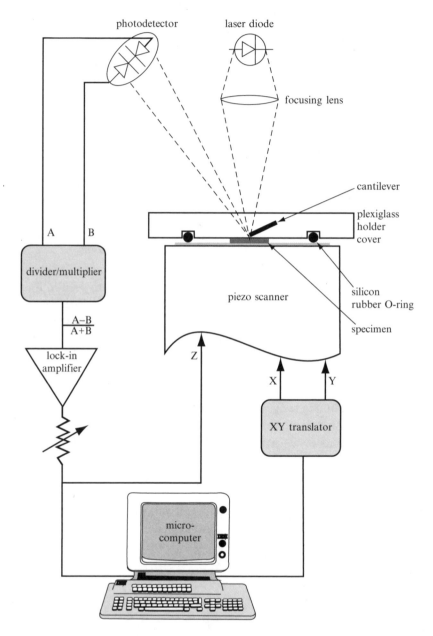

Figure 6.6   Highly schematic (and not-to-scale) illustration of the mode of operation of an atomic-force microscope, the deflection of the cantilever being measured through the optical magnification provided by reflection and detection of a laser beam. (Figure adapted from an original by J. Yang, L. K. Tamm, A. P. Somlyo and Z. Shao, *Journal of Microscopy* **171**: 183–198, 1993. Copyright 1993 by the Royal Microscopical Society.)

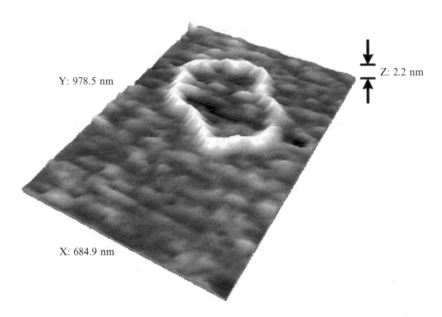

Y: 978.5 nm

Z: 2.2 nm

X: 684.9 nm

Figure 6.7 An atomic-force microscope image of a bacterial plasmid, deposited on a mica surface from a nickel chloride solution. Plasmids are circular double stranded DNA molecules, and their easy transfer from one bacterium to another, in addition to their tendency to mutate, is one cause of the large resistance to antibiotics displayed by some bacteria. (Image reproduced with the kind permission of Maria Holmberg.)

6.5 Optical Tweezers

The technique known as optical tweezers has its roots in the early 1970s, when Arthur Ashkin used light scattering to trap small glass beads. The basic mechanism relies on exploitation of the force associated with the change of momentum when a light beam is scattered. Together with Steven Chu, Ashkin subsequently extended the idea to trap things as large as living bacteria, and as small as atoms.

The underlying physics is reasonably straightforward. Let us start by considering the situation in which the size of the object is much smaller than the wavelength of the incident light. When the electric field E associated with a light beam impinges upon a piece of dielectric material, it induces an electric dipole d, where

$$d = \alpha \bullet E \tag{6.33}$$

and where α is the polarizability of the irradiated material. The interaction potential V_{int} acting between the induced dipole and the electric field is given by

$$V_{\text{int}} = -\boldsymbol{d} \bullet \boldsymbol{E} \qquad (6.34)$$

and the interaction gives rise to a force, $\mathcal{F}_{\text{light}}$, where

$$\mathcal{F}_{\text{light}} = -\nabla V_{\text{int}} \qquad (6.35)$$

Taken together, Equations (6.33), (6.34) and (6.35) indicate that

$$\mathcal{F}_{\text{light}} = \alpha \nabla E^2 = \alpha \nabla I \qquad (6.36)$$

where I is the intensity. So we see that the force is proportional to the gradient of the light's intensity. If there is spatial variation in that intensity, in the direction transverse to the beam's direction of propagation, the object will be pushed in that transverse direction, toward the region where the intensity is greater. If, at the same time, there is a sharp focussing of the beam, there will be an analogous force acting in the direction of the focal point because the intensity gradients all point toward that point. So the overall effect will be that the object is caught in a three-dimensional trap, provided that the above-discussed forces exceed those arising from light scattering. It is found that this latter condition is fulfilled only if the gradient in light intensity is of the order of magnitude produced by a microscope having high numerical aperture.

For cases in which the object is much larger than the wavelength, the analysis considers the changes of momentum implicit in the refraction of the beam, when it is incident on the object. It is a relatively simple matter to show that there again will be formed a three-dimensional trap, in which the object can be held indefinitely (see Figure 6.8). In both pictures, the lighter shade in the intensity profile indicates the greater light intensity, while the magnitudes of the forces are indicated by the widths of the grey arrows. The inset figures indicate the corresponding momenta of the incoming and outgoing rays, with the resultant momentum Δ indicated by the smaller grey arrow in each case. Because the overall system must conserve momentum, the change in direction of the momentum vector associated with the refraction must be compensated for by the additional momentum imparted to the diffracting object.

In practice, the object is neither very small nor very large compared with the wavelength, and the analysis for this intermediate case is far more complicated. One circumvents this difficulty by simply calibrating one's experimental set-up against a well-controlled situation. A popular example exploits the known physics of a particle moving in a viscous fluid. The particle (usually a small bead) trapped in the optical tweezers is surrounded by liquid that can be moved at a suitable speed by a piezoelectric stage. The latter typically has a resonance frequency around 4×10^2 Hz and a maximum amplitude around 10^{-4} m. A sinusoidal or saw-tooth frequency of up to 40 Hz is applied, with an amplitude

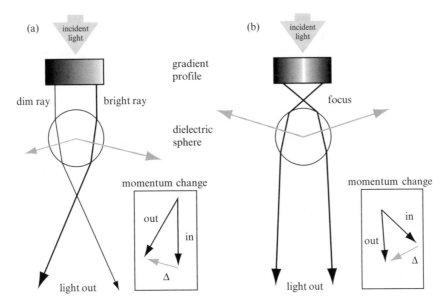

Figure 6.8 The origins of the forces in the parallel-light (a) and focussed-light (b) versions of the optical tweezer. In both pictures, the magnitudes of the forces are indicated by the widths of the grey arrows. The inset figures indicate the corresponding momenta of the incoming and outgoing rays, with the resultant momentum indicated by the smaller grey arrow in each case. (Figure adapted from an original by Karel Svoboda and Steven Block, *Annual Review of Biophysics of Biomolecular Structures* **23**: 247–285, 1994. Copyright 1994 by Annual Reviews Inc.)

in the range $(1-10) \times 10^{-6}$ m, and this produces a particle velocity of the order of 10^{-4} m s^{-1}. By Stokes Law (see Chapter 5), we have

$$\mathcal{F}_{\text{drag}} = 6\pi\eta a v = \gamma v \qquad (6.37)$$

where η is the viscosity of the liquid, a is the particle radius, v its velocity and γ is the drag coefficient. The motion of the particle in the optical trap is described by the Langevin equation (named for Paul Langevin – see Chapter 5):

$$m(dv/dt) = \mathcal{F}_{\text{thermal}}(t) - \gamma v - \kappa x = 0 \qquad (6.38)$$

where m is the particle's mass, κ_x is the effective lateral spring constant of the optical trap (the lateral direction lying along the x-axis), and $\mathcal{F}_{\text{thermal}}(t)$ is the thermal force, which naturally varies as a function of time in the usual Brownian fashion. The final equating to zero is justified by the fact that the inertial forces are negligible for the low Reynolds number (about 10^{-3}) that applies to the situation. (This is comparable to conditions that prevail for a bacterium in water, as is discussed in Chapter 10.)

A record of the temporal variation of the x-position (see Figure 6.9) typically displays a few changes of direction for each millisecond, and when this is Fourier transformed it produces a Lorentzian

$$S_x(f) = \frac{k_B T}{\pi^2 \gamma \left(f_c^2 - f^2\right)} \qquad (6.39)$$

where f denotes frequency, and where f_c is the so-called corner frequency (above which the Fourier transform drops away relatively rapidly – see Figure 6.10). This latter parameter permits one to calculate κ_x, because

$$f_c = \kappa_x / 2\pi\gamma \qquad (6.40)$$

In the experiment that gave rise to Figures 6.9 and 6.10, the bead diameter was 1.05×10^{-6} m, and the corner frequency of 420 Hz corresponds to a lateral spring constant of 2.6×10^{-5} N m^{-1}. This is about 40 times smaller than the spring constant of the cantilever in a typical AFM, as described above.

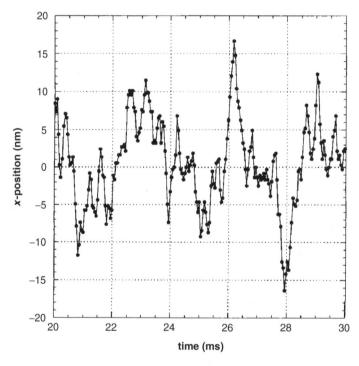

Figure 6.9 Temporal variations in the lateral position of a one micrometer polystyrene bead optically trapped five micrometers above a surface, the incident power of the laser being about 40 mW. The sampling frequency of 22 kHz was high enough to permit detection of the ballistic regime. (Data reproduced with the kind permission of Lene Oddershede, Sonia Grego, Simon Nørrelykke and Kirstine Berg-Sørensen.)

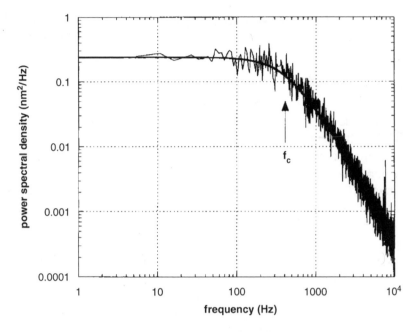

Figure 6.10 The power spectrum of the temporal characteristic shown in Figure 6.9. The thick line is the Lorentzian fit to the data, and it indicates a corner frequency (f_c) of 420 Hz. This corresponds to a trap stiffness of $2.6 \times 10^{-5}\,\mathrm{N\,m^{-1}}$. (Data reproduced with the kind permission of Lene Oddershede, Sonia Grego, Simon Nørrelykke and Kirstine Berg-Sørensen.)

Rather conveniently, it is possible to make an independent check on this result, because there is a simple relationship between the standard deviation σ of the histogram of the bead's axial position as a function of displacement and the spring constant. That relationship is

$$\sigma^2 = k_B T / \kappa_x \tag{6.41}$$

Figure 6.9 shows the results for the axial displacements obtained by Lene Oddershede, Sonia Grego, Simon Nørrelykke and Kirstine Berg-Sørensen, for the above-mentioned micron-sized bead. There is an added bonus from this type of analysis, because it is possible to derive the axial potential as a function of displacement, by inverting the displacement histogram and assuming that Boltzmann statistics apply (see Figure 6.10).

The optical tweezers technique is already proving to be very useful in the study of all manner of biological systems. It permits measurements of the forces relevant to individual molecules, be they located in the cytoplasm or in the membrane. One currently favoured trick is to attach a bead to a small stretch of biopolymer and then determine the micro-mechanics of the latter by subjecting it to a microscopic stress–strain test. The results obtained in these very important investigations will be discussed in later chapters of this book.

6.6 Patch Clamping

As will be discussed in a number of the later chapters in this book, the membranes of several different types of cell are dotted with myriad channels and receptors. These are all protein molecules, and they are embedded within the membranes, spanning its roughly 5 nm thickness. It had become apparent by the middle of the 20th century that these molecules are required by a cell if it is to possess the capacity for electrochemical excitability. Both the channels and the receptors display considerable variety, and it became an obvious goal of neuroscience to investigate the characteristics of individual members of these families of molecules. This presented a problem, however, because of the sheer variety of them present in a typical membrane; how could one separate the sheep from the goats, when there were also cows, pigs, horses, and a host of others to complicate the situation?

It was Erwin Neher and Bert Sakmann who, in the mid-1970s, realized that the necessary differentiation would become possible if one made measurements on an area of membrane so small that it contained only very few channels or receptors. They managed to achieve this ideal by studying the small area of membrane that can cover the circular opening at the end of a very narrow pipette. If the end of such a capillary pipette is very clean, and if the membrane surface is free of extraneous debris, the pipette can be made to adhere to the membrane, this usually requiring a light degree of suction, applied at the right moment. If the adhesion is sufficiently good, an electrical seal will have been formed whose resistance can be as high as 1 GΩ.

Several different versions of such patch clamping are employed, depending on what one is attempting to study. The cell-attached mode merely requires that the clamp be made to a cell's outer surface. If the (major) portion of the cell's surface that is not actually covering the end of the pipette is now removed, the remaining patch of membrane will have its normally inside surface facing outward, and vice versa. This scenario is used in the so-called inside-out mode. Alternatively, a brief application of sufficient suction, when the whole cell is attached, will rupture the patch covering the end of the pipette, whose interior will then be in direct contact with the cell's interior. This is known as the whole-cell mode. Finally, by gently applying a prolonged suction, in the whole-cell mode, the cell's membrane will gradually shrink, like a slowly deflating balloon, and the final situation will again be coverage of the end of the pipette by a small patch of membrane; but this time, one will have the outside-out mode. All these modes are illustrated in Figure 6.11.

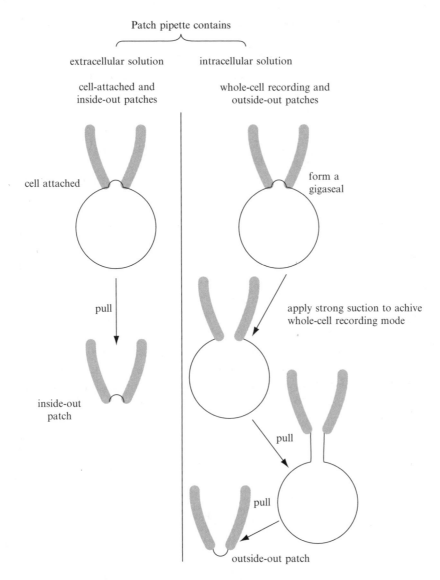

Patch pipette contains

extracellular solution   intracellular solution

cell-attached and   whole-cell recording and
inside-out patches   outside-out patches

cell attached   form a
                                                       gigaseal

pull   apply strong suction to achive
                                      whole-cell recording mode

inside-out
patch
                                                   pull

                                       pull

                                          outside-out patch

Figure 6.11   The different methods of producing cell-attached and inside-out (left) and whole-cell and outside-out (right) situations in a patch-clamp experiment. (Figure adapted from an original by David Ogden and Peter Stanfield, *Microelectrode Techniques: The Plymouth Workshop Handbook* (Cambridge: The Company of Biologists, 1987). Copyright 1987 by The Company of Biologists Ltd.)

Because the current through a channel is in the pA range, an amplification circuit is required in order to detect it, and measure it with sufficient precision. The key component in such a circuit is a resistor having a high resistance, R_f, across which the small current will produce a measurable voltage. It is referred to as a feedback resistor, and the circuit is called a high-gain current-to-voltage converter (see Figure 6.12). The output voltage, V_{out}, is measured with respect to a reference voltage, V_{ref}, which can be varied according to different temporal schedules, if desired. The latter might be, for example, a brief square-wave pulse. The various parameters are related by the following expression.

$$V_{out} = V_{ref} - i_p R_f \tag{6.42}$$

where i_p is the current flowing through the pipette.

One difficulty in such measurements arises from the thermal noise generated in the feedback resistor. Its variance, s^2, is given by

$$s^2 = 4k_B T f_c / R_f \tag{6.43}$$

where f_c is the bandwidth, that is to say the low pass filter setting. It is clear from this that the disturbing factor caused by noise can be reduced by increasing the feedback resistance. Some experimenters seek further noise reduction by cooling the resistor. For a resistance of 10 GΩ, the standard deviation of the noise is 0.04 pA at 20 °C, if the bandwidth is 10^3 Hz.

The phenomena of interest clearly have temporal components, and this forces one to take account of factors arising from capacity effects. The current will therefore consist of two components, one ionic and one capacitative. Let us consider the case of patch-clamp recording from a whole cell. The two contributions to the overall current will be

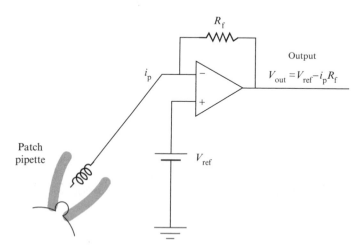

Figure 6.12   Schematic representation of the electrical circuit in a typical patch-clamp experiment. The gain (which equals V_{out}/i_p) is set by the magnitude of the feedback resistance R_f. (Figure adapted from an original by David Ogden and Peter Stanfield, *Microelectrode Techniques: The Plymouth Workshop Handbook* (Cambridge: The Company of Biologists, 1987). Copyright 1987 by The Company of Biologists Ltd.)

$$i_{ionic} = V_{cell}/R_{cell} \tag{6.44}$$

and

$$i_{cap} = C_{cell}dV_{cell}/dt \tag{6.45}$$

The temporal variation of the cell voltage will thus be described by

$$V_{cell}(t) = [V_{cell}(0) - V_{cell}(\infty)]\left(1 - e^{-t/\tau}\right) \tag{6.46}$$

where

$$\tau = C_{cell}R_{seal}R_{cell}/(R_{seal} + R_{cell}) \tag{6.47}$$

The cell resistance usually far exceeds the seal resistance, so we may simplify this to

$$\tau = C_{cell}R_{seal} \tag{6.48}$$

This time constant applies both to voltage variations imposed externally and currents emanating from the cell membrane, so the latter are effectively low-pass filtered. This provides a value for one of the parameters used above, for we thus have

$$f_c = 1/(2\pi\tau) \tag{6.49}$$

Patch clamping has become a staple of the physiologist's repertoire of techniques. It provides information about conductance of individual channels and about the kinetics of channel opening and closing (see Figure 6.13, for an example). It has also led to the discovery of hitherto unknown channels, and it

Figure 6.13   The endogenous channel activity patch-clamp recorded from a Sf9 (*Spodoptora frugiperda*) insect cell. The current trace is from an excised outside-out membrane patch with a bath containing $[Cl^-] = 150\,mM$, a pipette containing $[Cl^-] = 40\,mM$, and a pipette potential of $+40\,mV$. Several gating modes can be discerned. (Data reproduced with the kind permission of Claus Nielsen.)

can be used under circumstances in which it would not be possible to insert electrodes.

6.7 Molecular Dynamics

Having reviewed a number of experimental techniques for investigating bio-logical molecules and larger structures, let us turn to a somewhat different approach, namely computer simulation. Some have referred to this as the third basic scientific activity, because it is neither experimental nor theoretical; it has been called theoretical experimentation. There is hardly a branch of the scien-tific endeavour that has not gained from the advent of computer simulation. One sees it being used in such widely differing enterprises as weather forecasting, the mechanical performance of buildings and bridges, and predictions of stock-market fluctuations. Despite this impressive diversity, the underlying strategy is always the same: construct a theoretical model of one's system, simulate its behaviour under a given set of conditions, and then compare the results of the latter with actual observations.

The simulation technique known as molecular dynamics should not be con-fused with another branch of science that (legitimately) goes under the same name; it is distinct from the experimental investigation of the dynamics of molecules by spectroscopic techniques.

Molecular dynamics is remarkably simple in principle. If one has reliable information regarding the interactions between the particles in a many-particle system, one can solve Newton's equations (named for Isaac Newton, of course) and compute the trajectory of each and every particle as a function of time. For our purposes here, the particles are atoms, and the relevant interactions are described by analytical expressions for the interatomic potentials. If all the atoms were of the same element, there would be just the one potential energy function, $\mathscr{E}(r)$. The Cartesian coordinates of each and every atom are known, at any one instant during the simulation, so it is a straightforward matter to compute all the components of force on a given atom, both with respect to magnitude and direction. The prescription for finding each component is the familiar

$$\mathscr{F}(r_{ij}) = -\nabla \mathscr{E}(r_{ij}) \qquad (6.50)$$

where r_{ij} is the distance between the ith and the jth atoms. These force compon-ents are then added vectorially, to produce the net force on the atom in question. And this procedure is repeated for all the other atoms. Knowledge of all the individual forces enables one to compute how much each atom will move within a small increment of time, because one has the well-known relationship

$$\mathcal{F}_i = md^2r_i/dt^2 \tag{6.51}$$

$\mathcal{F}_i$ being the force on the ith atom. In practice, because r_i is itself a position vector, one calculates the individual Cartesian components of the force and then adds them vectorially. Knowing the acceleration of each atom, and its instantaneous direction of travel, one can compute where it will be at the end of the time increment. When this has been done for each and every atom, the entire procedure is repeated, and so on.

In effect, one is simulating the generally curved trajectory of each atom by a series of straight lines, and it is clear that this sets a severe constraint on the size of the time step. For example, the time taken for an atom in a typical large molecule to perform a single vibration about its mean position is approximately 10^{-12} s. This being the case, it would be obvious nonsense to attempt to carry out a simulation using a time step of the same size; the atom would fly off in a straight line instead of performing the oscillation. In practice, the optimal computational time step lies around 10^{-14} s.

As with all simulations, one benefit lies in being able to verify the validity of the original assumptions. In a molecular dynamics simulation, these will especially concern the assumed functional form of the interatomic potential. For example, one might wish to check whether purely central forces (see Chapter 3) are adequate. The other great benefit, which indeed provides such simulations with their prime motivation, is that one can examine the model at atomic resolution, in order to discover which are the critical movements underlying a given process, such as diffusion of an oxygen atom into or out of a myoglobin molecule.

Comparison of the simulation's thermodynamic variables with their experimental counterparts is straightforward. Because one has computed the positions of all the atoms as a function of time, one automatically has a record of their velocities. And in the classical limit, these velocities are simply related to the temperature T by

$$T = \frac{1}{3Nk_B} \left\langle \sum_{i=1}^{N} v_i^2 \right\rangle \tag{6.52}$$

where N is the total number of atoms and v_i is the velocity of the ith atom. The brackets indicate averaging over a suitable period of time, this being necessary because there will be inevitable fluctuations of the thermodynamic variables in a small system. Similarly, one can compute the pressure P through the expression

$$P = \rho k_B T - \frac{\rho}{6N} \left\langle \sum_{i=1}^{N} \sum_{j=1}^{N} r_{ij} \frac{\partial \mathcal{E}(r_{ij})}{\partial r_{ij}} \right\rangle \tag{6.53}$$

this following from the virial theorem, with ρ being the *number* density.

The situation at any instant in a multi-atom system is so complicated that an analytical expression for the underlying dynamics is out of the question. The equations of motion must therefore be solved numerically, and a number of different algorithms have been tested. These are all approximations, and the goal of the simulator is to minimize the influence of the round-off errors, the obvious constraint being that energy and momentum must be conserved. Because a typical simulation can involve hundreds of atoms, and thousands of time steps, one also strives for computational efficiency, while maximizing the time step. If the latter is too large, energy and momentum fails to be conserved and the system is observed to 'blow up'. With the advent of standard computer graphics routines, it is now easy to plot the positions of all the atoms as a function of time, and an entire simulation can then be replayed as an atomic-level movie. (One thus adopts the role of the proverbial Maxwell's demon.)

It is beyond the ambitions of the present section to give an exhaustive account of the different algorithms that have been employed, so suffice it to say that the following one, due to Loup Verlet, is particularly successful. To save space, the resolution of the position vector r into its Cartesian components will be simply implicit. Indicating the time step by Δt we have

$$r(t + \Delta t) = r(t) + v(t)\Delta t + (1/2)a(t)\Delta t^2 \tag{6.54}$$

$$v(t + \Delta t/2) = v(t) + (1/2)a(t)\Delta t \tag{6.55}$$

$$a(t + \Delta t) = -(1/m)\nabla \mathscr{E}\{r(t + \Delta t)\} \tag{6.56}$$

$$v(t + \Delta t) = v(t + \Delta t/2) + (1/2)a(t + \Delta t)\Delta t \tag{6.57}$$

where $a(t)$ is the acceleration. One advantage of this particular algorithm lies in its minimal memory requirements. There are $3N$ positional co-ordinates, and the same number of velocity and acceleration components, so $9N$ memory locations are required. However, there is no need to additionally store the sum of those same variables for the previous time step, as is the case for most other algorithms.

The above expressions are adequate if the simulation involves only independent atoms, but when molecules are to be simulated, complications arise from the fact that one needs to follow the dynamics of units that can rotate. This requires that Newton's equations be augmented by the Euler equations (named after Leonhard Euler). The three components of the torque on the ith molecule are given by

$$\Gamma_{\alpha, i} = I\dot{\omega}_{\alpha, i} - \omega_{\beta, i}\omega_{\gamma, i}(I_\beta - I_\gamma) \tag{6.58}$$

$$\Gamma_{\beta, i} = I\dot{\omega}_{\beta, i} - \omega_{\gamma, i}\omega_{\alpha, i}(I_\gamma - I_\alpha) \tag{6.59}$$

$$\Gamma_{\gamma, i} = I\dot{\omega}_{\gamma, i} - \omega_{\alpha, i}\omega_{\beta, i}(I_\alpha - I_\beta) \tag{6.60}$$

in which $I = (I_\alpha, I_\beta, I_\gamma)$ is the diagonalized moment of inertia tensor, and α, β and γ are the Eulerian angles. The angular velocity components are

$$\omega_{\alpha, i} = \dot{\alpha}_i \sin(\beta_i) \sin(\gamma_i) + \dot{\beta}_i \cos(\gamma_i) \tag{6.61}$$

$$\omega_{\beta, i} = \dot{\alpha}_i \sin(\beta_i) \cos(\gamma_i) - \dot{\beta}_i \sin(\gamma_i) \tag{6.62}$$

$$\omega_{\gamma, i} = \dot{\alpha}_i \cos(\beta_i) + \dot{\gamma}_i \tag{6.63}$$

Finally, we have the rotational component of the temperature, which is given by

$$T_{\text{rot}} = \frac{1}{2} \sum_{i=1}^{N} \omega_i \bullet I_i \bullet \omega_i \tag{6.64}$$

and in the classical limit one may expect equipartition to prevail between the translational and rotational components of the temperature.

Molecular dynamics is now a venerable science. It had its origins in the mid-1950s, when computers themselves were in their infancy, the pioneers being Berni Alder and Thomas Wainwright, who simulated the dynamics of hard spheres, and independently by George Vineyard and his colleagues, who were the first to employ realistic interatomic potential functions. Aneesur Rahman and Frank Stillinger carried out the first simulation of water, in 1972, and the first simulation of a protein was reported in 1977, by Martin Karplus and his colleagues, the object of their investigation being bovine pancreatic trypsin inhibitor.

6.8   Potential Energy Contour Tracing

Insofar as one is using a realistic analytical function to represent the inter-atomic and/or inter-molecular potential, a molecular dynamics simulation can be regarded as a reliable facsimile of the actual situation. One can also place considerable trust in the observed dynamical behaviour of the system being investigated. There is, nevertheless, a severe disadvantage, namely that of the brevity of the overall simulation. As was noted in the previous section, the optimal time step in molecular dynamics is typically about 10^{-14} s, so even a simulation spanning 10^6 time steps would cover only 10^{-8} s of real time, despite the fact that it would consume several hours on a 2001-vintage computer; and 10^{-8} s is only about a millionth of the time it takes a typical protein to fold into its operational conformation.

It is frequently the case that one is not so much interested in following the actual dynamics as in predicting the minimum-energy conformation of a large molecule such as a protein. This is not yet possible, at the time this book is being written, but some progress can be made in charting the energy land-scape in the high dimensionality required to adequately describe a molecule comprising many atoms. The point here is that each atom's position requires specification of three Cartesian coordinates, so a complete description of a molecule comprising N atoms requires one to take account of $3N$ coordinates. And because the potential energy is influenced by the position of each and every atom, finding the lowest-energy configuration of the molecule involves investigation of a potential-energy surface (or manifold) in $3N$ dimensions.

A molecular dynamics simulation does probe this potential-energy surface, of course, but it does so by exploring the energy barriers between local minima, in the haphazard fashion that characterizes thermal motions. The purpose of potential energy contour tracing (PECT), on the other hand, is to probe the topology of the energy manifold in a manner that is inaccessible to the thermal regime. A method is available which enables the state point to run around contours of constant potential energy, thus giving an indication of the local topology of the energy manifold. A simple analogy of these contours is seen in a geographical map, which depicts the contours of equal height above sea level. A constant potential energy contour is formally defined by

$$\mathscr{E}(r) = \mathscr{E}_0 \qquad\qquad (6.65)$$

and it is of course a $(3N - 1)$-dimensional manifold. The vector r has $3N$ individual Cartesian components, and it describes the total configuration of the molecule. This is often referred to as the state point in configuration space. For any such state point, the system can undergo an incremental change in $(3N - 1)$ mutually orthogonal directions but nevertheless comply with Equation (6.65), the one proviso being that the state point does not happen to coincide with a local minimum or local maximum of the potential energy manifold.

The PECT method is based on differential equations bearing a resemblance to those used in molecular dynamics, but they result in a trajectory consistent with Equation (6.65). The starting point for a PECT run is a set of position and velocity coordinates, just as in molecular dynamics. Indeed, it usually proves most convenient to start a contour study by initially simulating the normal dynamics, and then switching to PECT. The PECT trajectory will then initially lie close to the one that has been followed with the molecular dynamics. Let us assume that is in fact what has been done, and let the vectors describing the position and velocities at the time of the switch be r_0 and v_0 respectively. In the molecular dynamics simulation, as discussed in the previous section, these vectors will have changed to r_1 and v_1 a small time step Δt later.

In the case of molecular dynamics, Newton's Laws gave us the relationships

$$r_1 = r_0 + v_0\Delta t \tag{6.66}$$

and

$$v_1 = v_0 + \mathcal{F}\Delta t/m \tag{6.67}$$

where $\mathcal{F}$ is the force, as in the previous section. In the PECT approach, the prescription for finding the new configuration vector is just as it is in molecular dynamics, but the expression for the new velocity vector is rather different. We have, then,

$$r_1 = r_0 + v_0\Delta t \tag{6.68}$$

and

$$v_1 = v_0 - \text{grad}\mathcal{E}\,(\text{grad}\mathcal{E} \bullet v_0)/(|\text{grad}\mathcal{E}| \bullet |\text{grad}\mathcal{E}|) \tag{6.69}$$

which leads to

$$v_1 \bullet \text{grad}\mathcal{E} = 0 \tag{6.70}$$

As can be seen from Equation (6.70), the velocity vector has no component along any direction in which the potential energy is changing. It is thus tracing out a contour (that is to say, $(3N - 1)$-dimensional manifold) of constant potential energy, as desired.

These expressions still include the quantity Δt, though in the case of PECT it has no physical significance. However, it nevertheless plays the same computational role as it does in molecular dynamics (MD), and it still has the dimension of time, of course. One could thus call it a pseudo time step. The state point traces out a smooth trajectory in PECT, just as it does in MD, and the composite trajectory – MD before the switch, and PECT after – is continuous, but not necessarily differentiable at the switch-over point. Differentiability would prevail, however, if the switch-over occurred when the potential energy in the MD simulation was at a local minimum or local maximum value. In default of this, there will be a cusp, the break being of the minimum magnitude necessary to change the velocity from its MD value to a PECT-compatible value.

One must bear in mind that there is a $(3N - 1)$-dimensional manifold of possible velocities compatible with PECT at a point in $3N$-dimensional configuration space. So although this simulation method provides a unique way of probing the energy landscape, one must not underestimate the size of the task if one desires a complete picture of the energy manifold's overall topology. This is particularly so if one seeks to discover the point in configuration space corresponding to the global minimum of potential energy. One's hope would be that the topology of the manifold reasonably close to the global minimum would actually reflect that proximity. This would not be the case if the topology were analogous to that of a billiard table, with a moving ball being given no warning of its impending fall into one of the pockets. Those interested in the structure of proteins, for example, would naturally hope that the topology near the global energy minimum would be more like what is seen on a putting green. The PECT method was invented by Rodney Cotterill and Jens Ulrik Madsen in 1988, and it was first used to investigate protein structure 4 years later, by Barry Robson and his colleagues.

Exercises

6.1   Prove that the relationship

$$d_{hkl} = \left(\frac{h^2}{a^2} + \frac{k^2}{b^2} + \frac{l^2}{c^2}\right)^{-1/2}$$

is true, d_{hkl} being the inter-planar spacing between a set of planes with Miller indices h, k and l, while a, b and c are the side lengths of the unit cell. In the simple cubic system, what is the angle between the directions [110] and [112]? In the face-centred cubic system, what is the angle between the [111] and[110] directions?

6.2   What is the Larmor frequency of a proton in a magnetic field of 2 T, and what is the energy difference between the upper and lower spin states, given that the magnetogyric ratio of a proton is $2.675 \times 10^8 s^{-1} T^{-1}$? What will be the Boltzman factor for a system of protons at a temperature of 300 K?

6.3   Through what distance is the tip of an atomic force microscope's cantilever maximally deflected by a single interatomic bond of the Lennard-Jones type, given that the cantilever's spring constant is $10^{-3} Nm^{-1}$, the Lennard-Jones distance and energy parameters being 0.20 aJ and 0.15 nm, respectively?

6.4   Write a small computer program (in the language of your choice) that simulates the one-dimensional interaction between just two atoms, using the molecular dynamics technique. Assume that the interaction between the atoms is adequately described by the Lennard-Jones potential function, the distance and energy parameters being 0.50 aJ and 0.15 nm, respectively, and let the mass of each atom be that of an argon atom. Study the system for various initial separations between the atoms, and compute the velocity auto-correlation function, $\gamma(t) = \langle v(\tau + t) \bullet v(\tau) \rangle$, τ being some convenient starting time and the brackets indicating a system average. By taking the Fourier transform of $\gamma(t)$, compute the spectral density function for the system, under the conditions of your computer experiment.

Further reading

X-ray diffraction and molecular structure

Blundell, T. L. and Johnson, L. N., (1976). *Protein Crystallography*. Academic Press, New York.

Bragg, W. L., (1929). An optical method of representing the results of X-ray analysis. *Zeitschrift für Kristallographie* **70**, 475–492.

Buerger, M. J., (1960). *Crystal-Structure Analysis*. Wiley, New York.

Cochran, W., (1958). Structure factor relations and the phase problem. *Acta Crystallographica* **11**, 579–585.

Glusker, J. P. and Trueblood, K. N., (1985). *Crystal Structure Analysis*. Oxford University Press, New York.

Hauptman, H. A. and Karle, J., (1950). Relations among the crystal structure factors. *Physical Review* **80**, 244–248.

Hauptman, H. A., (1989). The phase problem of X-ray crystallography. *Physics Today* **42**(11), 24–29.

Klug, A., (1958). Joint probability distribution of structure factors and the phase problem. *Acta Crystallographica* **11**, 515–543.

Madsen, J. U. and Cotterill, R. M. J., (1978). Crystallographic phase determination from dark-field electron-microscope Moiré patterns. *Acta Crystallographica* **A34**, 378–384

Perutz, M. F., (1956). Isomorphous replacement and phase determination in non-centrosymmetric space groups. *Acta Crystallogaphica* **9**, 867–873.

West, J., (1930). A quantitative X-ray analysis of the structure of potassium dihydrogen phosphate. *Zeitschrift für Kristallographie* **74**, 306–332.

Wilson, A. J. C., (1970). *Elements of X-Ray Crystallography*. Addison-Wesley, Reading, MA.

Nuclear magnetic resonance

Abraham, R. J., Fisher, J. and Loftus, P., (1988). *Introduction to NMR Spectroscopy*. Wiley, Chichester.

Breitmeier, E., (1993). *Structure Elucidation by NMR in Organic Chemistry: A Practical Guide*. Wiley, Chichester.

Ernst, R. R., Bodenhausen, G. and Wokaun, A., (1987). *Principles of Nuclear Magnetic Resonance*. Clarendon Press, Oxford.

Gunther, H., ed., (1995). *NMR Spectroscopy*. Wiley, Chichester.

Neuhaus, D. and Williamson, M., (1989). *The Nuclear Overhauser Effect in Structural and Conformational Analysis*. VCH Publishers, New York.

Slichter, C. P., (1978). *Principles of Magnetic Resonance*. Springer-Verlag, Berlin.

Wüthrich, K., (1986). *NMR of Proteins and Nucleic Acids*. Wiley, New York.

Wüthrich, K., (1990). Protein structure determination in solution by NMR spectroscopy. *Journal of Biological Chemistry* **265**, 22059–22062.

Scanning tunnelling microscopy

Binnig, G. and Rohrer, H., (1982). Scanning tunnelling microscopy. *Helv. Phys. Acta.* **55**, 726–735.

Binnig, G. *et al.*, (1982). Surface studies by scanning tunnelling microscopy. *Phys. Rev. Lett.* **49**, 57–61.

Travaglini, G. *et al.*, (1987). Scanning tunnelling microscopy on biological matter. *Surface Science* **181**, 380–390.

Atomic force microscopy

Binnig, G., Quate, C. F. and Gerber, C., (1986). Atomic force microscope. *Phys. Rev. Lett.* **12**, 930–933.

Rademacher, M. *et al.*, (1992). From molecules to cells: imaging soft samples with the atomic force microscope. *Science* **257**, 1900–1905.

Schneider, S. W. *et al.*, (1999). Continuous detection of extracellular ATP on living cells by using atomic force microscopy. *Proc. Nat. Acad. Soc. USA.* **96**, 12180–12185.

Wiegräbe, W. *et al.*, (1991). Atomic force microscopy of a hydrated bacterial surface protein. *J. Microscopy* **163**, 79–84.

Worcester, D. L., Miller, R. G. and Bryant, P. J., (1988). Atomic force microscopy of purple membranes. *J. Microscopy* **152**, 817–821.

Optical tweezers

Ashkin, A., (1970). Acceleration and trapping of particles by radiation pressure. *Phys. Rev. Lett.* **24**, 156–159.

Chu, S., (1991). Laser manipulation of atoms and particles. *Science* **253**, 861–866.

Oddershede, L. *et al.*, (2001). Optical tweezers: probing biological surfaces. *Probe Microscopy* **2**, 129–137.

Sheetz, M. P., (1998). *Laser Tweezers in Cell Biology* (*Methods in Cell Biology, Volume 55*) Academic Press, San Diego.

Svoboda, K. and Block, S. M., (1994). Biological applications of optical forces. *Ann. Rev. Biophys. Biomol. Struct.* **23**, 247–285.

Patch clamp

Neher, E. and Sakmann, B., (1976). Single-channel currents recorded from membrane of denervated frog muscle fibres. *Nature* **260**, 779–802.

Hamill, O. P. *et al.*, (1981). Improved patch-clamp techniques for high-resolution current recording from cells and cell-free membrane patches. *Pflügers Arch.* **391**, 85–100.

Ogden, D. and Stanfield, P., (1987). Patch clamp techniques for single channel and whole-cell recording. In: *Microelectrode Techniques*, D. Ogden ed. The Company of Biologists Limited Cambridge.

Hille, B., (1992). *Ionic Channels of Excitable Membranes, Second Edition*. Sinauer, Sunderland, MA.

Neher, E. and Sakmann, B., (1995). *Single Channel Recording, Second Edition*. Plenum, New York.

Molecular dynamics

Alder, B. J. and Wainwright, T. E., (1957). Phase transition for a hard-sphere system. *Journal of Chemical Physics* **27**, 1208–1209.

Gibson, J. B. *et al.*, (1960). Dynamics of radiation damage. *Physical Review* **120**, 1229–1253.

Gould, H. and Tobochnik, J., (1988). *An Introduction to Computer Simulation Methods: Applications to Physical Systems (Parts 1 and 2)*. Addison-Wesley, Reading, MA.

Lykos, P., ed., (1978). *Computer Modelling of Matter*. American Chemical Society, Washington, DC.

McCammon, J. A., Gelin, B. R. and Karplus, M., (1977). Dynamics of folded proteins. *Nature* **267**, 585–590.

Stillinger, F. H. and Rahman, A., (1972). Molecular dynamics study of temperature effects on water structure and kinetics. *Journal of Chemical Physics* **57**(3), 1281–1292.

Verlet, L., (1972). Molecular motions in simple liquids. *Faraday Symposium of the Chemical Society* **3**, 116–121.

Potential energy contour tracing

Cotterill, R. M. J. and Madsen, J. U., (1990). Potential energy contour tracing: an efficient way of exploring configuration hyperspace. In: *Characterising Complex Systems*, ed. H. Bohr. World Scientific, Singapore, pp 177–191.

Li, J. *et al.*, (1992). Exploration of the phase space of molecular systems: assessment of established and new methods. *Biophysical Chemistry* **43**, 221–238.

7 Biological Polymers

The word polymer is, appropriately, a composite of the Greek word *poly*, which means many, and another Greek word *meros*, which means parts. The choice of this name stems from what has been found to be the defining characteristic of a polymer: it is a composite structure based on consolidation of many smaller units (monomers) into a single whole. The consolidation in question occurs in one topological dimension, to produce thread-like arrangements of the elementary units, but these threads can (and frequently do) become multiply folded, so that the final conformation extends in at least two, and usually three, dimensions.

Polymeric structures are not uncommon in the inorganic realm, asbestos being a good example, but we will here be interested only in the organic variety; and, indeed, only in a sub-class of the organic polymers, namely the biopolymers. Organic polymers not falling within the biological domain are interesting in their own right, and their investigation led to insights that were valuable when considering their biological counterparts. These organic non-biological polymers have structures in which the monomers tend to be of only one type, as in the numerous artificial polymers that have been synthesized. Nylon, polyethylene (commercially abbreviated to polythene), styrofoam and teflon are just a few of the names that readily come to mind.

Biopolymers, then, are simply polymers that occur in biological systems. They tend to be comprised of a variety of monomers, rather than of a single type, and this gives them physical properties that are far more sophisticated than those of their non-biological counterparts. It is this sophistication that enables them to imbue their parent tissues with the differentiated response repertoires typical of living structures. As William Cowper put it: *variety's the very spice of life*. Very broadly speaking, the biopolymers can be divided into four main classes: nucleic acids, proteins, lipids (and the related fatty acids) and polysaccharides. The latter are the main constituents of plant tissue (together with more complicated biopolymers known as lignins), and when they are present on the animal cell surface, they mediate cell–cell recognition. There will not be room in this book for a discussion of these molecules. The lipids must be included, however, and they will be dealt with in the next chapter, in the broader context of biological membranes. We will be concerned with the first two classes in this chapter, that is to say with the nucleic acids and the proteins.

7.1 Nucleic Acids

The importance of nucleic acids to heredity emerged only gradually. Gregor Mendel's discoveries in 1865, based on his observations on the breeding of peas, were followed 20 years later by August Weismann's hypothesis that the number of chromosomes must be constant. There is not a simple relationship between the number of chromosomes and biological sophistication, however, because the potato possesses more chromosomes than the human being (who has 23 pairs). In 1926, Thomas Morgan suggested that genes are not separated entities, but that they are grouped together in the chromosomes. Sewall Wright had already postulated that genes control the production of enzymes. In 1900, Hugo de Vries had noticed that changes in the colouring of primroses are not gradual, but occur abruptly as spontaneous mutations. Hermann Müller and L. G. Stadlerr discovered that the mutation rate can be greatly increased by exposing the gamete cells (that is to say the unfertilized egg cells) to X-rays, and George Beadle and Edward Tatum irradiated specimens of the common bread mould *Neurospora crassa* and discovered that it was then incapable of producing Vitamin B_6. Wright's gene-enzyme link had been proved correct.

Frederick Griffith showed that the pneumonia bacterium, *Diplococcus pneumoniae*, is virulent when its polysaccharide coat is intact. This is designated the S form, from the smooth appearance of the resulting colonies. Rough-looking colonies are formed by the variety that lacks its coat (the R form), this being a consequence of its not possessing the relevant enzyme. Oswald Avery, Colin MacLeod and Maclin McCarthy repeated the Griffith experiment with various parts of the S form removed. Prior removal of the polysaccharide coat had no effect, nor did removal of an underlying protein capsule. However, if the bacterium's deoxyrobonucleic acid (DNA) is removed or denatured, its lethal potential is destroyed. Avery and his colleagues had established that DNA is the repository of the genetic message.

Observed in an electron microscope, DNA appears as a long, thin thread, its diameter being a mere 2 nm. Maurice Wilkins, Rosalind Franklin and Raymond Gosling were surprised to find that X-ray diffraction patterns from such material did not display the expected variation with the species from which it had been extracted, and they were puzzled by the appearance of reflections corresponding to a distance of 3 nm, which is considerably larger than the size of a single nucleotide (that is to say, the relevant monomer). There are two types of nucleotide, namely the purines and the pyrimidines. These comprise identical sugar and phosphate groups, one of each indeed, whereas the atomic configuration of the additional ring structures differs in the two types of nucleotide.

Moreover, the purines come in just two varieties, namely adenine and guanine, and the same is true of the pirimidines, which are all either thymine or cytosine.

Erwin Chargaff and his colleagues investigated the chemistry of DNA and found roughly equal amounts of deoxyribose (the sugar) and phosphate. This was not surprising because there are equal amounts of these substances in each nucleotide monomer, as we have just seen. And they found variation in the amount of purine, for example, from species to species. This presumably had something to do with the hereditary message being stored in the molecule. The real fascination lay in certain regularities. They found that the proportions of pyrimidine and purine were equal. Even more remarkably, although the amounts of adenine (a purine) and guanine (a purine) varied from one species to another, the amount of adenine was always the same as the amount of thymine (a pyrimidine), and a similar equivalence was found between the amounts of guanine and cytosine (a pyrimidine).

Incorporation of these X-ray and biochemical facts into an acceptable model of DNA was accomplished by Francis Crick and James Watson in 1953. The pivotal discovery was the fact that a co-planar hydrogen-bonded adenine–thymine pairing produces a di-nucleotide having exactly the same overall length as a co-planar hydrogen-bonded guanine–cytosine pairing. They found that the molecule is an example of what polymer chemists call a ladder polymer, the rungs being of equal length because of the equality of these AT and GC pairings. The sides of the ladder are composed of regularly alternating sugar and phosphate units. Watson and Crick found that steric hindrance causes the two sugar-phosphate backbones to twist into a helical conformation: the celebrated double helix (see Figure 7.1). The degree of twist is such that successive base pairs are mutually offset by roughly 34°, so one complete turn of the double helix (that is to say, the pitch) comprises 10 such pairs. (In practice, the number is closer to 10.6.) This distance corresponds to 3.4 nm, and this was therefore the origin of the mysterious X-ray reflections. As one can see from Figure 7.1, the two strands which constitute the double helix are complementary, and this enables each of them to serve as a template for the formation of a new complementary strand (known as replication), following the separation of the two original strands in cell division.

Much has been discovered since that breakthrough. It is now well established, for example, that each gene codes for a specific protein, and that some proteins play structural rather than enzymatic roles. Moreover, the genetic message is not used directly in dictating protein structure. Instead, the message is first transferred to a strand of messenger-RNA (RNA being ribonucleic acid), this stage being known as transcription. Apart from containing a different form of sugar, RNA differs from DNA in the substitution of uracil for thymine. Vitally, uracil can form a base pair with adenine. The final translation stage involves triplets of bases (known as codons) being recognized by proteins known as ribosomes, each such act of recognition providing the cue for attachment of an

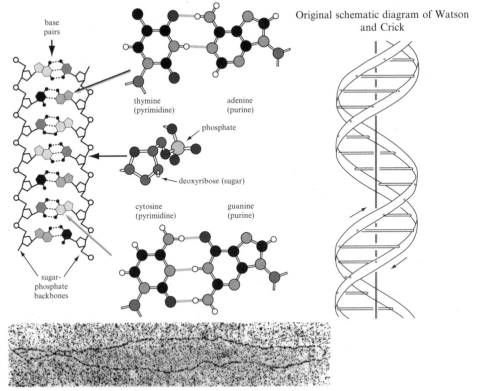

base pairs

thymine
(pyrimidine)

adenine
(purine)

phosphate

deoxyribose (sugar)

cytosine
(pyrimidine)

guanine
(purine)

sugar-
phosphate
backbones

Original schematic diagram of Watson
and Crick

Electron photomicrograph of DNA of simian virus SV40

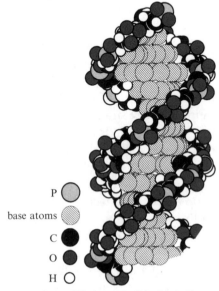

P

base atoms

C

O

H

Space-filling model of double helix

additional amino-acid residue (see below). The latter is the remnant of an amino-acid molecule, the correct variety of which has been selected by a strand of transfer-RNA, and transported to the appropriate site on the ribosome. The growing polypeptide chain increases its length by one amino-acid residue for each such event. There are 20 amino acids to choose from, so a genetic code consisting of pairs of bases would provide only 16 different possibilities, and would thus be inadequate, as first argued by George Gamow. A triplet code, on the other hand, would provide 64 possibilities. This is more than enough, and the code is in fact redundant, some amino acids being coded for by more than one triplet (see Table 7.1). (There are also three triplets that code for 'stop', namely UAA, UGA and UAG.)

There is a considerable amount of physics implicit in what has already been described. Most obviously, there is the hydrogen bonding between purine–pyrimidine base pairs: two bonds in an AT pair and three in a GC pair. Then

Table 7.1   The genetic code

		Second base					
		U	C	A	G		
F		phe	ser	tyr	cys	U	T
i		phe	ser	tyr	cys	C	h
r	U	leu	ser	stop	stop	A	i
s		leu	ser	stop	trp	G	r
t							d
		leu	pro	his	arg	U	
		leu	pro	his	arg	C	
	C	leu	pro	gln	arg	A	
b		leu	pro	gln	arg	G	b
a							a
s		ile	thr	asn	ser	U	s
e	A	ile	thr	asn	ser	C	e
		ile	thr	lys	arg	A	
		met	thr	lys	arg	G	
5′							3′
		val	ala	asp	gly	U	
e		val	ala	asp	gly	C	e
n	G	val	ala	glu	gly	A	n
d		val	ala	glu	gly	G	d

Figure 7.1   The ladder polymer deoxyribonucleic acid (DNA) consists of two sugar-phosphate backbones joined by pyrimidine-purine rungs. The upper left-hand diagram shows the molecule unwound and with the bases rotated into plan view. Thymine pairs only with adenine, and cytosine only with guanine, such base pairings producing rungs of equal length. The backbones are twisted into a double helix, as indicated in the original Watson-Crick diagram (upper right), while the basal planes that form the rungs lie at right angles to the helix axis (bottom picture). The electron photomicrograph of simian virus DNA (middle picture) does not resolve such details, and the molecule appears as a thin continuous line

there is the steric hindrance effect in the sugar-phosphate backbones. A third issue concerns the interaction between a strand of messenger-RNA and a ribosome molecule, and also the one that governs the interaction of the ribosome with the transfer-RNA. Then again, there are the intra-molecular forces that dictate the three-dimensional structure of a transfer-RNA molecule. This structure involves a sort of clover-leaf arrangement (see Figure 7.2), the various sub-structures being stabilized by hydrogen bonding between complementary bases on a *single* thread of RNA. Moreover, we earlier mentioned the denaturing of biopolymers, and when this is induced by a rise in temperature, it can be regarded as a special type of melting. Finally, there is radiation-induced mutation, which is reminiscent of what was discussed at the end of chapter 4.

The electronic digital computer is now an indispensable tool in the study of biopolymer structure and interaction. Pharmaceutical companies at the turn of the millennium were investing large amounts of money in what is referred to logically as computer-aided molecular design (CAMD), and the larger companies

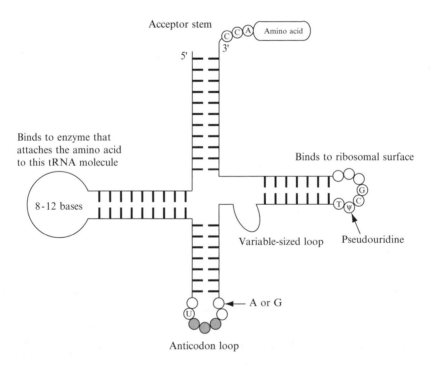

Figure 7.2 This planar view of an unwound transfer ribonucleic acid (tRNA) molecule indicates the cloverleaf structure formed by base-pairing, the unpaired bases (some unusual) producing four loops. The largest loop contains between eight and twelve bases, while two of the others have seven. The three-base anticodon is indicated by the shading. The native three-dimensional structure, determined by Alexander Rich and his colleagues, resembles a twisted L

have armies of theoretical scientists who sit at computer screens attempting to tailor molecules to fit medical requirements. We have not yet considered protein structure, so as a first example in this chapter of the application of physical principles let us consider the structure of nucleic acids.

7.2 Nucleic Acid Conformation: DNA

It seems safe to assume that the above-quoted pitch of the double helix of DNA corresponds to the free-energy minimum for that molecule. Within a cell, however, the scarcity of room forces the DNA into a strained condition. If the DNA in a bacterium such as *Escherichia coli* was stretched out into a line, it would measure almost 1 mm, which is about a thousand times the length of the creature. We must conclude that the DNA is forced into higher degrees of coiling, to make it more compact. Similarly, in the eucaryotic (that is to say, nucleus-containing) cell, the DNA in the chromosomes displays a considerable degree of such supercoiling.

We can compare the two sugar-phosphate backbones in the DNA molecule with the edges of a ribbon, which are twisted about each other and also about the helix axis (which is a straight line in the non-supercoiled state shown in Figure 7.1). The molecule often exists in the form of a closed loop, that is to say with no free ends. The complete topology of the molecule requires specification of three numbers: the linking number, the twist and the writhe. The linking number, L, is the number of times that one ribbon edge winds around the other. Because the molecule is a closed loop, the linking number is always an integer. If the linking number of our ribbon is 0, a cut along the entire length of the ribbon axis would produce two non-linked half-ribbons. If the linking number is 1, however, the cut would produce two linked half-ribbons; they cannot be separated from each other. (A Moebius strip has a linking number of 0.5, because the ribbon is rotated through only 180° before its ends are rejoined. It has only a *single* continuous edge. This is not possible in DNA, because the backbones have directionality, and they run in opposite directions. If that had not been the case, a linking number of 0.5 would have been topologically possible, but the genetic message would then become ambiguous.) The linking number obviously stays constant, unless the helix is cut, twisted, and then rejoined. This can happen, in fact, the responsible enzymes being known as a topoisomerases.

The twist, T, is the number of times either edge winds around the helix axis. It can vary from one place to another along the length of the molecule, and it need not be an integer. Depending on the sign of the twist, T can be positive or negative. It is clear that an approximate value of T can be obtained by dividing the number of base pairs by the above-quoted number of pairs per pitch of the helix, namely 10.6. So the DNA of the simian SV40 virus shown in Figure 7.1, which comprises about 5500 base pairs, has a T value of slightly less than 500. The writhe, W, is the number of turns that the helix makes around the supercoil axis, so it is automatically zero if there is no supercoiling. As in the case of T, W need not be an integer, and it too can be positive or negative. If the helix axis lies on a plane, W is zero. When W is non-zero, the helix axis itself winds up into a double helix. This can be left-handed or right-handed.

We see that L and T describe the relative positions of the ribbon edges, whereas W is related to the spatial trajectory traced out by the ribbon axis. It is not difficult to show that the three numbers are interdependent, because we have

$$L = W + T \tag{7.1}$$

Given the constancy of L, this means that

$$\Delta W = -\Delta T \tag{7.2}$$

The topology of the supercoil is biologically important. If the writhing is particularly pronounced, some of the base pairs can be prised apart, and this makes the individual bases available for interactions with other molecules. It is thus of importance to enquire as to the energetics of supercoiling, but the situation is complicated because of the multiplicity of energy contributions. These include elastic effects which stem from the underlying interatomic interactions, and there are also effects due to the surrounding aqueous solution.

One must also take thermal fluctuations into account. Body temperature is a relatively high temperature for DNA. Indeed, its melting point lies in the range 80–90°C. So we must think in terms of average values when considering the three numbers identified above. Let these be denoted by the small letters l, t and w, and let the corresponding relaxed values be l_0, t_0 and w_0. The energy contributions can then be written

$$\mathscr{E}_L = \alpha_L f_L(l, l_0) \tag{7.3}$$

$$\mathscr{E}_T = \alpha_T f_T(t, t_0) \tag{7.4}$$

$$\mathscr{E}_W = \alpha_W f_W(w, w_0) \tag{7.5}$$

The coefficients α_L, α_T and α_W include a temperature-dependent energy factor, while the functions f_L, f_T and f_W are purely numerical. We naturally need to know the nature of those functions.

The function that dictates the linking energy can be explored experimentally by arranging for the DNA to interact with an enzyme that cuts the molecule into fragments, and then with another enzyme that joins the resulting free ends, so as to produce smaller closed loops of various circumferences. These are then subjected to gel electrophoresis (invented by Arne Tiselius), in which the loops with the greater amount of writhing diffuse more rapidly, because they are wound up into clumps having the smaller overall diameters (we recall the formula derived by George Stokes, discussed in Chapter 5). It is found that the relative probability density for diffusion length $n(x)$ follows an Arrhenius distribution

$$\frac{n(x)}{n_0(x)} = A\mathrm{e}^{-\mathscr{E}(l-l_0)^2/k_B T} \tag{7.6}$$

In this equation, A is a constant and $n_0(x)$ is the total density of coils. The second moment of the probability distribution, $\langle (l - l_0)^2 \rangle$ is given by

$$\langle (l - l_0)^2 \rangle = k_B T/2\mathscr{E} \tag{7.7}$$

These results indicate that the unknown energy function for the linking number is well represented by a harmonic approximation (see Chapter 3). Moreover, the twist energy is the simplest component of the three, and our knowledge of elasticity (see Chapter 5) suggests that it too is amenable to an harmonic approximation. Thus we have

$$\mathscr{E}_L = \alpha_L(l - l_0)^2 \tag{7.8}$$

$$\mathscr{E}_T = \alpha_T(t - t_0)^2 \tag{7.9}$$

The energy of supercoil formation $\mathscr{E}_{SC}$ for a given value of L, is the sum of the twist and writhe energies, and it is the latter that is the most difficult to derive from fundamental principles. This is to be expected, given the complexity of a coiled coil. We have

$$\mathscr{E}_{SC} = \mathscr{E}_T + \mathscr{E}_W = \alpha_T(t - t_0)^2 + \alpha_W f_W \tag{7.10}$$

Moreover, because we are working with average values of the twist and writhe numbers, the supercoil energy must already be at its minimum with respect to changes in those parameters. So we have

$$\mathrm{d}\mathscr{E}_{SC} = 2\alpha_T(t - t_0)\mathrm{d}t + \alpha_W f_W' \mathrm{d}w = 0 \tag{7.11}$$

And because of the topological constraint expressed by Equation (7.2), namely that $dt = -dw$, we obtain

$$t - t_0 = (\alpha_W/2\alpha_T)f_W' \tag{7.12}$$

If we make a Taylor expansion for f_W, the zeroth and first-order terms vanish at $w = 0$ if the function has a zero-valued minimum at that position, which seems reasonable in view of the similar behaviour of the other two functions. The first-order and second-order terms are therefore $wf'_W(w)$ and $w^2f''_W(w)/2$, respectively. So Equation (7.12) is transformed into

$$t - t_0 = (\alpha_W/2\alpha_T)wf''_W(0) \qquad (7.13)$$

This expression can be inserted into Equation (7.10), and w can be written in terms of l and t, because of Equation (7.1). This leaves an expression for the supercoiling energy that contains only parameters that can be determined by experiment. Indeed, it is possible in principle to derive them by computer simulation, using analytical forms of the various interatomic interactions given in Chapter 3. In practice, such calculations are still very difficult.

Studies of the topology of DNA, under various conditions in the local environment have an important bearing on chromosome structure. In the chromosomes, a given stretch of DNA is found to lie in juxtaposition with a protein known as a histone, and it is important to know how the former wraps itself around the latter. For a linking number of 1 or 2, the minimum-energy form of a closed loop or DNA is a circle. When the linking number rises to 3, the DNA adopts a figure-of-eight form, and increasingly higher linkage numbers lead to increasing amounts of supercoiling. The DNA-histone scenario is a complicated one, and there has been the suggestion that one should introduce two additional topological numbers: the surface linking number, SL, and the around-the-surface winding number, Φ. These are related to the linking number by the equation $L = SL + \Phi$.

7.3 Nucleic Acid Conformation: RNA

Let us now move on to the issue of RNA structure. As was mentioned at the start of this chapter, a given type of transfer-RNA (tRNA for short) permits attachment of a specific amino acid, and transfers this to the ribosome. The tRNA base sequence includes a triplet of bases known as the anti-codon, which is recognized by the complementary codon on the messenger-RNA (mRNA). It is in this manner that the correct amino acid residue becomes attached to the appropriate end of the growing polypeptide.

As shown in Figure 7.2, the tRNA molecule is rather small. Phenylalanyl tRNA comprises 76 bases. In that particular variety, a total of 42 bases are so positioned in the sequence of bases that they can pair up with complementary bases, the resulting 21 base pairs naturally being stabilized by hydrogen bonds. This base pairing does not occur in a haphazard fashion. On the contrary, there are four regions of the molecule in which two complementary stretches of the sequence produce a contiguous series of base pairs (see Figure 7.2). And because the four resulting base-paired stretches are deployed in a roughly symmetrical manner, the outcome is a structure that resembles a cloverleaf. The extremities of the three leaves of the structure all have regions that are free of pairing, and this produces loops. The middle member of these three loops possesses the anti-codon, while one (un-paired) extremity of what could be called the stem is the site at which the amino acid becomes attached. A specific enzyme is required to catalyse that attachment.

Let us suppose that we wish to know the most favourable base-pairing state, and thus the overall structure of the molecule, as a function of temperature. This could, for example, be part of an effort to determine the melting temperature (that is to say, the temperature at which the tRNA becomes denatured). At the time this book is being written, it is not yet practicable to study the folding of a tRNA molecule by molecular dynamics (see Chapter 6), though that exciting development should become feasible within the next decade or so. Instead, the favoured approach has been to calculate the partition function. From this, as discussed in Chapter 4, all the other thermodynamic parameters can be calculated. We recall that the partition function, Z, is given by

$$Z = \sum_i e^{-\varepsilon_i/k_B T} \tag{7.14}$$

This would enable us to calculate the energy as a function of chain conformation, and then find the most likely conformation by minimizing with respect to the various parameters. The difficulty stems from the sheer multiplicity of possible base-pairing. As is so often the case in thermodynamics, one is forced to make approximations. A good approach is to work with a partition function in the form

$$Z(I) = \sum_{\mathscr{E}} g(\mathscr{E}) e^{-\mathscr{E}/k_B T} \tag{7.15}$$

The first term after the summation sign is the number of conformations that have energy $\mathscr{E}$. That density of states term is then multiplied by the Boltzmann weighting term. This turns the problem into one of counting states: a sort of statistical mechanical book-keeping exercise. However, the latter task is by no means straight-forward, and various counting strategies have been employed, the currently favoured one being that advocated by Shi-Jie Chen and Ken Dill (see the suggestions for further reading, at the end of this chapter). A particularly thorny issue is that of excluded volume, which leads to an over-counting of some density of state terms. In essence, this difficulty arises because certain impossible situations are included in the summation when they ought to be omitted.

It is usually practicable only to extend to second-order terms in the calculation of the conformational energy. This means that one includes base pairs and also adjacent *pairs* of base pairs, these latter being referred to as stacks. The energy $\mathscr{E}$ may then be written

$$\mathscr{E} = \sum_{\text{stacks}} \mathscr{E}_{\text{stack}}(i, j, h, k) + \sum_{\text{pairs}} \mathscr{E}_{\text{pair}}(i, j) \qquad (7.16)$$

It must be borne in mind, when making such summations, that there will be contributions from backbone steric hindrance effects. The bare hydrogen-bond energy is about 0.03 aJ, as was discussed in Chapter 3, so a base pair has an energy of about 0.06 aJ (for an AU pair) or 0.09 aJ (for a GC pair). Stacking energies are found to be typically 0.05 aJ to 0.10 aJ, this energy having to be added to the pair energies of the involved pairs, in each case. Analyses of this type are capable of predicting denaturation (melting) temperatures that are in good agreement with experimental values.

Although further details will not be given here, the underlying strategy in the determination of this temperature is straightforward. The internal energy and entropy terms in the expression for the free energy are in opposition, because the latter term is preceded by a negative sign (see Chapter 4). The entropy term will become progressively more important as the temperature increases, therefore, and it will ultimately dominate. In the tRNA molecule, each base pair will restrain the ability of the backbone to explore phase space, and it will thus decrease the entropy. This effect will be less important than the favourable lowering of the energy by base-pair formation, provided that the temperature is below that of the transition. However, above that temperature, it is the entropy term that dominates.

The structure of these molecules took on added significance when Sidney Altman and Thomas Cech discovered that some RNAs have autocatalytic abilities. This raised the possibility that perpetuation of the earliest forms of life was based on RNA, rather than on the more complicated interaction of DNA and proteins. There has been talk of an RNA world, around four billion years ago, based on RNA's dual ability to store information and act as the required enzyme. In this scenario, a genetics based on RNA was subsequently augmented, and ultimately replaced, by the DNA-protein mechanism. It is not out of the question that we will one day see this competition played out in a large-scale computer simulation.

7.4 Proteins

As we have already seen, the distinguishing characteristic of a given biopolymer is the chemical composition of its elemental (monomer) units. This can be as simple as the CH_2 (i.e. H—C—H) unit present in the chains in lipids to the more complicated amino-acid residue units present in proteins. The latter will provide us with our first example of these vital threads. An amino acid (see Figure 7.3) consists of three parts: an electrochemically basic amino group, NH_2, at one of its extremities; an acidic carboxyl group, COOH, at the other end; and, lying between these two ends, a single carbon atom, to which are attached a hydrogen atom and also a side group, usually denoted by the letter R. (The reader should be careful *not* to think of this letter as standing for the word *residue*.) That centrally-located carbon atom is referred to as the alpha-carbon (α-carbon), to distinguish it from the carboxyl carbon, and the great variety of possible protein structures stems from the fact that the side-group R can have any of approximately 20 different chemical compositions.

The different side groups interact with the surrounding environment (that is to say with the aqueous surroundings, with other parts of the protein itself, or with other substances such as lipids) in different ways, so the physical proper-ties of a given protein are determined by the sequence of its Rs.

There are two ways in which monomers can become joined together so as to produce a link in the polymer chain, namely either by what is known as addition or by the process called condensation. The former merely involves rearrange-ment of some of the interatomic bonds, so as produce the link (see Figure 7.4).

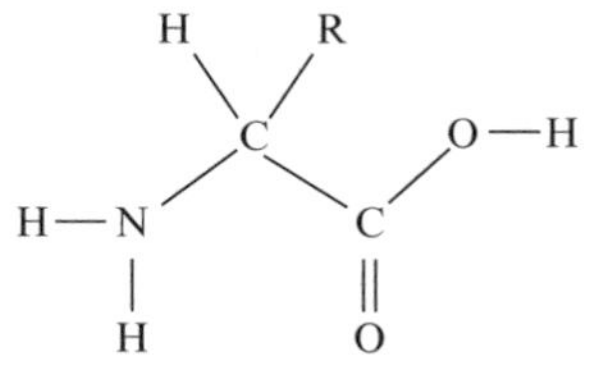

Figure 7.3 An amino acid molecule has amine (NH_2) and organic acid (COOH) terminals. The centrally located alpha-carbon provides the point of attachment for a lone hydrogen atom and the side group (R). There are twenty different forms of the latter, and this produces the great variety of protein structures

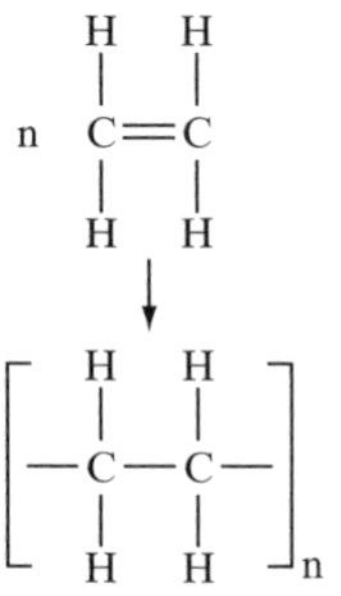

Figure 7.4 Addition polymerisation involves redistribution of certain bonds, as the con-solidation process proceeds. The process is conservative, producing no by-products

Such a process is conservative in that there are no by-products of the polymer-ization process. The latter, by way of contrast, depends upon the prerequisite removal of certain atoms from one or both of the consolidating units, so this process produces not only the link but also a by-product composed of those surplus atoms.

Proteins are produced by such a condensation process (see Figure 7.5). There is no redistribution of bonds within the molecule. The rearrangements exclu-sively occur at the termini and each step of the consolidation (that is to say, each addition of an amino-acid residue) produces a water molecule as a by-product. We see, therefore, that the monomers present in proteins are amino-acid molecules from which have been removed one oxygen atom and two hydrogen atoms; these monomers are therefore amino-acid residues, while the by-product is water. The word protein was coined by Jöns Berzelius, it stem-ming from the Greek term for *of the first rank*.

The 20 most common amino acids can be divided into two groups: those with side groups that are polar and those with side groups that are not. The side groups of members of the latter group do not interact with water, which is composed of polar molecules (each having two positive and two negative poles – see Chapter 2). There are nine non-polar amino acids, ranging from glycine (denoted by *Gly*, in the three-letter nomenclature, and by *G* in the one-letter system, see Table 7.2), with its side group consisting of a single hydrogen atom, to the quite complex tryptophan, the side group of which comprises nine carbon atoms, eight hydrogens and a nitrogen. In view of what is known about the electronegativities of the various elements, it comes as no surprise that none of the non-polar side groups contains an oxygen atom.

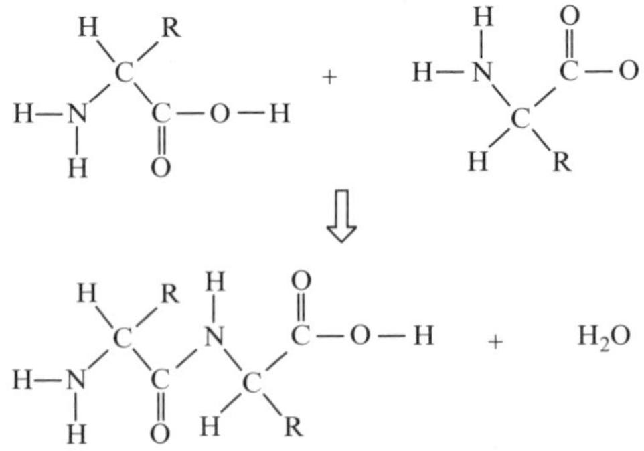

Figure 7.5 Production of a polypeptide from individual amino acid molecules is a con-densation process. No bonds are redistributed, the changes occurring exclusively at the termini, each consolidating step producing a water molecule as a by-product

Table 7.2 Nomenclatures for the various common amino acids

Amino acid	abbr	
Alanine	Ala	A
Arginine	Arg	R
Asparagine	Asn	N
Aspartic acid	Asp	D
Asparagine or aspartic acid	Asx	B
Cysteine	Cys	C
Glutamine	Gln	Q
Glutamic acid	Glu	E
Glutamine or glutamic acid	Glx	Z
Glycine	Gly	G
Histidine	His	H
Isoleucine	Ile	I
Leucine	Leu	L
Lysine	Lys	K
Methionine	Met	M
Phenylalanine	Phe	F
Proline	Pro	P
Serine	Ser	S
Threonine	Thr	T
Tryptophan	Trp	W
Tyrosine	Tyr	Y
Valine	Val	V

The other group, with polar side groups, can be further divided into three sub-groups. The first of these, comprising aspartate and glutamate, develop negative charges at physiological values of pH, and they are thus acidic. The members of the second sub-group are lysine, arginine and histidine, all of which develop positive charges at physiological pH; they are consequently basic. The final group comprises members such as glutamine and cysteine, which are all electrochemically neutral at physiological pH.

Cysteine is noteworthy because its side group contains a sulphur atom, which has great relevance for what is known as secondary bonding in proteins having the corresponding amino-acid residue. Another rather special side group is the one present in an amino acid that has not yet been mentioned, namely proline. It consists of a chain of three carbon atoms, and it also contains six hydrogen atoms. It is unique amongst the amino-acid side chains in that it is attached to the remainder of the amino acid molecule at two places, rather than the customary single site, the two attachments being at the α-carbon atom and at the nitrogen atom.

The upshot of this is that the single hydrogen atom that is usually attached to that nitrogen atom (see Figure 7.5) is in this case replaced by a CH_2 group, and the latter interacts more strongly with the next amino-acid residue in the sequence. If that residue involves a glycine side group, there will be considerable restrictions on the possible orientations that the two monomers can adopt, with respect to each other. This has important consequences for the structure of the protein in that vicinity. The pH value (which we remember is the negative of the logarithm to the base ten of the hydrogen-ion concentration, that is to say $-\log_{10}[H^+]$) in the surrounding aqueous medium is an important parameter in the determination of protein structure. Therefore, we ought to take a closer look at its influence. From a knowledge of the energies relevant for the water molecule, we can invoke the Boltzmann factor when determining the probability that one of the O—H bonds in that molecule will be broken, so as to produce the two ions OH^- and H^+. The expression for the probability is also known as the Arrhenius equation, namely

$$P = A \exp(-\Delta\mathscr{E}/k_B T) \tag{7.17}$$

where A is a constant (which, as we saw in Chapter 4, will include a vibration factor). This shows that there will always be a certain concentration of the above ions in solution.

The pH will, in turn, determine the state of ionization of any amino acid present (see Figure 7.6). Thus at low pH (that is, for a *high* concentration of hydrogen ions), the amino terminal will tend to acquire an extra hydrogen atom, and that end of the molecule will then terminate with an NH_3^+ unit. Conversely, at high pH (and therefore *low* concentration of hydrogen ions), the carboxyl end of the amino-acid molecule will tend to lose its hydrogen atom, and this will leave behind a COO^- unit. Finally, at physiological pH (that is, pH ~ 7), *both* ends of the molecule are found to be modified in the above manner, that is to say the termini will be NH_3^+ and COO^-, respectively. When the amino acid molecule is in this state, it is said to be zwitterionic.

It is important to note that the zwitterionic condition *does not* promote polymerization. In fact, as can be seen by comparing Figures 7.5 and 7.6, the tendency actually runs counter to what would be desirable for the condensation process; the NH_2 terminus has acquired an extra H, whereas the condensation process requires that it *lose* one of the Hs that it already has; and likewise, an O—H bond has been broken at the carboxyl end, whereas the condensation process needs this bond to remain intact, while a C—O bond is broken instead.

The system has a vested interest in this situation normally prevailing, of course, because it would be dangerous if polymerization could occur spontaneously and haphazardly. In practice, the condensation process is under the control of an *enzyme* (known as a ribosome).

For many of the side groups, analogously, the ionization state will depend upon the pH of the surrounding solution. It is therefore necessary to know the

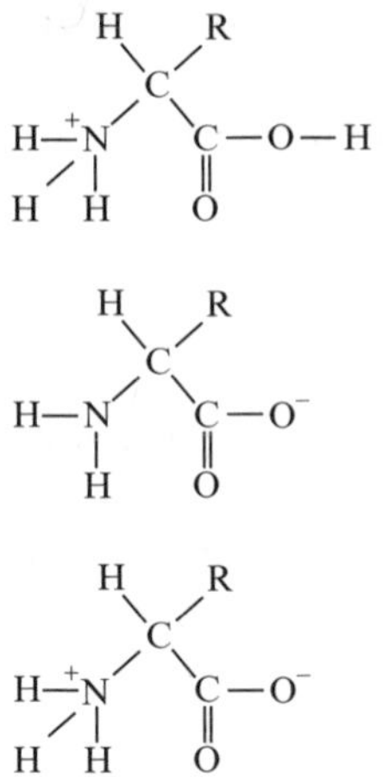

Figure 7.6 Availability of hydrogen ions (as measured by the pH) determines the probability that a terminal group of an amino acid will become charged. Their high concentration at low pH favours proton capture by the amino end, while the low concentration at high pH favours proton loss at the acid terminus. At medium pH, both processes occur with high probability, and the resulting zwitterionic form does not favour spontaneous amino acid polymerisation (which thus requires a suitable enzyme)

Table 7.3 Values for pK for some of the more common amino-acid side groups

Amino acid	pK (25°C)
Aspartate	3.9
Glutamine	4.3
Histidine	6.0
Cysteine	8.3
Tyrosine	10.9
Lysine	10.8
Arginine	12.5

degree of acidity at which the group will become polarized, because the charge state will have an important bearing on the overall structure of the protein. The pK value is defined as that value of pH for which there is a 50% probability that the side group is polarized (see Chapter 4). Values of pK, at 25°C, for some of the more common amino-acid side groups are shown in Table 7.3.

Before we go on to discuss the three-dimensional structure of proteins, we ought to note another way in which the amino acids can be divided into two groups, namely according to whether they are essential or non-essential, members of the latter group being those that the organism can produce for itself. The term *essential* is thus applied to those amino acids which must be present in the diet of the species in question. This division varies from species to species. In the plant kingdom, in which consumption of material originating externally is not common, the organism produces *all* the amino acids it requires, so there simply are no essential types.

It ought to be mentioned, however, that some plants actually *do* display a form of eating, namely the insect-consuming varieties. Each leaf of these plants is equipped with tiny hairs, which sense the presence of prey, and dispatch signals which cause the leaf to close around the victim. Enzymes are then secreted from the surface of the leaf, and when the insect has decomposed, amino acids and minerals are absorbed through appropriate routes. Such processes have their counterparts in the animal kingdom, of course.

Returning to the question of protein structure, and to the link produced by the condensation process, and recalling our earlier discussion of chemical binding, we might expect the bonds in the monomer to be arranged as shown in Figure 7.7(A), that is to say that the only pi-bond in the structure would be the one joining the carboxyl carbon with the oxygen atom. Now it transpires that a so-called resonance effect actually causes the bonding arrangement to perpetually flip between those shown in Figures 7.7(A) and 7.7(B).

This, in effect, causes a modified pi-bond to be stretched out from the nitrogen atom, across the carboxyl carbon, and over to the oxygen atom, as indicated in Figure 7.7(C). The resulting stretched-out orbital is known as the

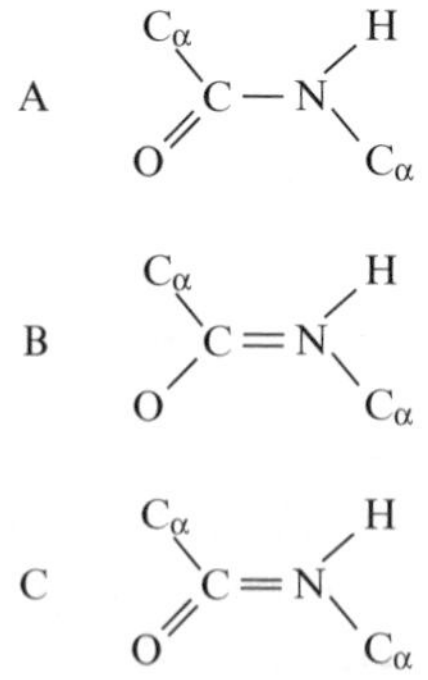

Figure 7.7 The conceptual association of valence with the "bonds" in simple molecular diagrams, such as these, suggests that form A is correct. But the pi-bond between the oxygen and carbon atoms resonates with the form shown in B, in which that bond has shifted across to the carbon-nitrogen link. The upshot is a situation in which the pi-bond can be regarded as stretching across all three atoms, thereby causing them to lie in a plane. Such an amide link is indicated in C

amide link. Moreover, because these three atoms are not in a straight line, their positions will actually define a plane, in which the N, C and O atoms are constrained to lie. This was first suggested by Linus Pauling, a pioneer in the study of chemical bonding. Further simplification of the situation derives from the fact that the alpha-carbon atoms and the hydrogen atom attached to the nitrogen *also* lie in this plane. So the job of working out how the atoms in the polypeptide are arranged with respect to one another is reduced to that of finding out how the resulting planar elements are juxtaposed (see Figure 7.8). Although the protein folding problem still awaits a satisfactory solution, it is clear that the juxtaposition of these planes, and of the side groups which are attached to the interplanar links, limits the possible arrangements in the intermediate-level motifs collectively referred to as the molecule's secondary structure.

The reader should make careful note of the fact that the elemental planes just discussed are *not* identical with the amino-acid residues (of which the protein is also composed). On the contrary, each such residue extends across parts of two adjacent elemental planes. Conversely, each elemental plane comprises remnants of two successive amino-acid residues. A polypeptide chain consisting of N amino-acid residues therefore possesses $(N - 1)$ amide links. The reader

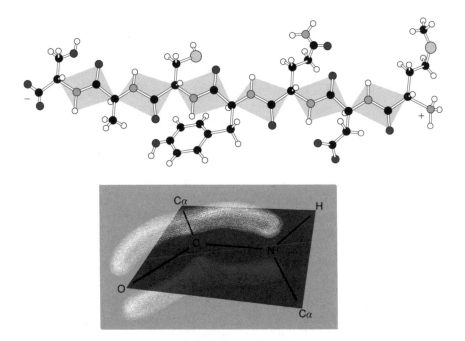

Figure 7.8   Most of the atoms in a polypeptide backbone are constrained to lie in planes defined by the amide links (below), and this limits the possible arrangements in the intermediate-level motifs, collectively referred to as the molecule's secondary structure. The amino-acid sequence shown here is Methionine-AsparticAcid-Glutamine-Tyrosine-Cysteine-Alanine-Serine

should also note that the side groups, being attached to the alpha-carbon atoms, dangle as it were from the points at which pairs of these elemental planes make mutual contact.

Relative movement between adjacent elemental planes is made possible by the fact that the $C-C_\alpha$ and $N-C_\alpha$ bonds are of the sigma-type. Indeed, were it not for the presence of a side group (and also a lone hydrogen atom) at this same inter-planar point, the mutual rotation of the elemental planes would meet with only minor resistance. As it is, there is considerable restriction to such movement, such resistance being an example of the steric hindrance that we encountered in Chapter 3.

Rotations around the above-mentioned two sigma bonds vary what are known as the dihedral angles of the situation (see Figure 7.9), these usually being denoted by the symbols φ and ψ. G. N. Ramachandran and his colleagues have calculated the probability of occurrence of the various possible $\varphi-\psi$ combinations. They used the Arrhenius equation (Equation (7.17), above), inserting in each case the corresponding energy, derived from a knowledge of the various interatomic distances, and following assumptions as to the form and parameters of the interatomic potentials.

The resulting plot of the configuration energy, as a function of φ and ψ is now referred to as a Ramachandran diagram, and it must be emphasized that there is a different diagram for each side group. Figure 7.10 shows the Ramachandran diagram for the case of (non-polar) alanine, which has a CH_3 side group. The contours on such a plot indicate the relative probability of finding the inter-planar link in the state defined by the two dihedral angles.

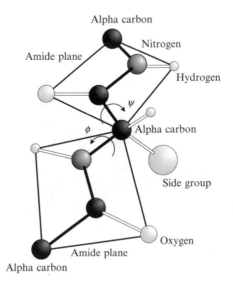

Figure 7.9 The relative orientation of two amide planes is completely defined by the two dihedral angles ϕ and ψ shown in this diagram

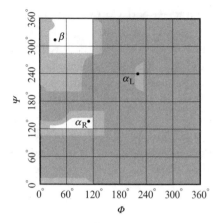

Figure 7.10 A Ramachandran diagram shows, by means of equal-energy contours, the probability of two adjacent amide planes being mutually oriented with a given pair of ϕ and ψ angles. In this simplified version, corresponding to an alanine side group, higher energy is indicated by increasing density of shading, and the $\phi-\psi$ values for the beta-sheet and right and left hand alpha-helix motifs are indicated

It must be kept in mind, however, that there is more to protein structure than merely working out the most favourable combination of φ and ψ for each side group. The final three-dimensional structure will be the result of a global minimization of the free energy, and this will usually correspond to a situation in which the various inter-planar links have φ and ψ angles that are somewhat removed from their individual minimum-energy values.

The structure of a protein can be described in a variety of different ways, and at different levels. The primary structure is determined by the sequence of amino-acid residues, that is to say by the sequence of the side groups, since the backbone units are identical. The primary structure therefore takes the form $R_1, R_2, \ldots, R_i, \ldots, R_n$, where R_i identifies the type of side group on the ith amino-acid residue. The primary structure unambiguously describes a given protein, because such biopolymers are never branched. The first determination of the primary structure of a protein was accomplished by Frederick Sanger, the protein being insulin.

The secondary structure is determined by the number, types and extents of local motifs, of which there are only a limited number of kinds. The most common of these are the alpha-helix (α-helix) and the beta-sheet (β-sheet). The alpha-helix, as its name implies, is a helical arrangement in which there are 3.6 amino-acid residues per turn (that is to say, per helical pitch). This gives a translation displacement of 0.15 nm along the axis of the helix for each successive residue, while the diameter of the helix is approximately 0.6 nm, if one ignores the side groups.

It was Linus Pauling and Robert Corey who first predicted the presence of this secondary-structure motif in proteins, 6 years before it became possible to resolve this atomic arrangement by X-ray diffraction. The key to the stability of the α-helix lies in the fact that the above-mentioned planes are mutually deployed so as to line up pairs of hydrogen and oxygen atoms (see Figure 7.11), thus permitting the formation of hydrogen bonds, that is to say arrangements of the type $=N—H\cdots O=C=$ with the dots denoting the bond.

The helix shown in Figure 7.11 is seen to involve such a hydrogen bond between the nth and the $(n-4)$th residue, for all values of n, this regular pattern being broken only at the extremities of the helical segments. Not all stretches of

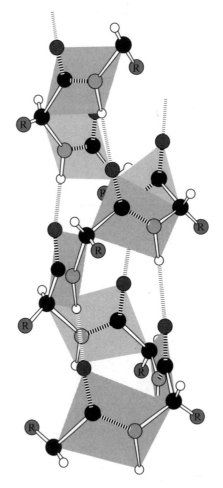

Figure 7.11 In the alpha-helix secondary-structure motif, predicted by Linus Pauling and Robert Corey, and subsequently verified by X-ray crystallography, there are 3.6 amide links per helical pitch. This arrangement favours the formation of two counter-directed hydrogen bonds per amino-acid residue, these providing the overall structure with its mechanical stability

the polypeptide find it energetically favourable to form an alpha-helix, of course, particularly if the stretch includes a proline, which is indeed said to be a helix breaker. Another possible factor working against the presence of a helix, at a given place in the amino-acid sequence, could be interactions between the local side groups.

The other common secondary-structure motif, the beta-sheet, is shown in Figure 7.12; its presence in proteins was also predicted by Pauling and Corey. In this case, the hydrogen bonds are established between adjacent polypeptide strands, with those strands running either parallel or anti-parallel to each other. This does not mean that the strands necessarily belong to different polypeptides, however, since a single one of these could close back on itself, in either a parallel or anti-parallel fashion. In the case of this arrangement, the displacement along the axis, for each amino-acid residue, is 0.347 nm. The β-sheet motif is not limited to two polypeptide chains; it is not uncommon for proteins to comprise sheets that are four, five, and even six strands wide. It should be noted, however, that there is another fundamental difference between the helix and sheet motifs in that the latter can (an invariably does) link amino-acid residues that are well separated in the primary structure. We will return to this point in the next section, when we consider protein folding.

The tertiary structure of a protein is the three-dimensional globular configuration of all its constituent atoms, and it will in general include a variety of secondary-structure motifs, of various sizes and mutual juxtapositions. The first globular protein for which the structure was determined was myoglobin,

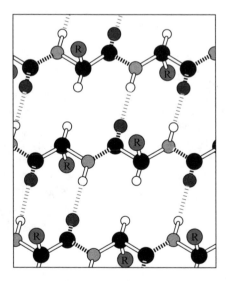

Figure 7.12 The beta-sheet arrangement is another common secondary-structure motif, and it too permits the formation of two counter-directed hydrogen bonds per amino-acid residue

this breakthrough being achieved by John Kendrew and his colleagues in 1962. It is special in that it comprises only alpha helices, interspersed with stretches of what is referred to as random coil. There are eight helices in myoglobin, and the tertiary structure is determined by their mutual packing. Figure 7.13 shows the structure of myoglobin, using the familiar ball-and-stick method of displaying the configuration of the C_α atoms in the backbone (the side groups having also been omitted, for the sake of clarity). When all the atoms of such a globular protein are shown, it becomes apparent that even a single water molecule would have a hard task in penetrating into the protein's interior.

The highest level of structure found in proteins is seen only in those which comprise more than one polypeptide chain, this level being that of the quaternary structure. The classic example is the oxygen-transporting molecule haemoglobin, the structure of which was determined by Max Perutz and his colleagues in 1962. It comprises four sub-units, each of which bears a striking resemblance to myoglobin, and each of which possesses a heme group, with an iron atom at its centre (myoglobin has just one of these units). The four sub-units pack neatly together, to give the molecule its globular three-dimensional structure. Perutz and Kendrew were colleagues.

As was discovered by Christian Bohr, the haemoglobin molecule displays a cooperative behaviour, in that the (temporary) linking of an oxygen molecule to one of the heme groups facilitates oxygen uptake by the remaining sub-units. This behaviour, known as the Bohr effect, reminds one of a principle enunciated in St. Matthew's gospel (...*to him that hath shall be given*...), and it stems from the fact that attachment of the first oxygen atom causes local

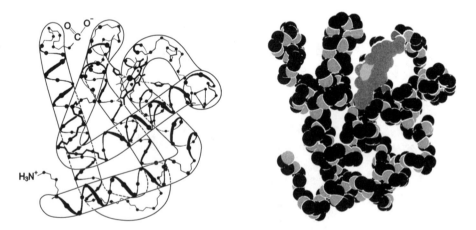

Figure 7.13 The tertiary structure of a protein is the three-dimensional arrangement of its constituent atoms. It is not yet possible to predict this conformation on the basis of the primary structure (the sequence of amino-acid residues). These two views of the tertiary structure of the oxygen-storing myoglobin molecule highlight the molecule's eight alpha helices (left) and all the carbon and nitrogen atoms in the molecule's backbone (right)

distortions in the sub-unit, which then produce distortions in the remaining sub-groups, those secondary distortions making further oxygen uptake easier.

One might speculate as to why these globular proteins have to be so large. The answer lies in the difficult tasks they must perform. For example, the iron atom in each heme group in the above molecules must attract an oxygen molecule, in order that it can be transported, but the binding between the iron and the oxygen must not be so strong that the oxygen atom cannot be released again when the target tissue is reached; the iron and oxygen must not be permitted to form rust! The globular molecule's shape guarantees formation of such a weak-but-sufficient bond, in a manner that could not have been achieved with a smaller protein.

Nature has gradually evolved these globular proteins, enabling them to perform a task whose necessity could easily be overlooked. The point is that the energy barriers encountered in the biological domain cannot normally be surmounted by the process discussed in Chapter 4, namely thermal activation. One could get the impression that it is simply a question of working out the corresponding Boltzmann factor, in order to calculate the temperature at which the enzyme function would modify (or break) the relevant interatomic bonds at an acceptable rate. This is not the case, however, because the temperature of the body is carefully controlled. Indeed, even a rise of a few degrees Celsius, that is to say a fever, can be a serious matter.

Instead, the enzyme *lowers* the energy barrier, by altering the shape of the molecule it is acting upon, and this brings the rate of the modification (or rupturing) process up to a useful level. The alteration to the shape of the acted-on molecule is achieved by the thermal motions of the atoms in the enzyme molecule, with each degree of freedom potentially contributing an energy equal to $k_B T/2$, as discussed in Chapter 4. In practice, however, not all the available degrees of freedom will be able to act in unison; much of the thermal effort will thus be wasted. This is why enzymes need to be fairly large molecules: they are not particularly efficient tools.

Another factor influencing enzyme size is the actual shape of the molecule being acted upon, a good example being that which underlies genetic engineering, through modification to DNA. As was discussed ealier in this chapter, DNA is a ladder polymer, twisted into a double helix by steric hindrance, the rungs of the ladder being the base pairs. The backbone chains, which are equivalent to the sides of the ladder, involve covalent bonds, while the base pairs are bound together by the much weaker hydrogen bonds. As shown in Figure 7.1, the structure of the DNA double helix is asymmetrical, because it has alternating wide and narrow grooves.

The enzyme that cuts the ladder into two pieces (see Figure 7.14) has regions on its surface which neatly fit into one or both of these grooves, enabling the thermal motion of its constituent atoms to break the correct bonds and thereby cleave the DNA molecule in two. It should be noted, in passing, that because

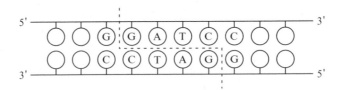

Figure 7.14 Some restriction enzymes cleave a few of the hydrogen bonds between the bases joining the two strands of DNA, as well as two of the covalent bonds in the backbones. The enzyme of this class known as *Bam H1*, found in *Bacillus amyloliquefaciens*, separates four adjacent base pairs in this manner

the base pairs are complementary, cuts like the one shown in Figure 7.14 will leave two so-called sticky ends, each of which will be able to combine only with another sticky end having the appropriate number and sequence of unpaired bases. It is these cleavage and recombination processes which underlie genetic engineering.

It is instructive to consider a break-through achieved by Kary Mullis. He was looking for a way to speed up the process by which the enzyme DNA-polymerase joins together nucleic-acid monomers so as to produce new strands of DNA, in the test-tube. It dawned on him that such a situation is not subject to the above-mentioned need to keep the temperature constant at 37°C, so a sufficiently large Boltzmann factor can be obtained merely by raising the temperature. This advance is now embodied in the technique called PCR (for polymerase chain reaction). One of this technique's most spectacular consequences was the production of sufficiently large samples of the DNA recovered from the putative remains of the last Russian royal family that a positive identification could be made. It was thus possible to give them a proper burial – 70 years after they were murdered.

Let us close this section with a look at some of the other important ways in which polypeptide chains can become linked together. The strongest of these is the cystine bond that is formed between two cysteine side groups. Each of the latter loses its terminal hydrogen atom, and the two sulphur atoms then become joined by a σ-bond, as shown in Figure 7.15. The process is not unlike that of vulcanization of certain types of rubber. Proteins with such disulphide bonds are amongst the most stable forms encountered, and they often serve functions in which molecular geometry is important. Thus immunoglobulin has 15 disulphide bonds, while ribonuclease has four. By way of contrast, myoglobin has none.

The next type, in terms of strength, is the hydrogen bond itself, which we have already discussed at length. We therefore pass on to what are known as salt bridges, an example of which is shown in Figure 7.16. This figure shows

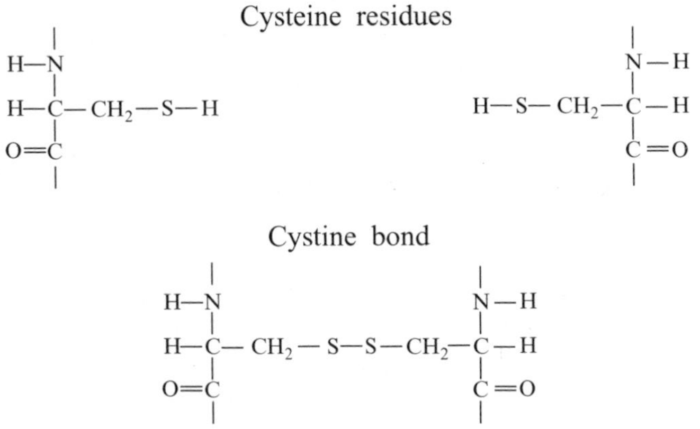

Cysteine residues

$$
\begin{array}{ll}
\text{H—N} & \text{N—H} \\
\ \ | & \ \ | \\
\text{H—C—CH}_2\text{—S—H} & \text{H—S—CH}_2\text{—C—H} \\
\ \ | & \ \ | \\
\text{O=C} & \text{C=O} \\
\ \ | & \ \ |
\end{array}
$$

Cystine bond

$$
\begin{array}{ll}
\text{H—N} & \text{N—H} \\
\ \ | & \ \ | \\
\text{H—C—CH}_2\text{—S—S—CH}_2\text{—C—H} \\
\ \ | & \ \ | \\
\text{O=C} & \text{C=O} \\
\ \ | & \ \ |
\end{array}
$$

Figure 7.15 The strongest type of bond between adjacent stretches of a polypeptide chain, or of two chains, is the cystine bond that is formed between two cysteine side groups. This involves the shedding of a hydrogen atom from each group, after which the two sulphur atoms join, in much the same way as is seen in vulcanised rubber

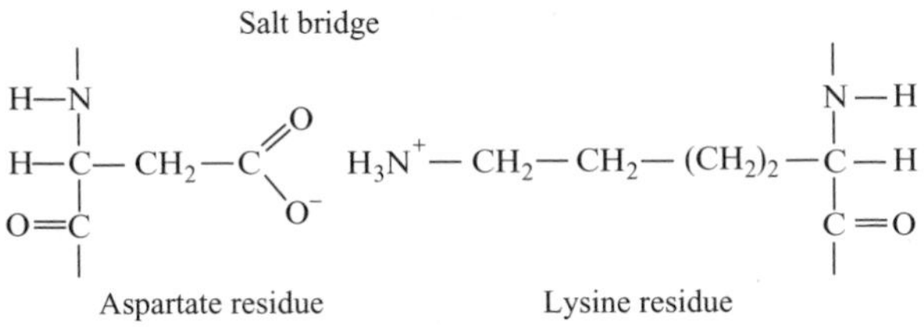

Salt bridge

$$
\begin{array}{ll}
\text{H—N} & \text{N—H} \\
\ \ | & \ \ | \\
\text{H—C—CH}_2\text{—C} \overset{\displaystyle O}{\underset{\displaystyle O^-}{<}} \quad \text{H}_3\text{N}^+\text{—CH}_2\text{—CH}_2\text{—(CH}_2)_2\text{—C—H} \\
\ \ | & \ \ | \\
\text{O=C} & \text{C=O} \\
\ \ | & \ \ |
\end{array}
$$

Aspartate residue   Lysine residue

Figure 7.16 A salt bridge can be formed between two amino-acid side groups if they both bear charges, and of opposite sign, as in the case of the aspartate and lysine side groups shown here, at approximately neutral pH

what happens when the side groups of an aspartate residue and a lysine residue approach each other, at the appropriate value of pH (i.e. pH $\sim$ 7). The COOH terminus of the aspartate looses its hydrogen atom, while the NH_2 terminus of the lysine acquires a hydrogen atom (refer back to Figure 7.6). The side groups thus have opposite electrical charges, and will therefore mutually attract. The energy of the bond can easily be calculated, using the expression for the Coulomb attraction given in Chapter 3 (see Equation (3.3)).

The one complication here is that we may not simply use the constant ϵ_0, the permittivity of free space, because the space between the two charged side groups is filled with water, not a vacuum. We must therefore replace ϵ_0 by $\epsilon_0 \epsilon_r$, ϵ_r being the relative permitivity. In other words, $\epsilon_r = 1$ for a vacuum. For bulk water, it is found that $\epsilon_r \sim 80$, but this applies only to the situation in which the water molecules are not subject to any constraints; that number is good only for bulk water. This would not be the case in the confined space between the side groups in a salt bridge, and in practice the relative permittivity for water in such cases is around 10. Salt bridges are much in evidence in the binding together of the different sub-units in the haemoglobin molecule.

Finally, there is a cross-link that is particularly important in providing the structural protein collagen with its remarkable combination of strength with flexibility. This is the aldol cross-link, and its geometry resembles that of a disulphide bridge. The link is provided by aldol condensation between pairs of lysine side-groups. Two adjacent lysine residues undergo, with the catalytic help of lysyl oxidase, aldehyde conversion. The structures thereby become aldehyde derivatives, and these are then linked by an aldol condensation (see Figure 7.17). As is usual in a condensation reaction, the by-product is water. The aldol condensation was discovered by Aleksandr Borodin, better known as the composer of the opera *Prince Igor*.

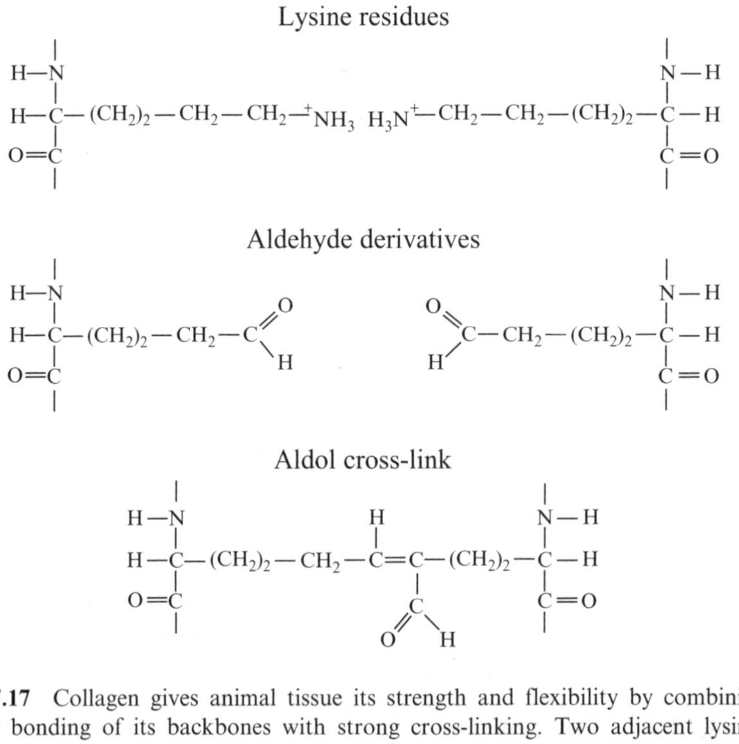

Figure 7.17 Collagen gives animal tissue its strength and flexibility by combining the covalent bonding of its backbones with strong cross-linking. Two adjacent lysine side groups (a) undergo aldehyde conversion, with the catalytic help of lysyl oxidase, and the resulting aldehyde derivatives (b) are then linked by the aldol condensation (c). As is usual in a condensation reaction, the by-product is water. The aldol condensation was discovered by Aleksandr Borodin (1833–1887), better known as the composer of the opera *Prince Igor*

7.5   Protein Folding

In 1940, Max Perutz intrigued his head of department, Lawrence Bragg, by writing a brief article with a most unusual title: *'Unboiling' an egg*. He had been inspired by certain experimental results obtained by Mortimer Anson and Alfred Mirsky, which indicated that the denaturing of a protein could be reversed, under favourable circumstances. As was discussed above, proteins are produced as strings of amino-acid residues, but it is not obvious that each successive residue automatically slots into its final position in the protein's overall three-dimensional structure. If that is not the case, the protein must fold up after its fabrication has been completed, and this will be equivalent to such reversal of denaturing. (The Perutz article is reproduced in Rainer Jaenicke's edited volume referred to at the end of this chapter.)

The question arose, however, as to what governs protein folding. Did the usual thermodynamic principles apply, even to such a complicated structure? This issue was addressed by Christian Anfinsen, who took as his starting point the work of Stanford Moore and William Stein on the connection between chemical structure and catalytic activity of the ribonuclease molecule. Ribonuclease consists of a single 124-residue chain, its structure being stabilized by four disulphide bridges. Anfinsen showed that this protein can be reversibly denatured in a solution of beta-mercaptoethanol and urea, the former disrupting the bridges. The urea apparently influences the conformation of the backbone, because its delayed removal from the denaturing solution allows the protein to adopt a scrambled arrangement in which bridges form between the wrong pairs of cysteine residues. The thermodynamically correct structure can be recovered from such a metastable state by gentle warming in a dilute solution of beta-mercaptoethanol. This permits the disulphide bonds to rearrange themselves. This indicated that the dominant role was indeed being played by thermodynamics. The three-dimensional structure of a native protein in its normal physiological environment, consisting of solvent at the correct acidity, ionic strength and temperature is the one for which the free energy is lowest.

Determination of the structures of many proteins, by the X-ray technique, showed these molecules to have interiors almost as compact as their non-biological organic counterparts, and that the hydrophilic side-chains tend to lie on the outside. This latter trend does not lead to an ideal hydrophobic–hydrophilic separation, however, for there is a certain amount of structural water inside many proteins, as Frederic Richards demonstrated. A molecule of bovine pancreatic tripsin inhibitor, for example, has four water molecules in its interior. This water occurs as single molecules, each with its electrostatic charges of both signs, so these are able to provide a neutralizing bridge between charged side-chains that happen to lie inside the protein. This suggests that highly charged solvents should be able to denature proteins, a fact established by Gerardus Mulder, and subsequently explained by Kaj Linderstrøm-Lang. This work can be regarded as complementing that of Max Perutz, whose studies of mutant forms of haemoglobin led him to stress the importance of tightly fitting *non*-polar contacts on the protein surface. These seal off the interior by providing a bulwark against intruding water.

Ab initio computer simulations of protein folding are confronted with a formidable challenge because of the multiplicity of interactions that have to be taken into account. Harold Scheraga has put much effort into calculating the magnitudes of the numerous parameters. As an illustration of what is involved, let us consider the expression for the total energy employed by Michael Levitt. It has six terms:

$$\mathcal{E}_{total} = \mathcal{E}_{bonds} + \mathcal{E}_{angles} + \mathcal{E}_{torsions} + \mathcal{E}_{non\text{-}bonded\ distances}$$
$$+ \mathcal{E}_{partial\ charges} + \mathcal{E}_{hydrogen\ bonds} \tag{7.18}$$

The individual terms are as follows:

$$\mathcal{E}_{bonds} = \sum_{bonds} K_b(b - b_0)^2 \tag{7.19}$$

$$\mathcal{E}_{angles} = \sum_{angles} K_\theta(\theta - \theta_0)^2 \tag{7.20}$$

$$\mathcal{E}_{torsions} = \sum_{torsions} K_\phi\{1 - \cos(n\phi + \delta)\} \tag{7.21}$$

$$\mathcal{E}_{non\text{-}bonded\ distances} = \sum_{\substack{non\text{-}bonded \\ distances}} \varepsilon\{(r_0/r)^{12} - 2(r_0/r)^6\} \tag{7.22}$$

$$\mathcal{E}_{partial\ charges} = \sum_{\substack{partial \\ charges}} q_i q_j / r \tag{7.23}$$

$$\mathcal{E}_{hydrogen\ bonds} = \sum_{\substack{hydrogen \\ bonds}} F(\theta, r) \tag{7.24}$$

This is reminiscent of the complexity we encountered in Chapter 3, when considering non-central forces. (Indeed, the reader would be well advised to glance back at Equation (3.14), which uses many of the same variables; and the fourth of the above terms is, of course, the Lennard-Jones function.) Levitt and Scheraga, and their respective colleagues adjusted the magnitudes of the parameters in these expressions by using the energy function to calculate such experimentally verifiable quantities as sublimation energies, unit cell dimensions in the corresponding crystals, equilibrium values of bond lengths and angles, vibration frequencies, and so on. In really ambitious examples of this type of approach, even the interactions with the surrounding solvent are taken into account.

Ideally, this approach would permit one to follow the dynamics of the molecule as it folds up from an initial unfolded state. However, that process takes anything up to a minute in practice, and simulating such a long time would be out of the question, given that the computational time step has to be not larger than about a hundredth of a picosecond. Nevertheless, computer programs have been written which do, at least, permit one to simulate the dynamics of the protein in its folded state, an accomplishment that is interesting in its own right. Teams led by Martin Karplus and Barry Robson have even made such ambitious computer programs commercially available.

The fundamental difficulty facing the would-be simulator of protein folding lies in the exquisite balance that exists between relatively large individual contributions to the free energy. Alan Fersht has been prominent among those who have determined the free energy changes associated with the various components. The largest stabilizing factor is due to hydrophobic effects, and for a typical 100-residue protein at 25°C this amounts to about 1.85 aJ. The formation of van der Waals bonds produces a comparable contribution, this producing a free-energy change of about 1.6 aJ. The part played by hydrogen bonds varies considerably with the type of protein, and the free-energy term consequently displays much variation; the approximate range is 0.35 aJ to 5 aJ. So we see from the latter that hydrogen bonding could even be decisive in certain proteins. On the debit side, that is to say the factors that tend to destabilize the protein, are the contributions from entropy and from the fact that a folding pattern globally favourable in terms of free energy might force certain side groups into *un*favourable situations. The latter contribution can be as high as 1.4 aJ. The entropic effect is potentially the most important of them all, because it can produce a change in free energy as large as 7 aJ, the value for an average protein being perhaps a third of this.

When we considered the denaturation of RNA, earlier in this chapter, we treated the phenomenon as a thermodynamic phase transition, which indeed it is, likening it to melting. We saw how the entropic term dominates at sufficiently high temperature. The same principles apply to proteins, and the delicate balance between the various free-energy terms manifests itself in melting temperatures that are actually comparable to that of the human body. Our proteins lead a precarious existence, close to the limits of their stability. Barnase, for example, melts at around 312 K, while for ribonuclease the figure is about 306 K. Figure 7.18 shows the free-energy differences between the total stabilizing and total unstabilizing factors, for three proteins, as

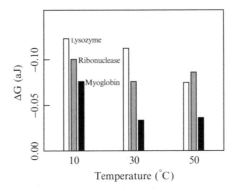

Figure 7.18   The free energy difference between the folded and unfolded states of three common proteins is shown here as a function of temperature, the negative values indicating that the folded form has the lower free energy in all cases. One sees that increasing the temperature does not always destabilise a protein, though this is generally the case. The important point made by these data (compiled by David Sheehan) is that stability is marginal, the free energy difference being about the same as that stored in four or five hydrogen bonds, or in a couple of molecules of ATP

a function of temperature. We should note the very small magnitudes of these differences, compared with the individual free-energy contributions discussed above.

The melting temperatures referred to above should more strictly be described as the temperatures at which the degree of denaturation is 50%. Recalling our discussion of equilibrium constants in Chapter 4, these half-denatured temperatures can be inserted into the expression derived by Jacobus van't Hoff (Equation (4.42) and (4.43)). From the van't Hoff Equation (4.43), taking logarithms and differentiating once with respect to T, we find

$$\Delta \mathscr{E} = RT^2 \frac{d\ln K}{dT} \qquad (7.25)$$

K being the equilibrium constant. Transition energies derived from Equation (7.25) are found to be in close agreement with those measured independently by calorimetry, so the assumption underlying Equation (7.25), namely that there are essentially just two states, folded and unfolded, appears justified. However, spectroscopic studies of pulse-heated and then rapidly-quenched molecules indicate that this assumption is safe only for proteins comprising less than about 100 residues. For larger molecules, one has to allow for the existence of short-lived intermediate configurations. Indeed, there is now ample evidence that these larger proteins can become temporarily trapped in meta-stable states.

This brings us to the issue of timing. If a denatured protein molecule is an example of what Paul Flory referred to as a free-flying chain – in which all the joints between the amide-link-stabilized planes (see above) are free to adopt any angles whatsoever – a molecule in the act of folding will be able to explore an astronomic number of different conformations, and this will take an inordinate amount of time. This problem is known as Levinthal's paradox (after Cyrus Levinthal, the first person to draw attention to it). Our knowledge of the fact that real problems fold in minutes, at the most, permits us to conclude that they are not confronted with this difficulty. We must conclude, therefore, that there must be a biasing factor, which permits the folding protein to explore phase space in a more efficient manner. It is not difficult to guess the origin of this factor, for it is surely the case that the various parts of the protein will always be influenced by the other parts, through the interatomic forces, to which we have attached so much importance in this book. Putting it in the language of Appendix C, the folding protein will follow a trajectory in configuration hyperspace, the topology of which is intimately connected to the interatomic forces.

We have already seen that entropy plays a very important role in protein folding, and it will thus have an important influence on the shape of the hyperspace manifold. This means that one should direct one's effort toward determining how many different folding patterns are consistent with the production of the compact three-dimensional structures proteins are known to possess (as discussed above). In the terms usually employed when addressing this issue, the task is to find out how many folding classes there are. Cyrus Chotia has risen to this challenge, and he studied the structures of the 400 or so structures that had been determined by 1992. (Such structures are available in the Brookhaven data bank, and the list grows steadily.) He was able to identify about 100 distinct folding classes, and by observing certain trends he ventured an estimate that about 1000 classes would ultimately be found. Henrik Bohr and Per-Anker Lindgård have tackled the problem theoretically, using a lattice model, and they arrived at an upper bound of about 4000 distinct folding classes. In view of the boldness of their assumptions, this figure can be said to be in reasonably good agreement with Chotia's empirical result.

The lattice approximation imagines the polypeptide backbone as being constrained to lie along the edges of a cubic lattice, no such edge being occupied by more than one backbone segment. Indeed, no *point* in the lattice is permitted to be traversed by more than one backbone. This approach is easier to visualize in two dimensions, the square lattice resembling the lines on a chess board. Starting at one corner, say, one traces out a path that haphazardly changes its direction by 90°, and never overlaps or crosses itself. The task is to find how many different patterns of path can be drawn, within these restrictions. Introduction of the third dimension greatly increases the possibilities. In Bohr and Lindgård's model, the individual line segments represent stretches of alpha-helix and beta-sheet that have already been formed.

A useful quantity that emerges from this type of analysis is the relative contact order. This is defined as the average separation along the amino-acid sequence of residues in physical contact in the folded protein, divided by the length of the protein. As David Baker reported in 2000, the logarithm of the rate of folding is inversely related to the relative contact order, this correlation holding over a million-fold range of folding rates (see Figure 7.19). Proteins with a large fraction of their contacts existing between residues closer in the amino-acid sequence tend to fold faster than those having a higher proportion of non-local contacts. Baker rationalized these observations in terms of the influence contact order has on a protein's entropy. The formation of contacts well spaced out in the sequence will greatly reduce the number of conformational options available to the stretch of polypeptide lying between the contacts; there is a relatively large entropic cost involved in forming contacts early in the folding process. A lower entropy of a folding intermediate means a higher free-energy barrier to be traversed, and hence a lower rate of folding. This is a dramatic result, for it indicates that the physics underlying folding may be considerably simpler than has previously been assumed.

When protein molecules are sufficiently large, they can require the help of another protein molcule to nudge them into their final native conformation. This appears to be the case for about 10% of large proteins, the mediating molecules being referred to as chaperones (also known as chaperonins). These proteins appear to exist for what could be called good housekeeping. Large molecules require more time to attain their final conformation, and there could be the possibility of infelicitous aggregation with other molecules of the same type, with possibly dire consequences for the host cell. The first chaperones to be discovered were the so-called heat shock proteins, these appearing as a consequence of brief heating. It seems likely that they exist in order to preclude

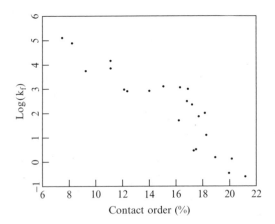

Figure 7.19 The logarithm of the rate of folding of a protein as a function of the contact order, the latter being defined as the average separation along the amino-acid sequence of residues in physical contact in the folded protein, divided by the protein's length. That such a plot reveals a systematic trend is clearly seen in these data compiled by David Baker

Prion protein

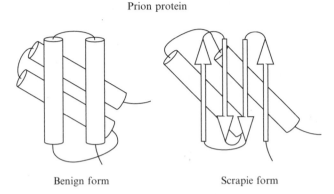

Benign form   Scrapie form

Figure 7.20 The existence of the prion protein was predicted by Stanley Prusiner. Its benign form comprises four alpha helices (here denoted by cylinders) and interconnecting random-coil segments. Single amino-acid substitutions can convert the molecule to its so-called scrapie form, in which there are two helices and two co-aligned stretches of beta sheet (each indicated by counter-directed arrows). This latter form is linked to the occurrence of bovine spongiform encephalopathy, commonly known as mad cow disease

aggregation between other proteins partially denatured because of the rise in temperature. Something akin to the chaperoning mechanism appears to under-lie the virulence of one form of prion. The existence of prions was postulated by Stanley Prusiner, who boldly predicted the existence of heredity molecules other than the nucleic acids (see Figure 7.20). The prion has a single primary sequence of amino-acid residues, but it comes in two different conformations, one innocuous and the other highly dangerous. The unfortunate thing is that the latter form can induce the former into adopting its pernicious conform-ation, and the transformation clearly has the capacity for what could be called a chain reaction. Prions occur naturally in the brain, and they have been shown to produce the devastating brain disease in cattle known as bovine spongiform encephalopathy (BSE), also known as mad cow disease. It is not clear whether the same type of molecule underlies the sheep disease scrapie, or indeed the rare human affliction Creutzfeld-Jacob disease (CJD).

Exercises

7.1 Using *before* and *after* diagrams, indicate what happens when three individual amino-acid molecules become consolidated into a single polypeptide molecule.

7.2 The following diagrams show the atomic positions in the DNA base pairs guanine–cytosine and thymine–adenine, with the hydrogen bonds indicated by broken lines.

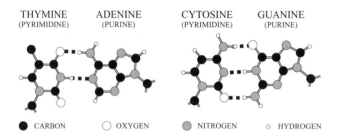

THYMINE   ADENINE   CYTOSINE   GUANINE
(PYRIMIDINE)   (PURINE)   (PYRIMIDINE)   (PURINE)

● CARBON   ○ OXYGEN   ● NITROGEN   ○ HYDROGEN

The restriction enzyme *Hind* III, which is used in genetic engineering, is able to cut the DNA molecule as indicated in the following diagram:

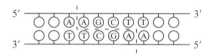

From your knowledge of the strengths of hydrogen and covalent bonds, and from the fact that an enzyme can cut a ribose-phosphate backbone by severing a single covalent bond, calculate the minimum number of atoms that must be acting co-operatively (at human body temperature) in the restriction enzyme, given that only an average of 0.1 of each degree of freedom can be effectively used.

7.3   A single nucleotide base in a DNA molecule is held in place by one covalent bond, with energy approximately 0.60 aJ, and at least two hydrogen bonds, each with energy approximately 0.03 aJ. A mutation involves the replacement of one nucleotide base by another. How much time will elapse, on average, between two successive thermally-activated spontaneous mutations in a given base in a DNA molecule, given that the body temperature is 37°C? Assume that the relevant atomic vibration frequency is 10^{12} Hz, and that all other pre–exponential factors can be set at unity.

7.4   A mutation of the type discussed in the previous question can be corrected by a suitable enzyme, which clearly must also be able to break the relevant interatomic bonds. Assuming that only one tenth of all degrees of freedom of the atoms in such an (three-dimensional) enzyme are available (i.e. can function cooperatively), calculate the minimum number of atoms the enzyme must be composed of.

Further reading

Alberts, B. *et al.*, (1998). *Essential Cell Biology: An Introduction to the Molecular Biology of the Cell.* Garland, New York.
Austin, R. H. *et al.*, (1997). Stretch genes. *Physics Today* **50**(2), 32–38.
Baker, D., (2000). A surprising simplicity to protein folding. *Nature* **405**, 39.

Chen, S. J. and Dill, K. A., (1998). Theory for the conformational changes of double-stranded chain molecules. *Journal of Chemical Physics* **109**, 4602.

Creighton, T. E., (1984). *Proteins: Structures and Molecular Properties*. Freeman, New York.

Creighton, T. E., ed., (1992). *Protein Folding* Freeman, New York.

Crick, F. H. C., (1968). The origin of the genetic code. *Journal of Molecular Biology* **38**, 367.

Darnell, J., Lodish, H. and Baltimore, D., (1986). *Molecular Cell Biology*. Scientific American Books, New York.

Dickerson, R. E. and Geis, I., (1969). *The Structure and Action of Proteins*. Harper and Row, New York.

Dill, K. A., (1990). Dominant forces in protein folding. *Biochemistry* **20**, 7133–7155.

Ferscht, A. R., (1999). *Structure and Mechanism in Protein Science: A Guide to Enzyme Catalysis and Protein Folding*. Freeman, New York.

Flory, P. J., (1969). *Statistical Mechanics of Chain Molecules*. Interscience, New York.

Garnier, J. and Robson, B., (1989). The GOR method for predicting secondary structures in proteins. In: *Prediction of Protein Structure and the Principles of Protein Conformation*, G. D. Fasman, ed. Plenum, Oxford.

Gesteland, R. F., Cech, T. R. and Atkins, J. F., eds., (1999). *The RNA World: the nature of modern RNA suggests a prebiotic RNA*. Cold Spring Harbour Press, New York.

Jaenicke, R., ed., (1980). *Protein Folding* Elsevier, Amsterdam.

Jaenicke, R. and Creighton, T. E., (1993). Junior chaperones. *Current Biology* **3**(4), 234–235.

Kellermayer, M. S. Z. *et al.*, (1998). Complete unfolding of the titin molecule under external force. *Journal of Structural Biology* **122**, 197–205.

Lindahl, T., (1993). Instability and decay of the primary structure of DNA. *Nature* **362**, 709–715.

Lindgård, P. A. and Bohr, H., (1996). Magic numbers in protein structures. *Physical Review Letters* **77**, 779–782.

Lindgård, P. A. and Bohr, H., (1997). Towards a systematic classification of protein folds. *Physical Review E* **56**, 4497–4515.

Lüthy, R., Bowie, J. U. and Eisenberg, D., (1992). Assessment of protein models with three-dimensional profiles. *Nature* **356**, 83–85.

Marszalek, P. E. *et al.*, (1999). Mechanical unfolding intermediates in titin molecules. *Nature* **402**, 100–103.

Nossal, R. and Lecar, H., (1991). *Molecular and Cell Biophysics* Addison-Wesley, Redwood City, CA.

Richards, F. M., (1991). The protein folding problem. *Scientific American* **264**(1), 34–41.

Robson, B. and Garnier, J., (1988). *Introduction to Proteins and Protein Engineering*. Elsevier, Amsterdam.

Scheraga, H. A., (1992). Some approaches to the multiple-minima problem in the calculation of polypeptide and protein structures. *International Journal of Quantum Chemistry* **42**, 1529–1536.

Sheehan, D., (2000). *Physical Biochemistry: Principles and Applications*. Wiley, Chichester.

Smith, D. A. and Radford, S. E., (2000). Protein folding: pulling back the frontiers. *Current Biology* **10**(18), R662-R664.

Stryer, L., (1981). *Biochemistry*. Freeman, San Francisco.

Tøstesen, E., Chen, J. S. and Dill, K. A., (1999). Designing RNA folding cooperativity. In: *Statistical Mechanics of Biocomplexity*, D. Reguera, J. M. Rubi and J. M. G. Vilar, eds. Springer, Berlin. pp. 56–61.

Watson, J. D. *et al.*, (1987). *Molecular Biology of the Gene*. Benjamin, Menlo Park, CA.

Wolynes, P. G., Onuchic, J. N. and Thirumalai, D., (1995). Navigating the folding routes. *Science* **267**, 1619–1620.

8 Biological Membranes

What is the largest organ in the human body? There are not many people who, if asked that question, would correctly reply: the *skin*. To many, the skin appears to provide a conveniently flexible and watertight, but otherwise inert, container. On the contrary, the skin is a dynamic structure, which supports a variety of essential processes. In much the same way, the skin that surrounds each and every one of an organism's constituent cells is a dynamic structure, without which the cell's interior would not be able to function. Moreover, biological membranes (or biomembranes), as these containers are called, also provide the surfaces of the organelles that lie within each cell, including the nucleus, if the cell possesses one. Amongst the tasks carried out by the typical biomembrane are the transport of ions into and out of the cell; communication with other cells, primarily through the sugar molecules that lie on the cell's external surface; and energy conversion, for example through the production of ATP molecules in the organelles known as mitochondria. In this chapter, we will consider the physics of the biological membrane.

8.1 Historical Background

In 1855, Carl Nägeli noted differences in the rates of penetration of pigments into damaged and undamaged plant cells and concluded that there must be an outer layer with its own special properties. He called it the plasma membrane. In 1897, Wilhelm Pfeffer demonstrated that the membrane is a universal barrier to the passage of water and solutes. Shortly thereafter, E. Charles Overton showed that this extreme view required modification. Although polar molecules certainly find it difficult to get through the membrane, non-polar groups such as alkyl chains and cholesterol have a relatively easy passage. Overton concluded that the membrane exercises selective control through its differential permeability and that it is composed of certain types of liquid crystals (or mesogens) known as lipids. Aleksander Oparin stated in his classic volume *The Origin of Life*, in the mid-1930s, that it is unlikely that there are living organisms, however primitive, that are not physically separated from the environment. Such an organism differs from its surroundings, by definition, so it must possess some structure that can maintain its special properties by providing a rampart against the outside world.

An important hint as to the actual molecular structure of the membrane was provided by Irving Langmuir's demonstration, in 1917, that lipid molecules tend to spread out into a layer one molecule thick, at air–water interfaces. Eight years later, Evert Gorter and F. Grendel measured both the surface area of an erythrocyte, commonly known as a red blood cell, and the area of a reconstructed film formed by the lipids extracted from its membrane. The ratio was approximately one-to-two, and they concluded that the membrane thickness spans two lipid molecules, placed end-to-end. In the mid-1930s, James Danielli, Hugh Davson, and E. Newton Harvey made accurate measurements of the surface tension of the plasma membrane and found this to be considerably lower than for most lipids. It was known that the addition of protein to oil (egg white to mackerel oil for example) lowers the surface tension, and this led Danielli, Davson and Harvey to propose a membrane model that is still regarded as essentially correct.

The Gorter–Grendel lipid bilayer was envisaged as being located at the centre of the membrane, while the proteins formed a thin film at the lipid–water interfaces. This model was revised by Davson and Danielli, in 1954. The hydrophobic parts of the lipid were conjectured to lie in the bilayer interior, the hydrophilic regions of these molecules facing the water at either side, in the gaps between the proteins. Davson and Danielli also proposed the existence of small pores through the membrane. Confirmation of the bilayer hypothesis was obtained in the mid-1950s, by J. D. Robertson. He found a way of staining membranes so as to permit their direct observation in an electron microscope. He was able to resolve two parallel lines, corresponding to the two layers of molecules, although no distinction between protein and lipid was observable. Robertson was led to elevate the model to the status of a universal structure, always comprising a protein–lipid–protein sandwich: the so-called unit membrane.

Proteins were discussed in the previous chapter, where it was shown that they too have hydrophobic and hydrophilic parts. The globular structure of a protein is generally attributed to the fact that a roughly spherical arrangement, with the hydrophobic parts hidden inside away from the water, corresponds to the lowest free energy. This is at variance with the Davson–Danielli–Robertson picture of unfolded proteins spread over the lipid exterior. In 1966, S. Jonathan Singer and Donald Wallach independently postulated a quite different model of protein–lipid architecture. They visualized normal globular proteins dotted around the lipid surface, invariably penetrating into it and sometimes spanning the entire 6 nm width of the bilayer. Singer suggested that the latter variety of protein might have two distinct hydrophilic regions, one being attracted to each lipid–water interface.

The lipid portion of biological membranes is not composed of a single meso-gen, and not all membrane molecules are mesogens. Of the three main constitu-ents, lipids, proteins, and sugars, only the first form liquid crystal phases. The relative amounts of the different components show considerable variation, depending on the part of the organism in which the tissue is located. In the myelin membranes that sheath nerve cells the ratio of lipid to protein is about 10 to one by weight, whereas in the membranes that surround the mitochondria, which provide each cell with energy, it is approximately one to three. The most common type of lipid in biological membranes is the phospholipid, the lesser members being glycolipids and cholesterol. It is appropriate to consider phos-pholipids because they hold the key to membrane structure. They are based on certain esters of the trihydric alcohol, glycerol. Attachment of a fatty-acid residue at each of the (three) hydroxyl positions on the glycerol molecule would produce a neutral lipid. Such lipids are present in the body as fat, stored in the adipose tissue. The fatty-acid chains are linear and can be defined by the number of carbon atoms they contain, together with the number of unsaturated bonds. The unsaturated (double) carbon–carbon bonds tend to be located roughly in the middle of the chains and give the latter a perceptible bend.

The neutral lipids, also known as triglycerides, do not form liquid crystal membranes; they lack a vital extra feature: a terminal polar group. The phos-pholipids differ from their neutral cousins in that one of the outer fatty-acid residues is replaced by one of a variety of groups, all of which incorporate a phosphate. It is clear that we are dealing with a multiplicity of families of molecules, in that there is a considerable degree of choice in all three terminal groups. It is usually the case that one of the two fatty-acid residues is saturated while the other is unsaturated, the latter invariably being located at the middle position of the glycerol backbone. Why this should be so is not clear, and neither is it understood why the membranes in the alveolar lining of the lung, with their two saturated palmitic chains, should be a prominent exception.

One end of a phospholipid molecule is charged and therefore hydrophilic, while the other extremity has two fatty-acid chains that are hydrophobic. How are these dual tendencies satisfied in a cell? It is neatly accomplished by the molecules arranging themselves as the above-mentioned bilayer, with the head groups in the outer layer facing the extra-cellular water and the head groups of the inner layer facing inwards towards the cell's aqueous medium, that is to say the cytoplasm. Since each phospholipid molecule is approximately 3 nm long, the bilayer is approximately 6 nm thick. Depending on the temperature, and also the chemical composition of the surrounding aqueous medium, the molecules in each of the two layers are either ordered or disordered. In the disordered state the composite structure is reminiscent of the smectic liquid crystal phase. In this state, the individual lipid molecules can readily migrate within their own layer.

Lateral movement of molecules within the membrane was detected by David Frye and Michael Edidin, in 1970. They studied cell–cell fusion induced by the Sendai virus, probing the subsequent distribution in the composite cell of proteins originally located on one or other of the participants. It was found that rapid mixing takes place at biological temperatures and, because this had occurred in the absence of an external energy source, Frye and Edidin concluded that it was due to lateral diffusion. Harden McConnell and Philippe Devaux, using what are known as spin labels, later established that lipids move around at a higher rate than proteins, as we should expect since the lipid molecules are smaller. It is found that the flipping of a molecule from one side of the bilayer to the other is a quite rare event. This is consistent with the fact that many membranes are quite asymmetric in their lipid composition. The bilayer offers a remarkably high resistance to the passage of metallic ions, and most substances require the help of other membrane-bound bodies to pass through the membrane.

Interesting facts emerge if we examine the fatty-acid composition profile in different species and different organs. The profile for the neutral triglycerides varies from one species to another. This is hardly surprising since diets of various species can differ widely. For the phospholipids in a given species, the composition profile varies from one organ to another, but if we compare the profile in a given organ, there is remarkably little variation from one species to another. In the brain, for example, the most common fatty acid is of the oleic type irrespective of whether the animal is a mouse, rat, rabbit, pig, horse, ox, or sheep, and in all cases palmitic acid and stearic acid are the next most common. This interesting observation is not particularly surprising if it is taken to indicate that the physical properties of a membrane are determined by its fatty-acid composition. The variation from one organ to another simply reflects the fact that there is a connection between physiological function and physical properties.

The fluidity of a biological membrane is influenced by its cholesterol content. The role of the latter substance appears to be the lowering of the order–disorder transition temperature by disrupting the arrangement of lipid molecules. Physiological function derives from the membrane-bound proteins. Some of these appear to be confined to one side, but it is also known that others span the entire width of the bilayer. As we would expect, the mobility of such proteins in the plane of the membrane is low when the lipid molecules are in the ordered state, and considerably higher when the lipid is disordered. Since function might in some cases depend on the mobility of the proteins, it can be important for an organ's function that the degree of lipid order lies within the desirable range. A dramatic example of this principle is seen in certain fish (said to be poikilothermic), which involuntarily adjust the composition of their membranes according to the water temperature. A similar thing is seen in some bacteria. S. Jonathan Singer and Garth Nicolson have synthesized the large amount of information on membranes, gained in recent years, into what is called the fluid-mosaic model. Its chief features are that the phospholipid serves both as a solvent for protein and as a permeability regulator, the latter by adopting the bilayer configuration; that lipid–protein interaction is possibly essential for protein function; and that membrane proteins are free to migrate within the bilayer, unless restricted by specific interactions, but cannot rotate from one side of the membrane to the other.

The Singer–Nicholson fluid-mosaic model of the biological membrane underwent considerable refinement. In 1977, J. N. Israelachvili noted that the lipids and proteins would have to adjust to each other, and that there would be several factors favouring a heterogeneous arrangement of the various membrane components. Also, Erich Sackmann emphasized the importance of the sugar-based structures (known as the glycocalyx) on the membrane's outer surface, as well as the (protein-based) cytoskeleton attached to the inner surface. Finally, mention should be made of Ole Mouritsen's mattress model, which acquires its name because the proteins embedded in the membrane often have a thickness somewhat less than that of the bilayer. This causes a local decrease in the membrane thickness, the overall effect resembling the dimples in a mattress. Olaf Andersen and his colleagues have observed this effect in connection with protein dimers, one member of which is located in each side of the membrane. Dimerization is possible only if such a dimple is formed.

In recent years, there has been much interest in membrane-bound units known as vesicles, or liposomes. Unlike cells, these structures often comprise several bilayers, arranged like the layers of an onion. Artificial liposomes were produced around 1960, by Alec Bangham. He studied how the lipid configuration varies with its concentration in water. Liposomes having a diameter as small as 30 nm can be formed by subjecting a lipid–water mixture to ultrasonic radiation. These structures are bound by a single lipid bilayer and are equivalent to diminutive cells with interiors devoid of everything but water, and their membranes purged of everything but lipid. They thus differ from the minimal configuration adopted by soap molecules: a single-layer sphere, with the tails inside and the polar heads facing the water, known as a micelle. Bangham and Robert Horne obtained cross-sectional pictures of liposomes in 1962, using electron microscopy, and resolved the individual layers. This showed that the bilayer itself is observable, even in the absence of protein. Liposomes have become the standard test vehicle for experiments in which individual phenomena are studied, without the complexities of the normal cellular environment. In 1974, they were put to novel use by Gerald Weissmann. Borrowing the Trojan Horse idea from Greek mythology, a team led by Weissmann used liposomes for packaging enzymes and delivering them to deficient cells.

8.2   Membrane Chemistry and Structure

In purely chemical terms, as mentioned above, biomembranes are based primarily on lipids, the predominant type being phospholipids. These have the general structure shown in Figure 8.1, and one sees that this is essentially a glycerol molecule to which have been attached fatty-acid chains (R_1 and R_2 in the figure) to two of the three available sites (via ester bonds), while the remaining site plays host to a phosphate group which, in turn, has a terminal group that can be any one of a variety of possibilities (designated by the letter X in the figure).

The chemical composition of the glycerol molecule itself is shown in Figure 8.2, while Figure 8.3 depicts one of the typical fatty acids found in phospholipids, namely palmitic acid, which comprises 16 carbon atoms and 31 hydrogen atoms. This fatty acid is said to be saturated in that the chain contains the maximum possible number of hydrogen atoms. We can imagine one hydrogen atom from each of two adjacent carbon atoms being removed, and the thereby-freed bonds joining up, so as to produce a double bond between those two adjacent carbon atoms. The fatty acid is then said to be unsaturated, and it is clear that various degrees of unsaturation are possible. Indeed, it is not uncommon for some of the fatty-acid chains in phospholipids to contain as many as

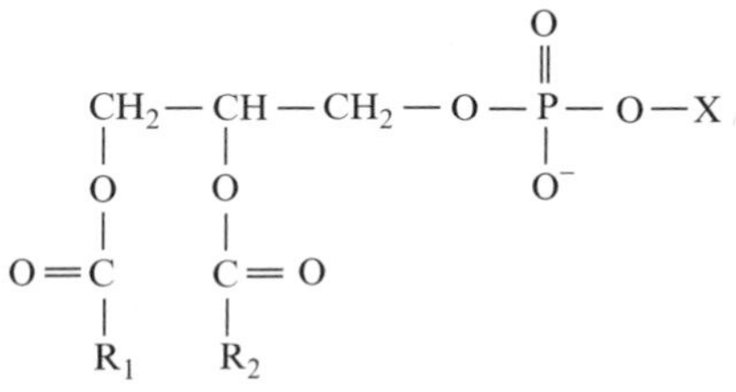

Figure 8.1   The phospholipid molecule

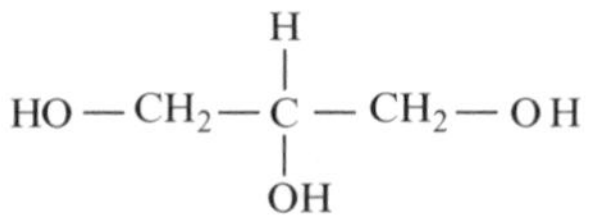

Figure 8.2   The glycerol molecule itself has a simplicity of structure that belies its great significance for the biological realm

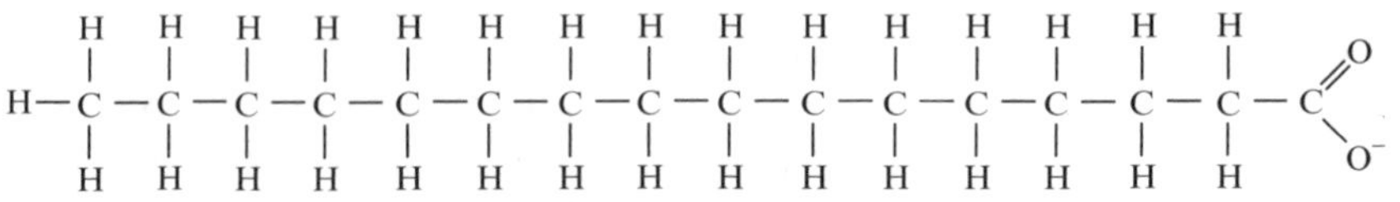

Figure 8.3   The palmitic acid chain

four of the double bonds (as is the case in arachidonic acid). It should also be borne in mind that the two chains in a given phospholipid molecule need not necessarily have the same number of carbon atoms, and in fact they frequently do not. The fatty acid chains which commonly occur in membrane lipids are listed in Table 8.1.

In a particularly common type of phospholipid, the group X is choline, the structure of which is shown in Figure 8.4. It is noteworthy that whereas one of the oxygen atoms attached to the phosphorus bears a negative charge, the choline group has a positive charge at the nitrogen atom. The entire terminal complex is thus an electric dipole. The nomenclature of the phospholipids is quite straightforward. For example, dipalmityl-phophotidyl-choline (also known as lecithin) comprises two palmitic acid chains (which therefore are of *equal* length, this being unusual), while the X group is a choline.

Another common type of membrane lipid is the glycolipid, which is based on the sphingosine molecule, rather than on glycerol. The structure of a typical glycolipid is shown in Figure 8.5, where R again denotes a fatty-acid chain, and where X is in this case a saccharide.

In both these classes of lipids, the constituent molecules have a non-polar part, namely the two fatty-acid chains, and a polar region, the latter often being referred to as the head group. These two parts, the positions of which are indicated schematically in Figure 8.6, are also referred to as the hydrophobic

$$HO-CH_2-CH_2-CH_2-N^+(CH_2)_3$$

Figure 8.4   When the X in Figure 8.1 stands for choline the terminal group is related to the molecule shown here

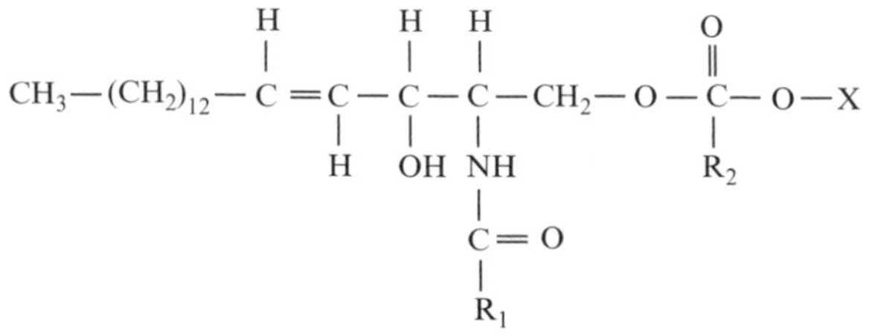

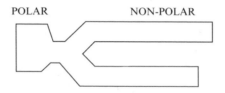

Figure 8.5   Structure of a typical glycolipid

POLAR   NON-POLAR

Figure 8.6   Schematic representation showing the polar and non-polar regions of a phospholipid or glycolipid molecule

Table 8.1 The fatty acid chains which commonly occur in the phospholipids of biological membranes, with both their common names and their chemical (systematic) designations

Symbol	Common Name	Systematic Name	Structure
Saturated Fatty Acids			
10:0	Capric	n-Decanoic	$CH_3(CH_2)_8COOH$
12:0	Lauric	n-Dodecanoic	$CH_3(CH_2)_{10}COOH$
14:0	Myristic	n-Tetradecanoic	$CH_3(CH_2)_{12}COOH$
16:0	Palmitic	n-Hexadecanoic	$CH_3(CH_2)_{14}COOH$
18:0	Stearic	n-Octadecanoic	$CH_3(CH_2)_{16}COOH$
20:0	Arachidic	n-Eicosanoic	$CH_3(CH_2)_{18}COOH$
22:0	Behenic	n-Docosanoic	$CH_3(CH_2)_{20}COOH$
24:0	Lignoceric	n-Tetracosanoic	$CH_3(CH_2)_{22}COOH$
Unsaturated Fatty Acids			
16:1(9)	Palmitoleic	9-Hexadecenoic	$CH_3(CH_2)_5CH=CH(CH_2)_7COOH$
18:1(9)	Oleic	cis-9-Octadecenoic	$CH_3(CH_2)_7CH=CH(CH_2)_7COOH$
18:1(9t)	Elaidic	trans-9-Octadecenoic	$CH_3(CH_2)_7CH=CH(CH_2)_7COOH$
18:2(9,12)	Linoleic	cis, cis-9,12-Octadecadienoic	$CH_3(CH_2)_4CH=CHCH_2CH=CH(CH_2)_7COOH$
18:3(9,12,15)	α-Linolenic	9,12,15-Octadecatrienoic	$CH_3CH_2CH=CHCH_2CH=CHCH_2CH=CH(CH_2)_7COOH$
18:3(6,9,12)	γ-Linolenic	6,9,12-Octadecatrienoic	$CH_3(CH_2)_4CH=CHCH_2CH=CHCH_2CH=CH(CH_2)_4COOH$
20:4(5,8,11,14)	Arachidonic	5,8,11,14-Eicosatetraenoic	$CH_3(CH_2)_4CH=CHCH_2CH=CHCH_2CH=CHCH_2CH=CH(CH_2)_3COO$

and hydrophilic regions of the molecule, respectively. The head group interacts with water, whereas the non-polar chains are repelled by water. That repulsion arises because the presence of the fatty-acid chains amongst an assembly of water molecules would decrease the latter's access to the otherwise available phase space. This would lower the overall entropy, increase the total free energy, and thus make the arrangement less likely. It is very important to note that hydrophobic effects are thereby associated with the *bulk* characteristics of water. It would be wrong to think of a fatty acid chain as being hydrophobic because it shuns water; it is hydrophobic because it is shunned *by* water. The fact that the fatty-acid chain region of a biomembrane is hydrophobic therefore does not imply that a single water molecule will be prevented from passing through this region. Indeed, such passage of water across a membrane is a common occurrence.

The polar head group of a phospholipid, on the other hand, will not interfere with the arrangement of the surrounding water molecules, because its polar nature enables it to masquerade as a sort of pseudo-water molecule. (The reader should recall from the previous chapter that similar considerations apply to the interactions of the various amino-acid side groups – some polar and some non-polar – with the water that commonly surrounds protein molecules.)

The different tendencies of the termini place phospholipids in the class of what are known as amphipathic molecules, and we have already seen that these disparate tendencies have a profound influence on the overall structure of the membrane. Before moving on to the details of that issue, we ought to consider just one more type of lipid, namely cholesterol and certain substances related to it. The chemical structure of this important molecule is shown in Figure 8.7, and one sees that it includes four ring structures, all of which share a side. This gives the molecule more rigidity than is seen in the phospholipids. Cholesterol also possesses a fatty-acid chain, though it is much shorter than those commonly found in the phospholipids. Finally, it should be noted that the OH terminal is polar, and weakly hydrophilic.

Biomembranes are remarkably heterogeneous structures. In addition to the lipids and saccharides mentioned above, the typical membrane also contains protein molecules, some of which span the entire 6 nm width of the membrane,

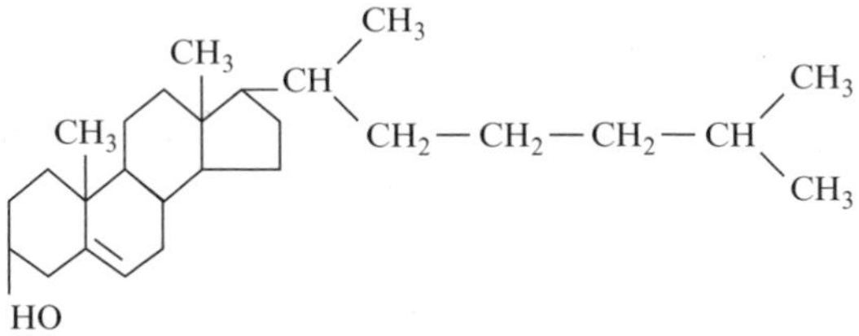

Figure 8.7 The cholesterol molecule

while others are confined to a single side. We see, therefore, that these dynamic bounding structures of cells contain three of the prominent components of our daily diet: fats, sugars and proteins. This should not surprise us; we are what we eat!

It is the actual chemical composition of a given membrane which determines its physical properties, and because this composition is different in different membranes a wide variation in physical behaviour is observed. Membranes with a high protein content tend to be rather stiff, for example, as is the case in the mitochondrion. That energy-converting organelle is involved in enzymatic and transport processes, both of which require proteins, so the roughly 3:1 protein:lipid ratio is perhaps not surprising. At the other extreme, the myelin membranes in the nervous system have a very low protein content (with a 1:10 protein:lipid ratio), and they have the high flexibility consistent with their role as – to put it simply – electrical insulators.

The structure of a membrane is also determined by the surrounding medium, of course, and we have already noted the importance of the water in determining the lipid configuration. Indeed, if this is replaced by alcohol, the membrane configuration rapidly dissolves away. With the water present, the dual tendencies of the amphipathic molecules can be satisfied if they form a double layer, with the head groups pointing outward, toward the water, and the hydrocarbon chains lying tail-to-tail on the inside. This arrangement, which is a special case of what is known as a smectic liquid crystal, is shown in Figure 8.8 (in which the cholesterol, sugars and proteins have been omitted, for the sake of clarity).

Figure 8.8 is highly schematic, however, and it cannot give the correct impression of the membrane's three-dimensional structure. The indicated density of packing of the phospholipid molecules is too high; as we will soon see, various factors produce a packing that is rather looser. Moreover, the presence of cholesterol makes for an even greater spacing between the phospholipid molecules, at least locally. Above all, Figure 8.8 is misleading with respect to the actual arrangement of the phospholipid molecules, because this is not crystalline, as could be concluded from the diagram, but more like a two-dimensional fluid.

Although it is true that the viscosity of this lipid fluid will be decreased as the temperature rises, this type of control is not available to organisms such as ourselves, with our carefully-controlled body temperatures. Instead, the system achieves the desired level of viscosity by varying the chemical composition, including the cholesterol content. Another useful variable, in this respect, is the degree of saturation in the various hydrocarbon chains in the phospholipid molecules. One of the previously-mentioned CC double bonds is found to cause a kink in the otherwise-straight chain, and this locally decreases the degree of packing between the molecules, and makes for lower viscosity.

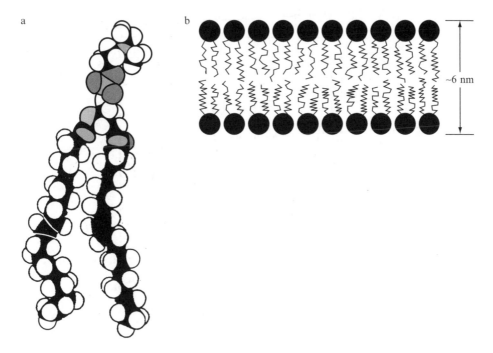

Figure 8.8   A typical phospholipid molecule (lecithin – left) and the bilayer configuration (right), which is a special type of smectic liquid crystal

As noted earlier, fish belonging to the so-called poikilothermic class, which are subject to large seasonal variations in water temperature, have membranes whose composition is automatically varied to compensate for these environmental fluctuations. We should also note, in passing, that different organs within the body of a member of a given species frequently have different lipid compositions, whereas remarkable constancy of composition is observed in the *same* organ in several *different* species (see Figure 8.9) The physical properties of a given phospholipid bilayer depend critically upon which fatty-acid residues are present in their 'tail' regions. It is frequently the case that the two 'tails' comprise different numbers of carbon atoms (denoted in the figure by the large numbers), and indeed different numbers of unsaturated bonds (denoted in the figure by the superscript numbers).

The typical situation is that the individual phospholipid molecules are reasonably free to migrate within that side of the lipid bilayer in which they are located, whereas the hydrophillic forces make it far less probable that a given molecule would flip from one side of the membrane to the other. Indeed, it is not uncommon for the two sides of the bilayer to have rather different compositions.

The mobility of the lipid molecules within the (flexible) plane of the membrane permits analogous mobility of the protein molecules, although this is

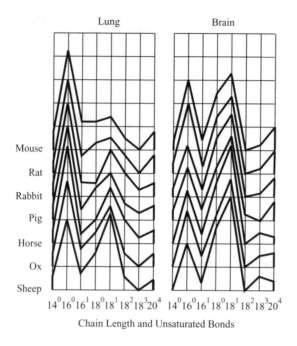

Figure 8.9 Lipid compositions for the lung and brain in different species

naturally more sluggish because of their much larger size. Such lateral mobility of the proteins can be demonstrated by the technique of freeze fracture. The water on either side of the membrane is deliberately frozen, and when the overall structure is subjected to a shearing force, as can be applied by a moving knife, it is found that fracture occurs preferentially along the plane separating the two halves of the bilayer. This is the case because the forces between the phospholipid polar head groups and the surrounding ice are much stronger than those between the ends of the two counter-facing sets of hydrocarbon chains. By cleaving membranes having different prior thermal histories, and by following the cleaving with suitable electron microscopy, it can readily be established that the protein molecules are indeed mobile.

The macroscopic shape of a membrane is determined both by its own composition and by that of the surrounding aqueous medium, and not the least by the latter's pH and ion content. Changes in these parameters could conceivably produce the type of shape change shown in Figure 8.10, although in practice such a modification would probably be stabilized by changes in certain protein molecules attached to the inside surface of the membrane, as indicated by Figure 8.11, which gives a more realistic picture of the true situation. Real biological membranes have a quite heterogeneous structure, with proteins of various types and sizes embedded in the phospholipids bilayer, and with saccharide molecules (not shown in the figure) sticking out from the surface. Some proteins span the entire width of the bilayer, whereas others tend to be located on one side. The former type is extremely important to the cells present in the nervous system.

We saw earlier that a single water molecule can pass from one side of the membrane to the other with relative ease. This might lead us to conclude that

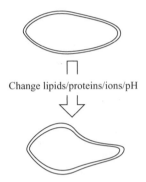

Change lipids/proteins/ions/pH

Figure 8.10 A vesicle is a closed envelope of phospholipid bilayer membrane. Its shape can be altered by changes in the membrane's lipid composition profile and/or protein content, and by changes in the ionic content and/or pH of the surrounding aqueous medium

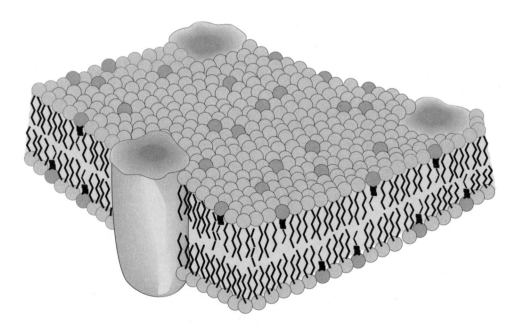

Figure 8.11 A more realistic representation of a biological membrane

ions like Na^+, K^+ and Cl^- would meet with even less resistance, in view of their smaller size. This is emphatically *not* the case, however. Such ions display a translocation probability (for passage across the membrane) that is a factor of about 10^9 lower than that observed for a water molecule. The reason for this lies in the fact that such an ion is usually surrounded by a cluster of water molecules, which are electrically attracted to the ion because of its relatively high surface charge density. In order to pass through the membrane, therefore, the ion would first have to break away from this retinue of water molecules.

The choice of the above ions, to illustrate the different diffusion regime, was not arbitrary; those same ions are the very ones on which signalling in the nervous system depends. As we will see in a Chapters 11 and 12, the proper functioning of that system is contingent upon there being gradients of the concentrations of these ions across the membrane. The gradients are maintained by specially-designed protein molecules (known as active pumps), which accomplish their task through expenditure of a commodity that we will be considering in the next chapter, namely biological energy as mediated by the ATP molecule. The first such active pump to be discovered was the enzyme Na^+, K^+-ATPase, found in crab muscles by Jens Skou in 1956 (see Chapter 9). About a quarter of all the energy we consume, through our food intake, is used to maintain the ionic concentration gradients in our nervous systems.

Another transport process that occurs at membranes is the one by which cells communicate with each other by chemical reactions. There are several versions of this general mechanism, which is known as exocytosis. The chemical molecules that are to be liberated from the interior of the cell are packaged into little membrane-bounded bags, identified above as vesicles, and these release their contents to the exterior by fusing with the cell's outer membrane.

8.3 Membrane Physics

As has been stressed above, the driving force for bilayer formation stems from the hydrophobic effect. The latter can be quantified through consideration of certain parameters we encountered in Chapter 4, and we naturally focus on the hydrocarbon chains of the lipid molecules. The chemical potential of hydrocarbon molecules in water is given by

$$\mu_{HC,w} = \mu^0_{HC,w} + RT \ln X_{HC,w} + RT \ln f_{HC,w} \tag{8.1}$$

Throughout this equation, the subscript HC, w stands for hydrocarbon in water, and the first term on the right-hand side is the chemical potential in the standard state, that is to say in infinite dilution in water. The symbols X and f stand for the mole fraction of hydrocarbon in water and the activity coefficient for the hydrocarbon in water at that concentration, respectively.

If all the subscripts HC, w in Equation (8.1) were replaced by HC, HC we would have the expression for the chemical potential of a hydrocarbon molecule in the purely hydrocarbon environment. Moreover, because the solubility of the typical hydrocarbon in water is very low, the mutual interaction between hydrocarbon molecules, when they are thus deployed, is negligible, so the activity coefficient is essentially unity (and its logarithm is thus zero). In a solubility experiment, with the pure hydrocarbon in direct contact with the aqueous solution, the former will naturally be in its standard state. The chemical potentials $\mu_{HC,w}$ and $\mu_{HC,HC}^0$ will then be equal, since equilibrium prevails. Noting also that $X_{HC,HC}$ is unity, we may thus write

$$\mu_{HC,HC}^0 - \mu_{HC,w}^0 = RT \ln X_{HC,w} \qquad (8.2)$$

Experimental determination of the solubility of hydrocarbon chains in water can thus yield values for the differences in chemical potential in the two standard states. C. McAuliffe has determined the solubility for a number of hydrocarbon families relevant to lipid bilayers, and he found

$$\ln X_{HC,w} = -4.11 - 1.49 n_C \qquad (8.3)$$

In this expression, n_C is the number of carbon atoms in the chain, and his results can readily be shown to indicate that the energy increases by 0.006 aJ for each additional methylene group in the hydrocarbon chain. This small increment, which is a mere 20% of the energy of a hydrogen bond, arises because each such new group will increase the number of water molecules forced to lie in contact with the hydrocarbon.

We are interested in the situation in which hydrocarbons of whatever variety are confronted with the choice of staying in aqueous solution or moving into the lipid bilayer, the subscript for the latter eventuality being HC,l. Because we may not assume that $X_{HC,l}$ is unity, unless the bilayer is composed of a single type of lipid and the candidate for admission to the bilayer is of the same variety, we will require the more general form of Equation (8.2), namely.

$$\mu_{HC,l}^0 - \mu_{HC,w}^0 = RT \ln(X_{HC,w}/X_{HC,l}) \qquad (8.4)$$

Once again, we have neglected the term involving the activity coefficients, because of our assumption of negligible interactions between the participating molecules. The argument of the logarithmic term in Equation (8.4) is simply the partition coefficient between water and the bilayer, of the hydrocarbon in question.

Until now, we have tacitly assumed that there are only two alternatives for a hydrocarbon molecule, when it is a lipid: either it can stay in aqueous solution or it can join a bilayer. The situation is more complicated in practice, because there are a number of other states of aggregation in which a lipid molecule can participate. Some of the most prominent examples are shown in Figure 8.12, and we should note that some of these configurations do *not* involve bilayers. In the micelle, the hydrophobic chains are separated from the water by a *single* sheet of head groups rather than by two such structures (as in the case of the flat bilayer or the vesicle). Let us consider the energetics of aggregation.

There will in general be two opposing tendencies. The driving force for aggregation will be countered by an entropy term, which will decrease with increasing aggregation, thus increasing the free energy, and also by electrostatic, steric and (lipid–lipid) repulsive factors. If all the aggregates comprise the same number m of lipid monomers, the chemical potential (per monomer) of the lipid molecules in micelles of that size will be

$$\mu_{\text{mic},m} = \mu^0_{\text{mic},m} + \frac{RT}{m} \ln \left(\frac{X_m}{m} \right) \tag{8.5}$$

and we see that the standard state of interest now is the one that applies to the micelle comprising m molecules. X_m is the mole fraction of lipid monomers in the m-fold micellar state. Equation (8.5) makes the rather sweeping assumption that the activity coefficient can be neglected because the micelles do not mutually interact. This assumption would be clearly dubious for the roll-like structure shown in Figure 8.12c.

When equilibrium prevails, we will have

$$\mu_{\text{mic},m} = \mu_{\text{w}} \tag{8.6}$$

for micelles of size m, and so by comparison with the above analysis we obtain

$$\mu^0_{\text{mic},m} - \mu^0_{\text{lip,w}} = RT \ln X_{\text{lip,w}} + RT \ln f_{\text{lip,w}} - \frac{RT}{m} \ln \left(\frac{X_m}{m} \right) \tag{8.7}$$

None of the terms in this equation needs special introduction, but we do note that the general suffix HC, w has now been replaced by one appropriate to the lipid, namely lip, w. The final term can usually be neglected when the micelles are large. As is apparent from this analysis, the driving force for aggregation stems primarily from the hydrophobic effect. If one were attempting the far more complicated task of determining which of the several arrangements shown in Figure 8.12 is the most stable, under a given set of conditions, Equation (8.7) would have to be augmented by the corresponding terms for the other configurations. This is certainly a most formidable proposition, and we will not consider it further.

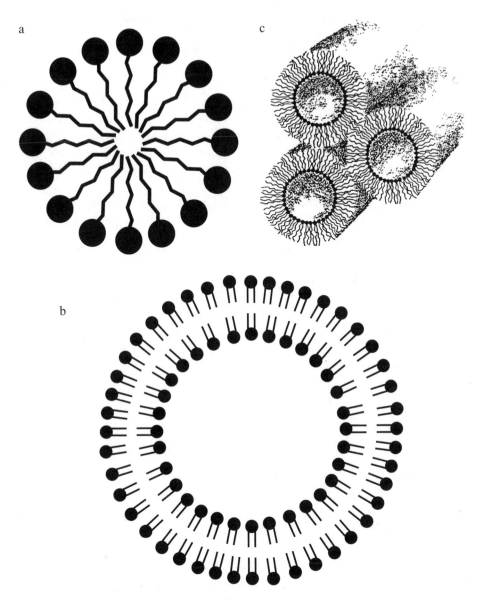

Figure 8.12 Depending upon the manner in which the phospholipid molecules are arranged, the resulting structure can be (a) a (three-dimensional) micelle, (b) a (three-dimensional) vesicle, or (c) a series of parallel rolls

Let us return, therefore, to the situation in which only micelles are present. What can be said about the size of the micelles that will be formed under given conditions? Assuming that the monomer activity coefficient in water really can be assumed to be unity, we can rearrange Equation (8.7) and take the antilogarithm. This produces

$$X_m = m X_{\text{lip, w}}^m \exp\left[\frac{-m\left(\mu_{\text{mic}, m}^0 - \mu_{\text{lip, w}}^0\right)}{RT}\right] \tag{8.8}$$

The final exponential component is, of course, a Boltzmann factor. In practice, the difference of chemical potential in the numerator of the exponent of this factor comprises a number of components. There is a term that arises from the hydrophobic effect, which has already been discussed, but there are also contributions from surface effects, curvature and packing of the lipid molecules.

Regarding surface effects, one should bear in mind that the negatively charged outer surface of a phospholipid layer will attract protons, and this will produce a local change in pH. It is found that

$$\text{pH}_{\text{surface}} = \text{pH}_{\text{bulk}} + \frac{q_e \Psi_0}{2.3 k_B T} \tag{8.9}$$

where q_e is the electronic charge and Ψ_0 is the surface potential.

Diffusion of solutes across a membrane is governed by the factors we considered in Chapter 5. The flux J of solute, for example, will be given by Fick's First Law. This can be written in the alternative form

$$J = -\frac{D\beta}{\ell}(C_1 - C_2) \tag{8.10}$$

where D is the diffusion coefficient and l is the membrane thickness. C_1 and C_2 are the concentrations of solute on the two sides of the membrane. In contrast to Equation (5.2), Equation (8.10) invokes the partition coefficient β. The quotient outside the parentheses on the right-hand side, $D\beta/l$, is known as the permeability coefficient, P.

As will be discussed at greater length in Chapter 11, the analysis becomes more complicated when the diffusing solute bears an electrical charge. If the gradient of the electrical potential were uniform throughout the membrane, the situation would be amenable to relatively simple treatment, but this is not usually the case. Instead, the electrical potential will have a maximum value V_{max} somewhere within the membrane, and the potential difference $V_{\text{max}} - V_1$ is the barrier which the diffusing solute will have surmount. The flux from side 1 to side 2 is then given by

$$J_{1 \to 2} = P C_1 \exp\left[\frac{-nF(V_{\text{max}} - V_1)}{RT}\right] \tag{8.11}$$

where F is the Faraday constant and n is the number of unit charges on each solute particle. Once again, we find a rate that is governed by a Boltzmann factor.

As we shall discuss in Chapters 11 and 12, diffusion through biological membranes occurs predominantly through the mediating agency of membrane-spanning proteins. It is thus of considerable interest to enquire how these molecules interact with the lipid bilayer. Is the latter merely a passive sea, within which the proteins carry out their work, or do the lipid molecules play an active role in the control of membrane function? The importance of this question is particularly stressed by the fact that membrane function can be altered by the presence of substances such as anaesthetics.

Robert Cantor has investigated the underlying physics of the situation, and he finds evidence that variations in membrane composition can induce changes in the transverse pressure profile across the bilayer. Moreover, that pressure can be surprisingly large, as we will soon learn. In analysing the situation, we see Equation (8.1) winning its spurs through its impressive generality, for we may regard the membrane-bound protein as a solute! Following Cantor, we simplify by assuming that the protein molecule in question can adopt one of only two conformations, characterized by the vectors (specifying the co-ordinates of all their constituent atoms) r and t. Each of these two conformations has associated with it a cross-sectional area $A(z)$ that varies with depth z in the membrane, and the gist of the matter is that this depth-dependent cross-section is changed by the transition from r to t, that is to say that $A_r(z)$ is not the same as $A_t(z)$.

Before getting down to a brief summary of Cantor's analysis of the situation, we should be aware of a possible confusion regarding dimensions. Bulk pressure p has the dimension of force per unit area, whereas lateral pressure π is a force per unit length. If $\delta\pi(z)$ is the lateral pressure acting within a thin slice δz of the bilayer, $p(z) = \delta\pi(z)/\delta z$ will be the lateral pressure density, and it will have the correct dimension of a bulk pressure. Now let us invoke Equation (8.1), and note that the chemical potential for the s state, at bulk pressure p, will be

$$\mu_s = \mu_s^0 + RT \ln [s] + V_s \Delta p \qquad (8.12)$$

in which V_s is the partial molar volume of the solute in state s (and not a voltage, as earlier in this section). The argument of the logarithm employs the standard symbol for a concentration, and we must bear in mind that we are here referring to a concentration (that is to say a probability of occurrence) of a *state*.

The total mechanical work dw associated with a change $dA(z)$ in the profile of the cross-sectional area is given by

$$dw = - \int p(z)dA(z)\delta z \qquad (8.13)$$

this being a sum over the contributions arising in the incremental slices at the various depths throughout the bilyaer. The negative sign corresponds to work being done by the protein, and we note that this work has the correct dimension of pressure multiplied by a change of volume. We may thus rewrite Equation (8.12) in the form

$$\mu_s = \mu_s^0 + RT \ln [s] + N_{Av} \int A_s(z) \Delta p(z) \delta z \qquad (8.14)$$

where N_{Av} is Avogadro's number.

There is a large excess of free energy associated with any oil/water interface, and in the case of the very thin lipid bilayer it is concentrated within this quite narrow region. This produces very high *local* lateral pressures. However, because the bilayer has nevertheless been produced by a process of self-assembly, there must be little or no *overall* pressure. So the local negative pressures at some depths must be effectively cancelled out by the local positive pressures at other depths. In particular, the large (surface) tension at the lipid/water interface must be compensated by the large compression (i.e. positive pressure) in the interior of the bilayer.

At equilibrium, the chemical potentials of the two states *r* and *t* must be equal. Moreover, we can eliminate the reference to the standard chemical potentials (denoted by the superscript zeros in Equations (8.12) and (8.14)) by equating those potentials at the two different pressures $p(z)$ and $p_0(z)$. We then have

$$\frac{[r]}{[t]} = \frac{[r]_0}{[t]_0} e^\alpha \qquad (8.15)$$

where

$$\alpha = \frac{1}{k_B T} \int \Delta A(z) \Delta p(z) \delta z \qquad (8.16)$$

The fraction F_t of the proteins that are in the state *t* can then be written

$$F_t = \frac{[t]}{([r] + [t])} = \frac{1}{(1 + y_0 e^\alpha)} \qquad (8.17)$$

where

$$y_0 = \frac{[r]_0}{[t]_0} \qquad (8.18)$$

Correspondingly, the situation for a pressure profile $p_0(z)$ will be characterized by

$$F_{t,0} = \frac{[t]_0}{([r]_0 + [t]_0)} = \frac{1}{(1 + y_0)} \qquad (8.19)$$

so we finally obtain the fraction f of protein in state t at $p(z)$ relative to that at $p_0(z)$, namely

$$f = \frac{F_t}{F_{t,0}} = \frac{(1 + y_0)}{(1 + y_0 e^\alpha)} \qquad (8.20)$$

From this analysis we see that the only way in which the protein's conformation could be unaffected by a change in pressure profile would be for ΔA to be independent of z, which seems unlikely even for the multiple-alpha-helix structures common in the cell membranes of the nervous system. In practice, therefore, the lateral expansion or contraction of the protein will vary as a function of depth in the bilayer. The sign of α is determined solely by the manner in which ΔA varies with z, whereas the average change of area has no influence. If the pressure increases near the lipid/water interface and decreases in the interior of the bilayer, α is positive, while the reverse situation corresponds to a negative α.

The phenomenon of greatest interest is switching of the activity state of a membrane-bound protein (from active to inactive, or vice-versa). Let conformation t correspond to the active state and conformation r to the inactive form. Because f varies exponentially with α, there will be only a negligible shift in the distribution of conformations until α exceeds $\ln(1/y_0)$, but thereafter the fraction in the active state will sharply decrease. It is not difficult to show that the effect is much stronger than the allosteric effect produced by the docking of a ligand (that is to say a small triggering molecule) to the protein. This is of great relevance to the functioning of nerve cell membranes, and not the least to those that mediate the sensation of pain. A ligand effect is produced when a neurotransmitter molecule docks with a receptor on the post-synaptic membrane (see Chapters 12 and 13). However, the more general influence arising from the lipid bilayer will be modified by any substance that acts within the membrane itself, such as an anaesthetic.

Given the obvious importance of Cantor's analysis, we must echo his disappointment over the scarcity of experimental data on $\Delta A(z)$ but, as he has noted, reasonable estimates can be made. The very important acetylcholine receptor has an average radius of 3.5 nm, and Cantor estimates that the change in radius accompanying the $t - r$ transition could be as much as 0.5 nm. Using a typical experimental value for the lateral pressure at an oil/water interface, namely

$0.05\,\mathrm{N\,m^{-1}}$, and noting that the thickness of the hydrophobic interior of the membrane is about 3 nm, Cantor found that the average pressure density can be as large as 300 atm, and that overwhelming shifts in the distribution of states are a viable proposition.

Apart from changes induced by the above-indicated chemical heterogeneity, the structure of the lipid bilayer can also be altered by a change of temperature. Indeed, a bilayer can undergo a melting transition, or rather, a gel-to-fluid transition. (It is for this reason that the term mesogen is favoured for the substance forming the bilayer, because one might otherwise erroneously assume that a melted liquid crystal is no longer a liquid crystal.) Differential scanning calorimetry is used to study such a bilayer transition, because this is accompanied by a heat change (that is to say, an enthalpy of fusion). This change will be given by

$$\Delta H_{\text{fusion}} = \int_{T_i}^{T_f} \Delta C_P dT \qquad (8.21)$$

The calorimetry produces data on the excess heat capacity (at constant pressure) over the *range* of temperature (T_i to T_f) in which the transition is observed to occur (see Figure 8.13), and integration produces the desired enthalpy. We note, therefore, that the transition is not a sharp one, as it would be if we were considering the melting of a piece of noble metal such as gold. Assuming that the melting of the bilayer is nevertheless a first-order transition, the change in free energy will be zero, and this enables us to find the transition entropy, namely

$$\Delta S_{\text{fusion}} = \frac{\Delta H_{\text{fusion}}}{T_{\text{fusion}}} \qquad (8.22)$$

The fusion temperature may be taken to be the arithmetic mean of T_i and T_f.

Ole Mouritsen and Paavo Kinnunen have noted that the fluctuation-dissipation theorem can be invoked to give a value for the heat capacity, in terms of the fluctuations in internal energy $\mathscr{E}_{\text{int}}$. The useful relationship is

$$C_P = \left(\frac{\partial \mathscr{E}_{\text{int}}}{\partial T}\right)_P = \frac{1}{(k_B T)^2}\left[\left\langle \mathscr{E}_{\text{int}}^2 \right\rangle - \left\langle \mathscr{E}_{\text{int}} \right\rangle^2\right] \qquad (8.23)$$

Through a similar analysis, they show that the isothermal area compressibility modulus is given by

$$K_T = k_B T A \left[\left\langle A^2 \right\rangle - \left\langle A \right\rangle^2\right]^{-1} \qquad (8.24)$$

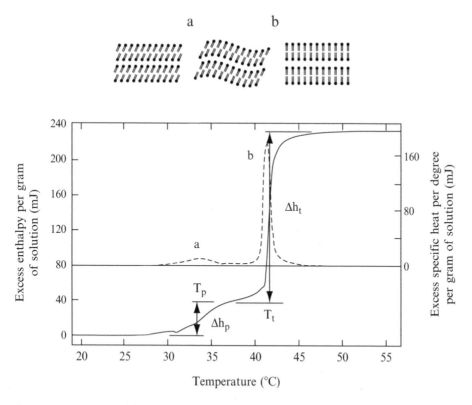

Figure 8.13 Phospholipid bilayer structure is temperature sensitive. Lecithin undergoes two structural transitions (a and b) when transforming from gel to fluid. (Calorimetric data recorded by Hinz and Sturtevant, 1972)

Finally, they show that lateral density fluctuations require the isothermal bending rigidity modulus κ_T to be multiplied by a factor $(1 - \chi^2/K_T)$, where χ is a materials constant that determines the strength of coupling between the density fluctuations and the local bilayer mean curvature.

Erich Sackmann has analysed the situation in which a vesicle and a flat surface of membrane approach each other (as happens at the presynaptic membrane – see Chapters 12 and 13), with their mutual fusion as a possible outcome. As a preliminary, one has to note that the typical membrane, consisting of a lipid bilayer with included proteins, has an extremely low isothermal bending rigidity modulus, which can be reduced to a few k_BT by the addition of small amounts of the appropriate solutes. Because of this, membranes can be excited into bending undulations, which lead to entropy-driven repulsion when two such membranes interact. The approximate form of the repulsion was found, by Wolfgang Helfrich, to be given by

$$\mathcal{E}_{\text{repulsion}}(r) = \frac{(k_B T)^2}{\kappa r^2} \tag{8.25}$$

and this can largely compensate for the van der Waals attraction through which the membranes are otherwise found to interact.

Turning to that possibility of fusion, Sackmann showed that the change of the Gibbs free energy due to the mutual adhesion would be

$$\Delta G = \int p dV - W A_c + \gamma \int dA + \frac{1}{2} \kappa \oint \left(\frac{1}{R_1} + \frac{1}{R_2} - \frac{1}{R_0} \right)^2 dS \tag{8.26}$$

where p is now the osmotic pressure inside the vesicle, W is the adhesion energy per unit area, A_c is the contact area, γ is the membrane tension, and R_1 and R_2 are the principal radii of curvature of the vesicle. R_0 is the spontaneous radius of curvature of the membrane that arises from the previously-mentioned asymmetry of the lipid composition on the two sides of the bilayer. In the case of no such asymmetry being present, R_0 would be infinite, and the final term in the last integral would disappear. As indicated by the special sign $\oint$, that integral is to be made over the entire surface of the vesicle.

Equation (8.26) offers ample scope for varying the Gibbs free energy through changes to the membrane characteristics, these being provoked by alterations in the local physical conditions, including the availability of substances which can actually enter the lipid bilayer. By using the appropriate parts of Equation (8.26), Helfrich was able to undertake a remarkably successful analysis of the dimpled shape of the typical red blood cell.

A special example of exocytosis occurs at the presynaptic membrane of a nerve ending, the process taking place only during the arrival of a nerve signal (see Figure 8.14). Given that the duration of the nerve impulse is about a millisecond (see Chapter 12), one sees that this particular release process is rather finely timed. One might guess that the electrical charges on the head groups of the phospholipid molecules would lead to electrical repulsion between the surfaces of the vesicles and the cell's outer membrane, and this is indeed the case. In fact, such repulsion is important because it prevents exocytosis happening in a haphazard manner. In practice, another factor is required,

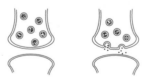

Figure 8.14   The neurotransmitter molecules contained in the membrane-bounded vesicles are released into the synaptic cleft when these temporary containers fuse with the presynaptic membrane

in order to trigger the fusion of the two membranes. That additional factor is the presence of Ca^{2+} ions, and in the case of the process at the presynaptic membranes in the nervous system, they diffuse into the cell immediately after the arrival of the nerve impulse.

Exercises

8.1 If the surface potential of a membrane is 0.05 V, what change in the surface pH will be observed when the temperature is raised from 0°C to 100°C, if the presence of a buffer in the surrounding bulk aqueous medium prevents any change in the bulk pH, the electronic charge being 1.6×10^{-19} Coulombs?

8.2 A membrane is found to undergo a melting transition at 45°C, the enthalpy of fusion being 0.02 aJ. What is the entropy change associated with the transition?

8.3 The resting potential (see Chapter 11) across the typical nerve cell membrane is 0.1 V, and the membrane thickness is 5 nm. Calculate the electrical field that acts across the membrane and, by referring to an appropriate tabulation of experimental data, establish how this compares with the electrical breakdown field strength for lipid material.

8.4 Write an essay on membrane structure, emphasizing how the lipid composition profile relates to physiological function, and also highlighting the role of cholesterol.

Further reading

Cantor, R. S., (1997). The lateral pressure profile in membranes: a physical mechanism of general anesthesia. *Biochemistry* **36**, 2339–2344.

Cantor, R. S., (1997). Lateral pressure in cell membranes: a mechanism for modulation of protein function. *J. Phys. Chem B.* **101**, 1723–1725.

Cevc, G. and Marsh, D., (1987). *Phospholipid Bilayers: Physical Principles and Models* Wiley, Chichester.

Danielli, J. F. and Davson, H., (1935). A contribution to the theory of permability of films. *J. Cellular Comp. Physiol.* **7**, 393–408.

Gorter, E. and Grendel, F., (1925). On bimolecular layers of lipoids on chromatocytes of blood. *J. Exp. Medicine* **41**, 439–443.

Hames, B. D. and Hooper, N. M., (2000). *Biochemistry, Second Edition.* BIOS Scientific, Oxford.

Israelachvili, J. N., (1977). Refinements of the fluid-mosaic model of membrane structure. *Biochim. Biophys. Acta* **469**, 221–225.

Merz, K. M. and Roux, B., eds., (1996). *Biological Membranes: A Molecular Perpective from Computation and Experiment* Birkhäuser, Boston.

Mouritsen, O. G. and Kinnunen, P. K. J., (1996). Role of lipid organization and dynamics for membrane functionality. In *Biological Membranes: A Molecular Perpective from*

Computation and Experiment K. M. Merz and B. Roux, eds., Birkhäuser Boston, pp 463–502.

Mouritsen, O. G. and Andersen, O. S., eds., (1998). *In Search of a New Biomembrane Model.* Munksgaard, Copenhagen.

New, R. R. C., ed., (1990). *Liposomes: a Practical Approach* IRL Press, Oxford.

Quinn, P. J., (1976). *The Molecular Biology of Cell Membranes* Macmillan, London.

Robertson, J. D., (1966). Granulo-fibrillar and globular sub-structure in unit membranes. *Ann. N. Y. Acad. Sci.* **137**, 421–440.

Sackmann, E., (1995). Biological membranes: architecture and function. In: *Handbook of Biological Physics*, Vol. 1 A, R. Lipowsky and E. Sackmann, eds. Elsevier Amsterdam, pp 1–63.

Singer, S. and Nicolson, G. L., (1972). The fluid mosaic model of cell membranes. *Science* **172**, 720–730.

9 Biological Energy

The goal of this chapter is to account for the energy changes in organisms, across the range of sizes from the macroscopic right down to the atomic. We wish to see how energy is gained from the environment, and how it is then utilized by the organism in order to provide the driving force for the various cellular and sub-cellular functions underlying its biological viability. We will note that only plants, algae and certain types of bacteria are able to harvest the energy present in sunlight, whereas animals must obtain their energy indirectly, in the form of food. In all these organisms, the energy thereby made available provides fuel for processes that occur at the molecular level, including all those that contribute to the maintenance of tissue. Amongst the other examples of such processes that we will consider later in this book are those relevant for movement in the animal kingdom, and for control of such movement. In addition, it will be interesting to see how the typical energies encountered in those processes compare with those that apply to the very important mechanism of photosynthesis in plants, algae and some bacteria. We will be looking at the energies involved in metabolism, both with respect to catabolic processes (breaking down) and anabolic processes (building up).

9.1 Energy Consumption

The unit of energy usually employed in the dietary domain is the calorie (symbol: cal), or the dietician's capitalized Calorie (symbol: Cal), which is equal to one kilo-calorie (i.e. 1 Cal = 1000 cal = 1 kilo-calorie = 1 kcal). In 1798, Count Rumford demonstrated the equivalence between thermal and mechanical energy, the unit of the latter being the Joule (symbol: J), and it was subsequently established that 1 cal = 4.1868 J.

Let us consider some typical amounts of energy in the biological domain. An adult human uses about 300 kcal per hour when walking sprightly, and about 800 kcal per hour when briskly ascending stairs. When sitting and quietly reading, we use a mere 25 kcal per hour, and this figure is doubled when we are eating. Regarding that latter activity, it may be noted that a 20 g piece of bread gives about 50 kcal, while a 200 ml glass of full fat milk gives around 140 kcal. The well-fed adult consumes a total of about 2400 kcal per day.

The daily consumption corresponds to about two 60 W lamps burning round the clock, that is to say continuously throughout the 24 hours of each day. We have earlier noted that this total energy consumption also corresponds to about a two-hundredth of the total energy stored in the body's interatomic bonds. Because we achieve our mature weight only after a couple of decades, this shows that the energy is also used for other purposes, such as maintaining body temperature and keeping the nervous system primed for action.

9.2   Respiration

Our first major task is to get an overall view of how these energy-gaining and energy-consuming processes take place. In Chapter 4, we learned that the energy of vibration that is present in all matter above the absolute zero of temperature is not generally available for bond-rearranging tasks because it is too diffusely dispersed. The exception to this rule is seen in the molecules of enzymes, in which the vibration energy becomes focussed at certain positions in their structures, thereby permitting them to carry out their remarkable cata-bolic and anabolic tasks. The heat we feel from incident sunlight does not produce interatomic rearrangements because it is indeed dispersed; it is pro-duced by light with wavelengths lying in the infra-red region of the electromag-netic spectrum, and these are too long to cause bond-breaking events. The shorter wavelength ultraviolet light photons do have the requisite energy, and the atomic-scale events they provoke produce the change of skin colour we call tanning. This process gives no useful energy in the service of metabolism, however, because our bodies do not produce the one substance known to be capable of harvesting light energy, namely chlorophyll. That substance is present in plants, algae and certain bacteria, and in plants and algae it is contained in structures known as chloroplasts. The energy-harvesting process is, of course, known as photosynthesis.

When light photons of sufficiently short wavelength (that is to say, suffi-ciently high energy) are incident on chlorophyll molecules, they cause ioniza-tion. This process was considered in Chapter 2, and in more detail in Chapter 4. We learned that the highest ionization energy is encountered in the case of neon, its magnitude being 3.454 aJ. In the isolated atoms of all heavier elem-ents, the ionization energy is lower despite the increased nuclear charge, because the electrons lie further away from the nucleus and the nuclear charge is partially screened by the complete inner shells of electrons (see Table 2.1). In the case of atoms which are members of a molecule, there will be additional quantum effects, which may lead to a stronger overall attachment for some of the electrons. However, the ionization energy is nevertheless not as large as it is for the isolated neon atom, for example. Indeed, it usually lies well below the ionization energy for hydrogen, which is 2.178 aJ.

The reason what could be called 'the photosynthetic machinery' is relatively complicated is connected with the need for preventing the reverse reaction, in which the electron is reunited with the positive ion. If that happened to any great extent, the energy-harvesting process would have been wasted. Instead, there are other molecules of various types waiting to grasp the liberated electron, as it were, and pass it amongst themselves. At some of these electron-transferring steps, anabolic processes take place that produce new molecules in which a portion of the original ionization energy has been stored. Those molecules are subsequently available as sources of the energy required by the plant, or by the animal which has eaten it. Such energy-liberating processes are collectively referred to as respiration. One should be careful not to be confused by this term, because it has two components, only one of which is in colloquial use. There is the external respiration by which the necessary bulk oxygen is acquired by the organism, as in the breathing of animals. The other component is the internal respiration, in which the acquired oxygen molecules participate in actual chemical reactions.

The summary equation for photosynthesis does not reflect the chain of molecular-level events which actually underlie the energy-harvesting process. It nevertheless provides a useful overview of what is involved. The equation reads

$$6CO_2 \ + \ 6H_2O \ + \ \text{light energy} \ \overset{\text{chlorophyll}}{\longrightarrow} \ C_6H_{12}O_6 \ + \ 6O_2$$

carbon dioxide   water   glucose   oxygen

$$(9.1)$$

the light energy in question coming from the incident photons. The process is said to produce a quantum yield of almost 100%, this parameter being a measure of the fraction of the incident photons actually contributing to the reaction. The overall efficiency of the entire set of energy-harvesting processes is, not unexpectedly, lower than this, but it is nevertheless an impressive 40%. Glucose is of course a carbohydrate, and the amount of carbohydrate produced at the surface of the Earth by photosynthesis is about 10 billion tons annually. This is equivalent to eight times mankind's energy consumption in 1990.

Turning to internal respiration, we find that it comes in two distinct forms, aerobic respiration and anaerobic respiration. As in the case of photosynthesis, either form involves a chain of chemical reactions, each of which is a series of oxidation processes. The summary equation for the aerobic process is

$$\underset{\text{glucose}}{C_6H_{12}O_6} \quad + \quad \underset{\text{oxygen}}{6O_2} \quad \overset{\text{enzymes}}{\longrightarrow} \quad \underset{\text{carbon dioxide}}{6CO_2} \quad + \quad \underset{\text{water}}{6H_2O} \quad + \quad \text{metabolic energy} \tag{9.2}$$

Taken at face value, this equation would seem to indicate that aerobic respiration is simply the reverse of photosynthesis. That is emphatically not the case, however, because the individual steps in the two chains of reactions involve quite different enzymes and atomic rearrangements.

The corresponding summary equation for the anaerobic process is

$$\underset{\text{glucose}}{C_6H_{12}O_6} \quad \overset{\text{enzymes}}{\longrightarrow} \quad \underset{\text{ethyl alcohol}}{2C_2H_5OH} \quad + \quad \underset{\text{carbon dioxide}}{2CO_2} \quad + \quad \text{metabolic energy} \tag{9.3}$$

The energy thereby produced is much lower than is the case for the aerobic reaction, and the ethyl alcohol is toxic to most living tissues. Organisms which are capable of surviving on relatively small amounts of energy, and in an alcoholic environment, are clearly rather specialized. An example is yeast, which is of course used in brewing (for the alcohol) and baking (for the carbon dioxide). Metabolically active tissue deprived of adequate oxygen, such as overworked muscles, will be forced to respire anaerobically. The same is true of the inner regions of large apples, in which the inward diffusive flow of oxygen is inadequate because of the sheer bulk of respiring tissue.

9.3 Photosynthesis

Let us now consider photosynthesis in more detail. It will be useful to start with a brief description of the relevant components in the typical plant leaf, which is of course the site of energy production. Working from the left extremity of Equation (9.1), the carbon dioxide enters at the leaf's under side, through pores known as stomata (singular: stoma), while the water supply is through the xylem vessels, which somewhat resemble the arteries in our own bodies. As mentioned earlier, the chlorophyll is present in organelles known as chloroplasts. These contain structures known as grana (singular: granum), which resemble neatly piled coins, and also less densely packed structures called intergranal lamellae. The grana are composed of flattened membrane vesicles called thylakoids, and these provide the physical environment for the chlorophyll molecules (see Figure 9.1). The intergranal space is known as the stroma, while the region of cytoplasm enclosed within a thylakoid is referred to as the lumen. Let us pass over the next item in Equation (9.1), namely glucose,

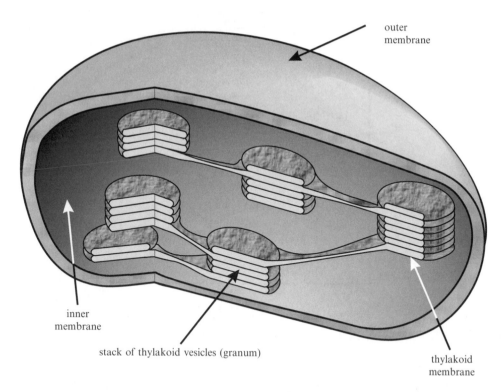

outer
membrane

inner
membrane

stack of thylakoid vesicles (granum)

thylakoid
membrane

Figure 9.1 Schematic cross-sectional view of a chloroplast, showing the form and position of the thylakoid membranes

the production of whose molecules we will consider later. The final item, oxygen is liberated into the atmosphere through the above-mentioned stomata.

There are actually several different types of chlorophyll. Higher plants possess chlorophyll *a* and chlorophyll *b*, while brown algae have chlorophyll *c* and red algae chlorophyll *d*. There is a similar range of bacteriochlorophylls. Each type has its characteristic light absorption maxima. Chlorophyll *a*, for example, absorbs maximally at the two wavelengths 435 nm and 675 nm (see Figure 9.2), while the two peaks for chlorophyll *b* lie at 480 nm and 650 nm. When light having wavelengths in the range of 435–480 nm falls on the human retina, it produces the sensation we associate with the colour blue, and we similarly associate wavelengths in the range 650–675 nm with the colours orange and red. When white light falls on a leaf, photons with wavelengths lying in these ranges will be absorbed, while photons having other wavelengths will be reflected. It is for this reason that we perceive leaves to have a green colour. Using the same principles, we are not surprised to find that the absorption bands for red algae span the range 490–575 nm, this region corresponding to our perception of blue and green.

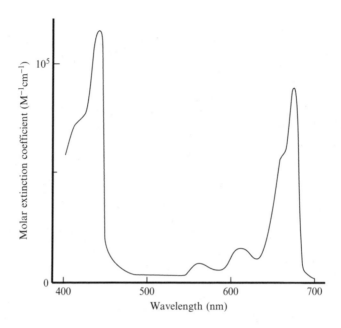

Figure 9.2   The absorption spectrum of chlorophyll *a*

We should note, in passing, that these wavelengths are considerably larger than the diameter of the typical molecule of biological interest. Indeed, the disparity is about a factor of 10. So it would be wrong to imagine a single atom in a chlorophyll molecule as being hit by a photon. On the contrary, the more reliable picture has the entire molecule being simultaneously 'bathed' in the photon's radiation.

Photon absorption will occur only if there are ground and excited energy levels having the appropriate magnitudes, and this is found to be the case for systems which have what are known as conjugated double bonds. These are produced when there is resonance between different, but equivalent, arrangements of double bonds, the important examples being those found in the chromophores of cytochrome, myoglobin and haemoglobin, and chlorophyll. In the latter (see Figure 9.3), the prosthetic group contains a pyrrole ring of nitrogen atoms, comprising extended pi-orbitals, and with a magnesium atom at its centre. (In haem, this is replaced by an atom of iron.)

When the chlorophyll-containing molecule has been excited by the incoming photon, several different processes could be envisaged as taking place. If there was a continuum of allowed electron energy levels, and if these were sufficiently strongly coupled to the vibration states of the molecule, the excitation energy would rapidly leak away into the surrounding medium, and dissipate itself in the form of heat. In the chromophores in question, there is no such continuity, the minimum energy of the lowest excited electronic state being separated

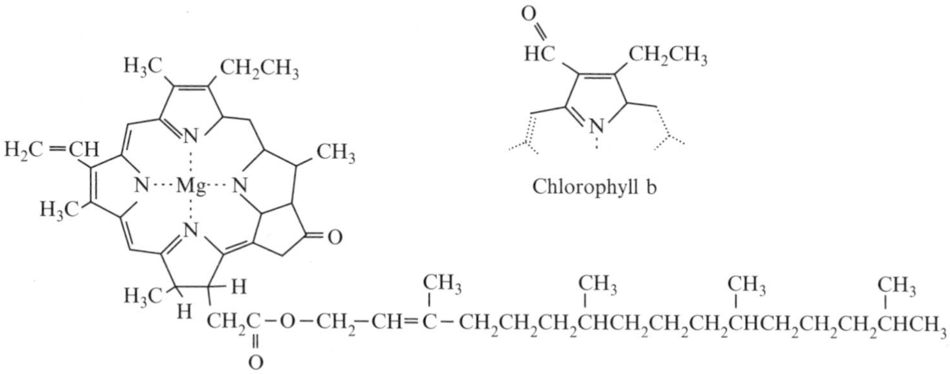

Figure 9.3 The atomic configuration in a molecule of chlorophyll *a*; the inset shows details of the slightly different arrangement in chlorophyll *b*

from the maximum energy of the electronic ground state by an energy gap, not unlike that which exists in the typical semiconductor. The upshot is that the preferred route for dissipating the energy of excitation is through transfer of the excited electron to another molecule. Indeed, and as we will shortly discover, photosynthesis (and also respiration) involves a series of such intermolecular electron transfers.

Whether or not electron transfer will take place depends upon a quantity known as the redox potential (see Chapter 4), and its entry into the story means that we should pause for a few definitions. A molecule is said to be reduced when it acquires an electron, while a molecule which loses an electron is said to be oxidized. In the aqueous environment typical of cell interiors and exteriors, a reduced molecule will be able to capture a proton (at a rate which will clearly depend upon the pH), so reduction is equivalent to the acquisition of hydrogen. An oxidized molecule, on the other hand, can readily capture an oxygen atom because of the latter's strong electronegativity. In the absence of water, however, the hydrogen and oxygen capture processes are not possible, and the energy transfer is purely electronic. It is referred to as an oxidation–reduction process, or a redox process, and again by analogy with what is seen in semiconductors one can think of the electron as being transferred from a donor D to an acceptor A. We may formally write

$$D + A \rightarrow D^+ + A^- \tag{9.4}$$

D becoming oxidized by the reaction while A is reduced.

Now because reduction is sometimes associated with hydrogen acquisition, we can think of oxidation as being equivalent to the opposite process, namely loss of a hydrogen, and simultaneously of an electron. This is important in the case of the early part of the photosynthetic process because the end result of a series of redox reactions is precisely that: oxidation of a hydrogen donor, to produce a relatively strong reducer (also called a reducing agent). The latter is then used to reduce CO_2 to a sugar. The reducer in question is nicotinamide adenine dinucleotide phosphate hydride, or NADPH for short, the reaction by which it is produced being

$$NADP^+ + H_2O + \text{light energy} \;\rightarrow\; NADPH + H^+ + 0.5O_2 \qquad (9.5)$$

It turns out that the difference between the redox potentials of $NADP^+$ and O_2 is approximately 1.1 V, and this is larger than the energy that is absorbed from a 675 nm wavelength photon. The reaction shown in Equation (9.5) cannot proceed, therefore, unless there is simultaneous absorption of more than one photon (see Figure 9.4). The higher plants thus have the molecular machinery necessary for double absorption, the key structural feature being the presence of two co-operating chlorophyll molecules.

This provision, and also the one required to ensure against electron re-capture at the original site of excitation, is served by what is known as the photosynthetic reaction centre. Its overall design places the redox-participating molecules in close mutual proximity, while keeping them well away from substances that would cause premature oxidation or reduction. This is achieved by having the molecules embedded in a large composite protein, which is itself embedded in the membrane. Determining the structure of the protein was made difficult by the fact that it would be denatured if the surrounding membrane lipids were removed. Luckily, Hartmut Michel devised a method by which a sufficient amount of lipid was retained around each protein molecule to permit crystallization without denaturation. In fact, the subsequent determination of the structure of the photosynthetic protein, by Michel, in collaboration with Johann Deisenhofer and Robert Huber, was the first ever for such a membrane-bound protein.

The reaction centre consists of four sub-units (see Figure 9.5). Two of them, designated the L and M sub-units, each comprises five membrane-spanning alpha helices. They are in intimate contact with each other, and it is they which contain the photochemically active groups. A third sub-unit, H, is located at the inner surface of the membrane, and it is anchored to the L–M dimer by a single alpha helix. It contains no active sites. Finally, the fourth sub-unit is a cytochrome molecule, which binds to the L–M dimer at the outer membrane surface. It contains four haem sites. The photochemically active chain has two dimerized chlorophyll molecules, two monomeric chlorophylls, two pheophytin molecules (these resembling the chlorophylls, except that they lack the latter's central magnesium ion), two quinone molecules (Q_A located in the L sub-unit and Q_B in M), and finally a single iron ion, this being located on the two-fold axis of symmetry of the overall structure (as is the above-mentioned chlorophyll dimer).

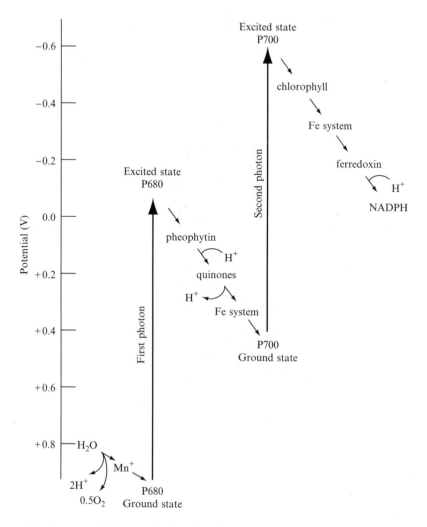

Figure 9.4 Because of the magnitudes of the various redox potentials, the photosynthetic mechanism will not function unless two photons are absorbed at slightly different wavelengths

   The determination of the three-dimensional structure of the reaction centre, in 1984, when taken together with the results of earlier kinetic and spectroscopic investigations, provided a detailed picture of the primary photosynthetic reactions. The initial photon-induced excitation of the chlorophyll dimer makes the latter strongly reducing (see Figure 9.6). This excitation is followed, 3×10^{-12} s later, by charge separation, an electron being transferred to the pheophytin in the L sub-unit. Within 2×10^{-10} s, the electron has been passed on to the quinone Q_A, and when the time has reached 3×10^{-7} s, the chlorophyll dimer has been re-reduced by the bound cytochrome. The first chain of

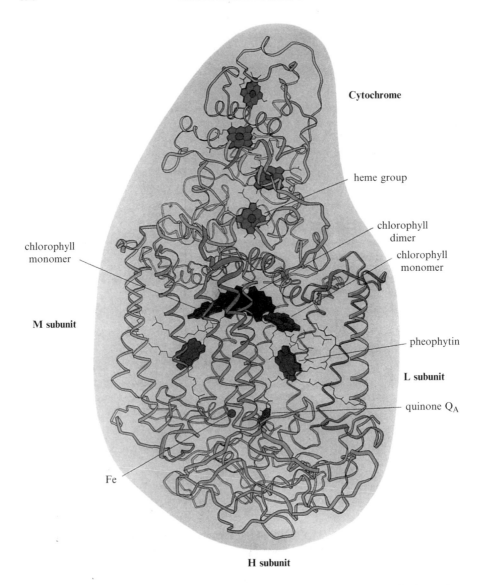

Figure 9.5 Schematic picture of a photosynthetic reaction centre from the bacterium *Rhodopseudomonas viridis*; the polypeptide chains are drawn as ribbons of different shades of grey for the four different protein sub-units

reactions is completed at the 1×10^{-4} s mark, when the electron is transferred from Q_A to Q_B, thereby reducing it. The subsequent absorption of a second quantum of light energy (that is to say, a second photon) produces a similar chain of reactions, again via the active sites in the L sub-unit, but this time the end product is reduction of Q_B to hydroquinone. The latter leaves the reaction

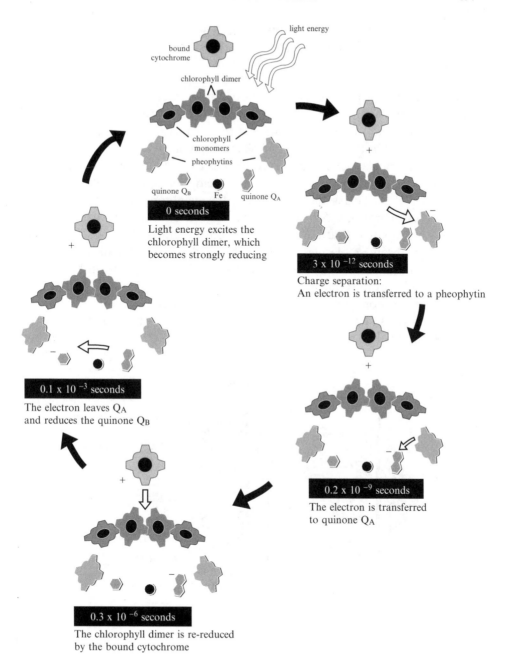

Figure 9.6   The various steps in the photosynthetic primary reaction

centre, and it is replaced by a new Q_B molecule from the surrounding membrane. The electrons are ultimately returned to the cytochrome region of the reaction centre by other electron carriers.

The electron transfer processes just invoked are examples of quantum mechanical tunnelling (see Appendix A), and it is possible to account for the rates at
which they occur if one knows the positions of all the nearby atoms in the
reaction centre, and can adequately describe the electron-atom interactions. In
practice this is a formidable task, and the quantum mechanical calculations are
necessarily rather approximate. The above-cited breakthrough in the structural
determination was nevertheless sufficient to explain why it is that only the L
sub-unit is used in the two chains of reactions: tunnelling is somewhat easier
along that particular route. The electron-transfer reactions are coupled to the
translocation of hydrogen ions across the membrane, just as Peter Mitchell
predicted in 1958. This is made possible by the membrane-spanning molecules
of bound cytochrome, which becomes the other end station of what could be
called a microscopic shuttle service based on the mobile hydroquinone and
soluble cytochrome molecules, the latter being identical to the molecules that
form one of each reaction centre's sub-units (see Figure 9.7). The upshot of the
hydrogen ion (proton) translocation is the setting up of a gradient of pH (that
is to say, a proton gradient), and it is this which subsequently drives the
production of ATP, the universal energy-transporting compound. We thus
see that the final step in the generation of biological energy is mediated by

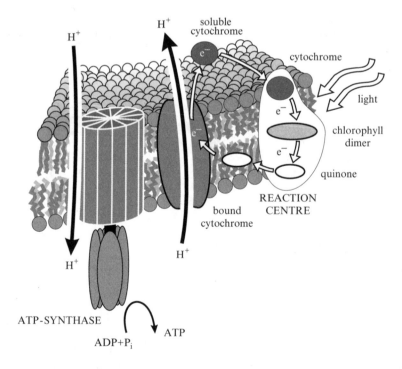

Figure 9.7 The light-induced electron transport and ATP synthesis in a photosynthetic
bacterium

protons rather than electrons, and Mitchell coined the term proticity, by analogy to electricity. His term for proton translocation, and the resultant setting up of a pH gradient, is chemiosmosis.

The production of ATP molecules is not limited to photosynthesis; it is also a vital feature of respiration. Because of its generality, we should defer further discussion of that process until after we have rounded off the discussion of photosynthesis itself. The final step is the production of glucose, and this requires a chain of chemical reactions, each of which relies on the presence of a suitable enzyme. A very important collective feature of these reactions, and indeed of all such reaction chains, is that there should be no net consumption or production of the molecules which participate as intermediaries. These reaction chains thus take the form of cycles, the one of relevance here being the Calvin cycle (named for its discoverer, Melvin Calvin). (The chemical details of an analogous cycle which operates during glucose-consuming respiration were elucidated by Hans Krebs.) In contrast to the reactions we have considered thus far, it does not require the contribution of light energy, and it is consequently known as the dark stage. It involves the fixation of CO_2 (see Equation (9.1) above), each molecule of which is incorporated into a five-carbon molecule (of ribulose bisphosphate, the reaction being catalysed by the enzyme ribulose bisphosphate carboxylase). This produces an unstable six-carbon molecule, which immediately breaks up into two three-carbon molecules (of 3-phosphoglyceric acid). Each of the latter is then phosphorylated (by ATP, which thereby becomes ADP, the enzyme being phosphoglyceric acid kinase) and reduced (by NADPH, which thereby becomes $NADP^+$, the enzyme being triose phosphate dehydroxykinase), so as to produce a molecule of glyceraldehyde 3-phosphate.

Only one molecule in every six of the latter is then converted into glucose (through a further series of catalysed reactions), whereas the other five participate in ATP-mediated phosphorylation processes which ultimately produce ribulose bisphosphate, thereby permitting the chemical books to be balanced, as noted above. The overall book-keeping is thus seen to yield one glucose molecule through the use of six ribulose bisphosphate molecules, there being formed 12 molecules of 3-phosphoglyceric acid. These latter therefore need a total of 12 ATP and 12 NADPH molecules. The ten 3-phosphoglyceric acid molecules not producing glucose require six further ATP molecules, *en route* to their production of six molecules of ribulose bisphosphate. The overall process thus requires 12 NADPH molecules and 18 ATP molecules.

9.4 ATP Synthesis

So much for the Calvin cycle, and indeed for photosynthesis itself. We must now move on to the fascinating mechanism whereby molecules of ATP are produced from molecules of ADP and inorganic phosphate, P_i. This very important reaction is catalysed by ATP synthase, and in 1964 Paul Boyer postulated that the process involves a structural change in that enzyme. Now as we will be discussing later, the ATP molecule is actually metastable; there is an energy barrier which prevents its spontaneous fission into ADP and P_i, with the release of the stored energy. About a decade later, Boyer himself demonstrated that the energy-requiring step in the synthesis involves the release of a complete ATP molecule from the enzyme concurrently with the simultaneous binding of an ADP molecule and a P_i group.

Electron microscopy had earlier revealed that part of the enzyme (now known as F_0) is embedded in the membrane, whereas a somewhat larger portion (F_1) sticks out into the surrounding cytoplasm. Ephraim Racker succeeded in isolating the F_1 fragment in 1961, and 20 years later John Walker determined the DNA sequences of the genes that code for the entire enzyme. Together with his colleagues, Walker went on to determine the three-dimensional structure of F_1, in full atomic detail, the task being completed in 1994. It turns out that F_1 is composed of no less than seven units, the proteins of three of which (designated α) are identical both with respect to primary and tertiary structure. The proteins of three of the remaining four units (designated β) are also identical with respect to primary structure, but not to tertiary structure, there being three distinct conformations: β_T – for *tight*; β_L – for *loose*; and β_O – for *open*. There is a cylindrical arrangement of six of these, the sequence running $\alpha\beta_T\alpha\beta_L\alpha\beta_O$ when running clockwise and viewing from above, toward the membrane (see Figure 9.8). The seventh, and final, unit (designated γ) is located on the axis of the cylinder, and because of its asymmetrical shape it causes changes in the conformations of the β sub-units as it rotates.

F_0 is also a composite structure, its units being designated a, b and c. The first two of these are anchored in the membrane, and the b unit is also attached to the under-side of one of the β units of F_1, thereby preventing the movement of the latter. Finally, several c units are arranged in a cylinder (within the plane of the membrane) and because their mutual centre is attached to the γ unit of F_1, they can rotate it when they themselves rotate. This causes the various β units of F_1 to cycle between the three possible conformations.

When a β sub-unit is in the loose configuration, an open slit on its outer surface admits entry of both an ADP molecule and one of P_i. When the γ unit rotates through 120° (because of the corresponding rotation of the multiple c units to which it is bound), it converts that particular β sub-unit from the loose to the tight configuration, thereby squeezing these two molecules so closely together that there is an atomic rearrangement, resulting in the production of an ATP molecule. A further rotation of the γ sub-unit by 120° converts the β unit in question from the tight to the open configuration, and this allows the newly formed ATP molecule to escape. Now because there are three β sub-units in the overall structure, three ATP molecules can be formed for each complete 360° rotation of the γ unit (see Figure 9.9). In the native

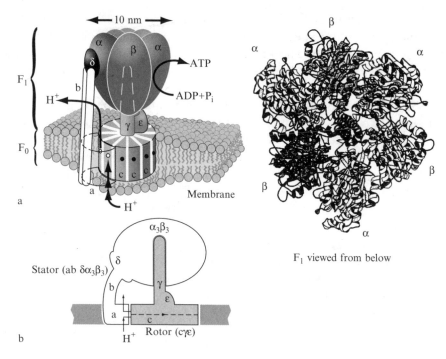

Figure 9.8   The structure of the F_1 unit of ATP synthase is known to atomic resolution, while that of the F_0 unit is still the subject of investigation

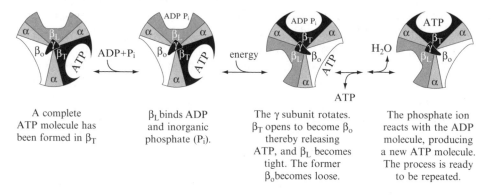

| A complete ATP molecule has been formed in β_T | β_L binds ADP and inorganic phosphate (P_i). | The γ subunit rotates. β_T opens to become β_o thereby releasing ATP, and β_L becomes tight. The former β_o becomes loose. | The phosphate ion reacts with the ADP molecule, producing a new ATP molecule. The process is ready to be repeated. |

Figure 9.9   Schematic representation of four consecutive stages in the synthesis of ATP molecules

biological state of ATP synthase, the γ unit rotates at an impressive 100 rev/s so the theoretical productive capacity of a single ATP synthase complex is 300 ATP molecules/s. This assumes, however, that there is an ample supply of ADP and P_i molecules, waiting to jump into the β sub-units when they are in the β_L configuration. We will return to this productive capacity shortly.

Meanwhile, let us now pass on to the atomic level within these molecules. The energy consumed in our food must be converted into a form that can be used by the body's molecules. In effect, this means getting it to bring about changes at the molecular level. As we have seen, the important molecule in question is adenosine triphosphate, or ATP. It consists of five distinct parts (as can be seen from Figure 9.10), of which three are phosphate groups. The remaining two parts are familiar components in the nucleic acid adenosine. This should not surprise us; the system strives to get good mileage from its various components.

The utility of the ATP molecule stems from the fact that each phosphate unit is bound to the rest of the molecule by an energy that is greater than that of the barrier that prevents its spontaneous breaking away. The situation is thus the familiar one of *metastability*, as is indicated in Figure 9.11. A small amount of energy must be expended, in order to liberate the larger amount stored in the repulsion between the terminal phosphate group and the rest of the molecule.

We can get a better appreciation of the underlying physics by considering the analogous situation in the pyrophosphate molecule, which is depicted in Figure 9.12, for neutral pH. The four OH bonds in this molecule have pK values of 0.85, 1.96, 6.54 and 8.44, respectively, so it is understandable that three of the termini bear negative charges at physiological pH.

The chemical rearrangement accompanying the hydrolysis of this molecule is indicated in Figure 9.13, which reveals the very important fact that the reaction involves no net change in the number of covalent bonds. This is a vital consideration, because the energy stored in such a bond actually exceeds that which is liberated by the hydrolysis. If this had not been the case, the process would have been useless as far as energy transduction is concerned.

Now it is found that the free energy liberated during the hydrolysis of ATP is about 0.06 aJ, a value which is so important that the reader ought to commit it to memory (noting also that it is about twice the energy in a hydrogen bond). Part of its source is revealed by Figure 9.12, namely the ion–dipole interactions

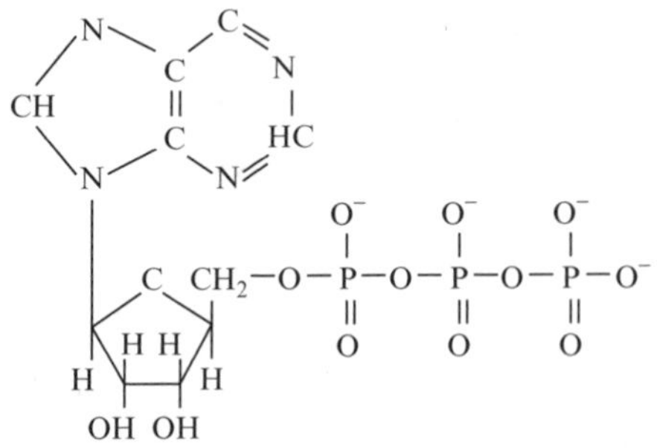

Figure 9.10   The adenosine triphosphate (ATP) molecule

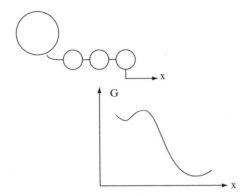

Figure 9.11 In this highly schematic picture of ATP, the large circle represents the adenosine while the smaller circles represent the phosphate groups; the terminal phosphate is metastable with respect to the rest of the molecule, as can be seen from the carefully-positioned plot beneath the molecule, this showing how the free energy varies with the position of the terminal phosphate

$$
\begin{array}{ccc}
\text{O} && \text{O} \\
\| && \| \\
\text{HO} - \text{P} - \text{O} - \text{P} - \text{O}^- \\
| && | \\
\text{O}^- && \text{O}^-
\end{array}
$$

Figure 9.12 The pyrophosphate molecule

$$
\begin{array}{cccc}
\text{O} & \text{O} & & \text{H} \\
\| & \| & & | \\
\text{HO} - \text{P} - \text{O} - \text{P} - \text{O}^- & + & 2\,\text{O} & \Rightarrow \\
| & | & & | \\
\text{O}^- & \text{O}^- & & \text{H}
\end{array}
$$

$$
\begin{array}{ccc}
\text{O} && \text{O} \\
\| && \| \\
\text{HO} - \text{P} - \text{O}^- & + & \text{HO} - \text{P} - \text{O}^- & + & \text{H}_3\text{O}^+ \\
| && | \\
\text{O}^- && \text{O}^-
\end{array}
$$

Figure 9.13 Hydrolysis of a pyrophosphate molecule

that the phosphate units enter into with the water molecules, because the final state comprises more electrical charges than there were originally. One would also expect there to be a change in the overall entropy of the system, because of the rearrangement of some of the atoms.

The water molecules themselves have four electron orbitals which project outward from the oxygen nucleus, toward the corners of an imaginary tetrahedron, as indicated in Figure 9.14. Two of these orbitals are rather short, each bearing a net fifth of an electronic charge (i.e. $-0.2e$), while the other two lobes are considerably longer, each linking up with a hydrogen nucleus, and shielding all but a fifth of its unit positive charge; the net charge at each of these lobes is therefore $+0.2e$.

Then again, there are subtle resonance effects, not unlike those we saw in connection with the amide link discussed in chapter 7; the double bond in the terminal phosphate group perpetually flipping from one side of the phosphorus atom to the other, as shown in Figure 9.15. However, the overwhelmingly

0.2e

−0.2e

H$_2$O molecule

−0.2e

0.2e

Figure 9.14 Schematic diagram of the electron probability lobes in the water molecule, with no attempt to illustrate their tetrahedral arrangement

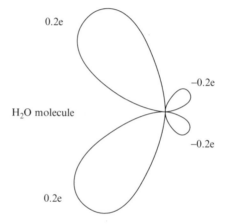

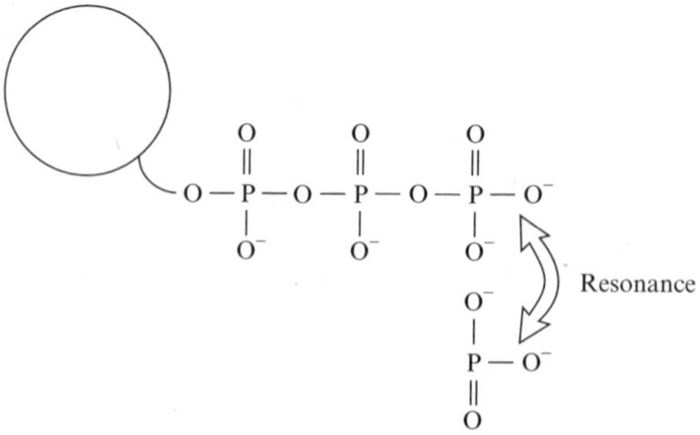

Resonance

Figure 9.15 A resonance effect causes the double bond in the terminal phosphate group to flip from one side to the other, but the major factor arises from the repulsion of two of the negatively-charged oxygen atoms

major factor arises from the fact that two of the negatively charged oxygen atoms, which strongly repel each other in the ATP molecule, become more widely separated because of the transition ATP $\rightarrow$ ADP $+$ P$_i$. This is most clearly seen from Figures 9.11 and 9.13.

Because this repulsion is of the Coulombic type, it is readily amenable to calculation, using the analytical form of that potential given in Chapter 3. We have

$$
\begin{aligned}
\mathscr{E}(r) &= \frac{1}{4\pi\epsilon} \times \frac{q_1 q_2}{r} \\
&= \frac{1}{4\pi\epsilon} \times \frac{\left(1.6 \times 10^{-19}\right)^2}{5 \times 10^{-10}} \\
&= \frac{4.1 \times 10^{-30}}{\epsilon}
\end{aligned}
\tag{9.6}
$$

where ϵ is the permittivity (and $\epsilon = \epsilon_0 \epsilon_r$, ϵ_0 being the permittivity of free space while ϵ_r is the relative permitivity, as discussed in Chapter 3). q_1 and q_2 are the respective (and equal) charges on the oxygen atoms, and r is the distance between them.

Given that the ATP molecule is three-dimensional, Figure 9.10 suffers from the disadvantage that usually plagues two-dimensional representations: it presents a distorted impression of the associated distances. Nevertheless, it is not difficult to persuade oneself that r is about 0.5 nm. Because we already know that $\mathscr{E}(r)$ is 0.06 aJ (assuming that the final separation of the relevant oxygen atoms is so large that the energy falls to zero), we can use Equation (9.6) to calculate the effective value of ϵ. This is found to be $\epsilon = 6.8 \times 10^{-11}\, C^2 N^{-1} m^{-2}$ and, because we know that $\epsilon_0 = 8.85 \times 10^{-12}\, C^2 N^{-1} m^{-2}$, this gives a final value of ϵ_r of approximately 8 for the region lying between those two oxygen atoms.

The relative permittivity of bulk water is around 80, so we see from the above result that we may *not* regard any water lying between the two oxygen atoms as being typical of bulk water. This is hardly surprising, given the confined space between those two atoms. On the other hand, the derived relative permittivity is well above the value of unity, which it would have had if the space between the oxygen atoms was essentially a vacuum.

The ATP molecules are rapidly used following their fabrication, that is to say following the consolidation ADP $+$ P$_i$ $\rightarrow$ ATP. We again use the parameters relevant for the typical adult. The daily intake is, as we have seen, 2400 kcal. Assuming that this all goes to the above fabrication process, at least in the first instance, the corresponding number of ATP molecules is 1.7×10^{26}. Now there are 86 400 s in a day, and because a single ATP synthase complex can in principle manufacture 300 ATP molecules in each second, its daily capacity will lie around the 25 000 000 mark. The daily energy intake can thus be totally converted to ATP if there are about 6.8×10^{18} ATP synthase molecules in the body working flat out, around the clock. Given that the average adult human body comprises about 10^{12} cells, this means that each cell would have to contain about seven million ATP synthase complexes, distributed amongst its mitochondria.

Let us now turn to the issue of weight. Knowing the atomic weights of the various atoms, it is not difficult to calculate the weight of a single ATP molecule, namely 8×10^{-25} kg. The total weight of the ATP molecules produced by one day's energy intake is thus about 140 kg, which is well above the average weight of an adult. We see, therefore, that if the energy we consume at a single meal was immediately all converted to ATP molecules, we would hardly be able to raise ourselves from our chairs after eating! In reality, the various enzymes ensure that only just a sufficient number of ATP molecules are made available at any one time. That number is, in fact, equivalent to the amount of energy that can be derived from about 2 g of sugar.

As we have seen, the production of ATP molecules is carried out in the mitochondria, where a number of different types of enzyme act at various stages of the (catabolic) break-down of the molecular remnants of ingested food. As we have also seen, some of these steps involve reduction by the transport of oxygen or electrons. Enzymes that perform this transport are known as electron carriers, and we noted that many of them form sequentially-acting chains, the net outcome of their efforts being the translocation of protons across the mitochondrial membrane.

We can make an estimate of the total number of protons being transported at any given instant, because we have already seen that this corresponds to the power being used by those two 60 W light bulbs. As we will learn in Chapter 11, there is a resting potential of about 0.1 V across the membrane, so the total current is about 1200 A. Since 1A corresponds to 6.0×10^{18} fundamental electrical charges/s, the total number of protons being translocated across mitochondrial membranes at any instant is 7.2×10^{21}/s. This gives about 6.2×10^{26} protons per day. Comparing this with the above-calculated number of ATP molecules per day, we arrive at the interesting result that the fabrication of each ATP molecule requires the translocation of 3–4 protons. This seems like a reasonable number, because it indicates that somewhere between nine and 12 protons pass through the membrane for each complete rotation of the γ sub-unit in the ATP synthase complex. The latter appears to be a rather efficient little molecular machine.

Exercises

9.1   Assuming that the electrical potential acting across a mitochondrial membrane is 0.1 V, and that this effectively produces an energy barrier that electrons would have to surmount in order to cross the membrane, calculate the magnitude of the barrier in units of $k_B T$ at body temperature. Use this ratio to find the equivalent number of pH units acting across the membrane [Hint: use Equation (8.9)].

9.2   The daily consumption of energy for an adult human is approximately 2400 kcal. The atomic weights (relative to hydrogen) of the elements that comprise ATP are as follows: Hydrogen 1; Oxygen 16; Nitrogen 14; and Phosphorus 31. Assuming that the total energy intake is exclusively used for creating ATP molecules, what is the total weight of these that are formed within any 24 h period?

Further reading

Abrahams, J. P. *et al.*, (1994). Structure at 2.8Å resolution of F_1-ATPase from bovine heart mitochondria. *Nature* **370**, 621–628.

Boyer, P. D., (1993). The binding change mechanism for ATP synthase – some probabilities and possibilities. *Biochimica et Biophysica Acta* **1140**, 215–250.

Cerdonio, M. and Noble, R. W., (1986). *Introductory Biophysics* World Scientific, Singapore.

Deisenhofer, J. *et al.*, (1984). X-ray structure analysis of a membrane protein complex. Electron density map at 3Å resolution of a model of the chromophores of the photosynthetic reaction center from *Rhodopseudomonas viridis*. *Journal of Molecular Biology* **180**, 385–398.

Deisenhofer, J. *et al.*, (1986). Structure of the protein subunits in the photosynthetic reaction center of *Rhodopseudomonas viridis* at 3Å resolution. *Nature* **318**, 618–624.

Dyson, R. D., (1974). *Cell Biology: A Molecular Approach* Allyn and Bacon, Boston.

Elston, T., Wang, H. and Oster, G., (1998). Energy transduction in ATP synthase. *Nature* **391**, 510–513.

Kolbe, M. *et al.*, (2000). Structure of the light-driven chloride pump Halorhodopsin at 1.8 Å resolution. *Science* **288**, 1390–1396.

Loewy, A. G. and Siekevitz, P., (1974). *Cell Structure and Function* Holt, Rinehart and Winston, London.

Mitchell, P., (1976). Possible molecular mechanisms of the protonmotive function of cytochrome systems. *Journal of Theoretical Biology* **62**, 327–367.

Moore, W. J., (1972). *Physical Chemistry* Longman, Englewood Cliffs.

Ray, P. M., (1972). *The Living Plant* Holt, Rinehart and Winston, New York.

Rhee, K. H. *et al.*, (1998). Three-dimensional structure of the plant photosystem II reaction centre at 8Å resolution. *Nature* **396**, 283–286.

Skou, J. C., (1957). The influence of some cations on an adenosine triphosphatase from peripheral nerves. *Biochimica et Biophysica Acta* **23**, 394–401.

Stryer, L., (1981). *Biochemistry* Freeman, San Francisco.

Sybesma, C., (1989). *Biophysics: An Introduction* Kluwer, Dordrecht.

Verkhovsky, M. I. *et al.*, (1999). Proton translocation by cytochrome c oxidase. *Nature* **400**, 480–483.

Yasuda, R. *et al.*, (2001). Resolution of distinct rotational substeps by submillisecond kinetic analysis of F_1-ATPase. *Nature* **410**, 898–904.

10 Movement of Organisms

Because plants possess chlorophyll, they do not need to move around in order to acquire energy; the sun provides for them *in situ*. However, this does not mean that there is no movement *within* a plant. The growth of the plant will involve cell divisions, for example, and these are mediated by a molecular mechanism that is related to the one underlying muscle contraction in animals. Moreover, there are other processes whereby plants undergo shape changes, such as those which permit them to continually face the direction of the incident sunlight. Our concern here, however, is with the movement of entire organisms, and these are the province of the animal kingdom. We will be considering the relatively primitive mechanisms by which bacteria propel themselves, and at the other end of the scale we will be taking a look at muscular movements in advanced animals such as ourselves.

10.1 Bacterial Motion

The small sizes of single-cellular organisms such as bacteria make for difficulties of movement. In contrast to the cases of swimming fish and humans, for example, viscosity plays a dominant role. Conversely, inertia is irrelevant; the typical bacterium cannot coast. The situation is governed by what is known as the Reynolds number, R (see Chapter 5), where

$$R = v\ell\rho/\eta \qquad (10.1)$$

and where ℓ is the characteristic length, v is the speed, ρ the fluid density and η the fluid (dynamical) viscosity, these final two parameters being $10^3 \, \mathrm{kg \, m^{-3}}$ and $10^{-3} \, \mathrm{kg \, m^{-1} \, s^{-1}}$ respectively for the case of water at 20 °C.

For a bacterium of characteristic size $2 \, \mu\mathrm{m}$, swimming at a speed of $25 \, \mu\mathrm{m \, s^{-1}}$ the Reynolds number is 5×10^{-3}, and it can coast for a mere $10^{-10} \, \mathrm{m}$, which is less than the diameter of a water molecule! The coasting distance is proportional to the Reynolds number, which for a human swimming at $1 \, \mathrm{ms^{-1}}$ is about 2×10^6. A human's inertia will thus permit coasting over a distance of about $1 \, \mathrm{m}$. It is interesting to note that the bacterium can move through a distance corresponding to about 10 times its body length in 1 s, and that a human can take about five strides in the same time, when running at full speed. The cadences of their motions are thus comparable, and it is the markedly differing Reynolds numbers which make all the difference when it comes to coasting.

A bacterium swimming in water is subject to approximately the same amount of viscous drag as a human attempting to swim in molasses. Even in water, a human trying to swim with arms stretched out to the sides, and without bending the arms, would make no progress at all, because any advancement achieved by sweeping the arms in one direction would be counterbalanced by the regress during the reverse sweep. It is for this reason, of course, that the breast-stroke *does* involve bending of the arms at one moment during each cycle. (The problem is solved in a different manner in the crawl, each arm being lifted clear of the water once during each cycle.) Some organisms, similarly, bend their flagella at one moment during each cycle of movement, so as to produce a wavelike motion (see Figure 10.1), whereas others avoid the problem by simply rotating them in a manner reminiscent of a ship's propeller.

Given the great difficulty that confronts a bacterium when it tries to swim, one might wonder why it even attempts to do so. This question has been examined in great detail by Howard Berg and his colleagues, and they were able to eliminate several possibilities. The bacterium's motion does not enable it to increase the number of potentially useful molecules striking its surface in a given period, for example, because simple diffusion accomplishes this more efficiently. Similarly, the stirring effect caused by the motion of the flagella does not lead to renewal of its environment at a rate that would be significantly greater than that caused by diffusion. Berg and his colleagues found the answer by studying the tracks of the seemingly erratic movement of individual bacteria, using an optical microscope. They discovered that these follow what could be called a biased random walk (see Figure 10.2(a)), which is to say stochastic changes of direction, at random epochs, but with the probability,

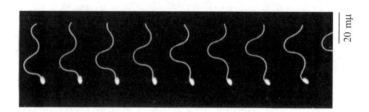

Figure 10.1 Propagation of a flagellar bending wave provoked by a solution of Mg-ATP in a spermatozoon of the tunicate *Ciona*, following the removal of the membrane by detergent (Courtesy of Charles J. Brokaw, California Institute of Technology)

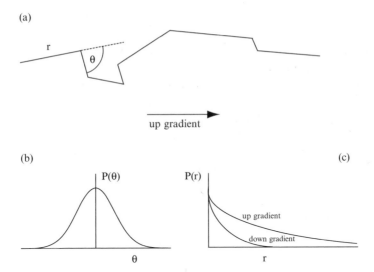

Figure 10.2 (a) A schematic (and two-dimensional) representation of the (three-dimensional) biased random walk performed by a swimming bacterium; (b) the distribution of the bacterium's turning angles under the influence of a gradient of chemoattractant; (c) the step-length response of a bacterium to the same chemoattractant, showing the greater tendency to move towards the source

$P(\theta)$, of a given change of direction, θ, sharply decreasing with the magnitude of that change (see Figure 10.2(b)).

The rationale for a bacterium's motion was revealed when Berg and his colleagues examined the probability distribution, $P(r)$, of straight-run segments of length r, which turned out to be reasonably well described by

$$P(r) = \lambda e^{-\lambda r} \tag{10.2}$$

λ being a decay constant whose value depended upon whether the concentration of a potentially useful molecular species was increasing or decreasing (see Figure 10.2(c)). Such biased motion, which enables a bacterium to travel in the direction of a chemical gradient, is known as chemotaxis.

We will shortly turn to the mechanism whereby the gradient can influence $P(r)$. Meanwhile, let us enquire as to whether the bacterium's effort is worthwhile, in terms of the energy it has to expend in order to move about. In 1856, George Stokes showed that the viscous drag, $\mathcal{F}_{drag}$ (which is of course a force), on a sphere with radius a moving with a speed v through a medium with viscosity η is given by

$$\mathcal{F}_{drag} = 6\pi a \eta v \tag{10.3}$$

The power the bacterium must expend, in order to achieve this speed, is simply $\mathscr{F}_{drag}$ v, and assuming that the creature's radius is about $2\,\mu m$, which is typical, the above-quoted speed of $25\,\mu m\,s^{-1}$ gives a power consumption of $2.5 \times 10^{-17}\,J\,s^{-1}$. This rate of energy expenditure must be supplied by the hydrolysis of ATP molecules, of course, the fission of a single one of these producing $0.06\,aJ$. In other words, the bacterium needs to be able to hydrolyse about 375 ATP molecules/s. As we saw in Chapter 9, the typical cell is capable of fabricating about 10^7 ATP molecules/s, so a bacterium can easily produce the power needed to propel itself through the water.

The development of the transmission electron microscope, and the associated specimen preparation techniques, have permitted investigation of the microstructure of cellular components which mediate motion. This has revealed details of the macromolecular assemblies which form the cilia and flagella that eucaryotic cells use to generate motion, and also of the way in which they are attached to the cell membrane. The core of the cilium, for example, is a cylinder roughly 200 nm in diameter, and it contains an impressively complicated arrangement of microtubules, which are, of course, composed of protein. In what is known as the axoneme (see Figure 10.3(a)), there is a circular arrangement of nine microtubule doublets around two singlet microtubules. The doublets are each composed of one complete microtubule (an *A* microtubule – made up of 13 protofilaments) and one incomplete microtubule (a *B* microtubule – made up of 11 protofilaments). It should be noted that the entire axoneme is encased within a membrane, and that the concentrations of ions in the cytoplasm which surrounds the axoneme are thus under the control of ion pumps and ion channels located in that membrane.

In addition to these major components, there are structures which form periodic cross-links between the microtubules, and it is they which provide the basis of movement. Particularly prominent amongst the latter are the dynein arms, which act as circumferential bridges between the microtubule doublets (see Figure 10.3(b)), and which actually supply the force that bends the cilium. The nexin links also form bridges around the perimeter, and their job is to hold the entire structure in place, as would the hoops of a barrel. Finally, there are the radial spokes and central sheaths, whose interactions with the two centrally-located microtubule singlets control the actual mode of cilary beating. In terms of the underlying molecular content, we may note that the arm of a radial spoke and its head portion are each composed of six distinct polypeptides. As a whole, the axoneme possesses neither mirror nor rotational symmetry. The relative positions of the *A* and *B* type microtubules, and of the dynein arms, is such as to produce a clear handedness; viewed from the base of the axoneme, the dynein arms point in a clockwise direction. It is this structural directionality which underlies the directionality of movement.

The actual bending movement of a cilium is caused by the mutual sliding of adjacent microtubule doublets, and by the fact that these are nevertheless

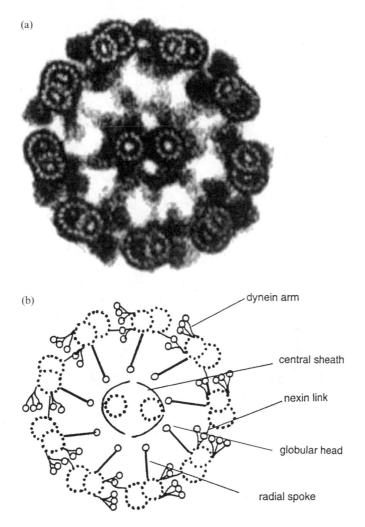

Figure 10.3   (a) Cross-section of a cilary axoneme, observed by transmission electron microscopy, showing the 9 + 2 arrangement of the microtubules within which the individual protofilaments are just visible; (b) a schematic representation of the same axoneme with the various components identified (Upper part courtesy of Lewis Tilney, University of Pennsylvania)

anchored at the base of the cilium (see Figure 10.4). The mutual sliding is caused by conformational changes in the dynein arms, each of which is a high-molecular-weight ATPase; it changes its shape as a concomitant of splitting ATP into ADP and inorganic phosphate, P_i, as was discussed in Chapter 9. The periodic shape changes of each dynein arm lead to a cycle of attachments and detachments between it and the appropriate microtubule doublet, the upshot

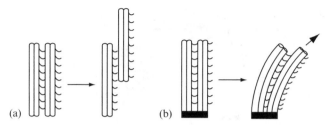

Figure 10.4 A schematic diagram (following Dennis Bray) showing how the mutual sliding of adjacent microtubule doublets (a) causes bending of the cillium if (b) the doublets are anchored at the base

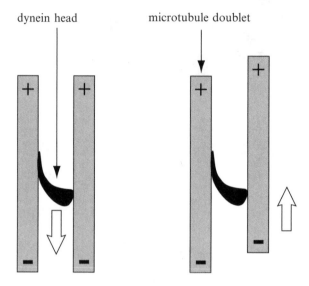

Figure 10.5 The negatively-directed movement of ciliary dynein ultimately causes a microtubule to move in the positive direction

being the 'walking' of the arm along the surface of the doublet (see Figure 10.5). This, of course, is the origin of the above-mentioned mutual sliding. The overall effect bears a striking resemblance to the mutual movements of the myosin and actin molecules in striated muscle, which we will move on to shortly.

These elemental movements do not, of themselves, account for the rhythmic motion of the entire cilium. Indeed, if every dynein arm was simultaneously activated, the cilium would twist itself up into a helix, because of the above-mentioned handedness. The additional factor is supplied by that centrally-located pair of microtubule singlets, and by their associated spokes and sheath. A study of the surface cilia of the unicellular organism *Paramecium* revealed that this central complex undergoes one complete 360° rotation for each cycle of beating. This causes the spokes to be displaced in a systematic manner, resulting in a similarly systematic activation of the dynein units.

The motions of the cilia are themselves ultimately caused by the environment, of course, the overall goal being to move the organism to a location possessing a more favourable distribution of its needs. The latter may include sunlight, and certainly will include substances present in the surrounding aqueous medium. On this issue too, it was investigation of *Paramecium* which adumbrated important principles. That organism will avoid a physical barrier, when swimming, by reversing the beat direction of its cilia for about 2 s or so. It was found that this reversal was related to a change in its membrane potential (a subject that will be our primary concern in Chapters 11 and 12). The normal resting potential of this creature's bounding membrane is about -30 mV, and the obstacle-provoking change drives it in the positive direction, that is to say it causes depolarization, this being associated with an influx of Ca^{2+} ions across the membrane. If, on the other hand, one taps *Paramecium* on its posterior, conceivably mimicking what could happen when it is being chased by a predator, the membrane potential becomes more negative (that is to say it is hyperpolarized) and the beat maintains its normal direction and actually speeds up. In this case, there is an efflux of K^+ ions across the membrane.

We ought to pause at this point and consider an interesting philosophical issue which has arisen in connection with speculations about consciousness. Some scientists have seriously suggested that consciousness could be present in single-celled organisms such as *Paramecium*. Their motivation for having this surprising attitude appears to stem from their linking of consciousness to the presence of microtubules, and we have seen that these molecules form part of the cilia. They are also present in the creature's cytoplasm. In order to bolster their case, these scientists point to the seemingly clever way in which obstacles and predators are avoided, by taking the appropriate evasive action. In the view of the present author, however, it would be quite extraordinary if consciousness were to be possible in a single-celled organism, which, by definition, cannot possess a nervous system. In any event, the apparently clever behaviour of *Paramecium* can readily be explained, without the need to invoke anything high-flown. For the simple fact is that the encountering of an obstacle causes Ca^{2+} ions to enter the cell, through the appropriate type of channel, these ions thereby moving in the direction of decreasing concentration. When the internal concentration exceeds the threshold value of about 10^{-6} M, the beat of each cilium is reversed, as has been demonstrated by experiments *in vitro*. Once the creature has backed away from the obstacle, the Ca^{2+} pumps rapidly expel some of these ions from the cytoplasm, bringing the concentration below the threshold, and restoring normal forward motion. These remarkably well-orchestrated changes all occur automatically, and they are driven by atomic-scale processes; there is nothing *clever* about the phenomenon.

The movement of organisms in response to concentration gradients in the surrounding aqueous medium is known as chemotaxis. In procaryotes (that is to say, bacteria and blue-green algae), such propulsion is mediated by the rotation of a rigid helical flagellum, this being driven by a tiny molecular motor situated at its base. Well-known organisms which exploit this mode of travel include *Salmonella* and *Escherichia coli*. In contrast to what is seen in eucaryotes, the procaryotic flagellum is not encased in a membrane; it is in direct contact with the surrounding liquid. This essentially precludes a Ca^{2+}-mediated beating mechanism, of the type seen in eucaryotic cilia and flagella, because the ionic concentration in the surrounding liquid would not be sufficiently homogeneous, over the relevant distance scale; it is difficult to imagine a systematic rhythm being supportable under such conditions.

Procaryotic flegella are smooth cylinders, about 14 nm in diameter, with the above-mentioned helical conformation, the length of which varies around a typical mean of about 10 μm (see Figure 10.6). These cylinders are composed of a sheath of parallel and contiguous flagellin molecules, surrounding a hollow lumen. The molecular motor via which they are bound to the organism's double membrane is an impressive structure (see Figure 10.7). Its inner portion, which spans the width of the plasma membrane, bears a strong resemblance to the F_0 portion of the ATP synthase complex which we encountered in Chapter 9. This is hardly surprising, given that both structures are motors which are driven by a flux of protons. Indeed, the speeds of rotation of the two motors is about the same, namely approximately 100 rev/s. However, the motor is somewhat more intricate in the procaryotic flagellum, because of the presence of the additional outer membrane. Nature's solution to this complication follows what could be called good engineering practice: it adds a bearing, which takes the form of a grooved ferrule, as can be seen in Figure 10.7.

The flagella of both *Salmonella* and *E. coli* are left-handed helices, so an anticlockwise rotation of the motor unit, as viewed from the other extremity of

5 μm

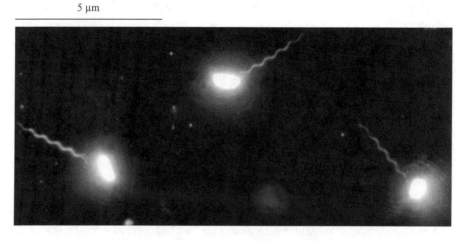

Figure 10.6 The helical flagella of bacteria revealed by dark-field microscopy (Courtesy of Hirokazu Hotani, Nagoya University)

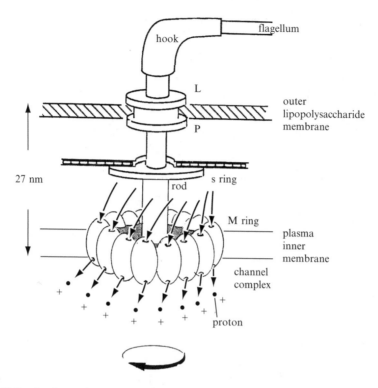

Figure 10.7   A schematic representation of the flagellar motor of a procaryote; the overall diameter of the motor is about 30 nm

the flagellum, will propel the organism forward. A clockwise rotation, on the other hand, will counteract this smooth progress. These two types of result are referred to as running and tumbling, respectively, and a bacterium typically alternates between these modes when swimming. As we noted earlier, the abrupt changes cannot be the result of thoughtful evaluation on the part of the organism, since it does not even possess a nervous system. The changes are imposed upon it by the environment, and the agencies in question are known as chemattractants and chemrepellents. *E. coli* can distinguish between about 30 such substances, which is rather impressive, given that the basis for discrimination must be molecular and purely deterministic.

The actual discrimination is mediated, initially, by receptors located on the bacterium's inner membrane, and thereafter by sequences of chemical reactions, which are ultimately believed to express themselves through conformational changes in some of the motor proteins (see Figure 10.8). Although the necessary structure determinations have still to be carried out, one could speculate that the flagellum's so-called M ring (see Figure 10.7) is caused to rotate by the local passage of a proton, as would the out-of-gear propeller of an aeroplane by the passing air molecules, during a sufficiently strong wind. The

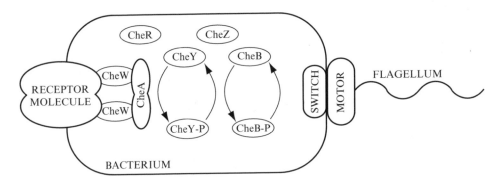

Figure 10.8 The behaviour-determining components in the *E. coli* bacterium are indicated in this highly schematic diagram in which no effort has been made to reflect the relative sizes faithfully; the direction of rotation of the creature's flagellum is determined by the concentration of the various molecular species, these concentrations mediating a primitive form of memory as the bacterium gains information about its surroundings through its own movements

conformational changes would then be akin to reversing the pitch of the aeroplane's propeller (except that in the case of the bacterium, the change occurs in one of the *stationary* components). The above-mentioned sequences of reactions are known to involve protein phosphorylations, and the final adaptation is similarly known to involve protein methylation. The genes responsible for chemotaxis have been identified. Indeed, this phenomenon is one of the best documented of all biological processes. It is impressive that Nature was producing these intricate pieces of nanoengineering aeons before we human beings were achieving our own Industrial Revolution.

10.2   Chemical Memory in Primitive Organisms

A bacterium like *E. coli* is unable to gauge the spatial variation of nutrients in its environment, at any instant, so it procures information by probing its surroundings, through the propulsion provided by its flagellum. In effect, this creature integrates incoming chemical signals during a few-second period of its travels, and adjusts its direction of advance accordingly.

The integration is achieved through temporary chemical modifications to molecules located in the bounding membrane, which transfer nutrients to the cytoplasm, and also through changes to certain other molecules in the interior of the cell (see Figure 10.8). Such integration is essentially a short-term memory mechanism. The stimulus in this microscopic expedition is thus the motor movement, while the response is that of the impinging substances. This is just the opposite of a reflex.

This mode of behaviour is so fundamental that it is worth considering the details, as they are seen in *E coli*. The cell's membranes are equipped with receptor molecules (not to be confused with the receptor *cells* discussed in the later chapters of this book) each of which has an extra-cellular domain that binds amino-acid ligands directly (or sugar ligands indirectly) when they are attached to a specific small protein (a ligand being a small molecule, or part of a small molecule, which interacts with a larger molecule, such as a protein molecule). The inner region of the membrane-spanning receptor molecule favours interaction with the coupling protein designated *CheW*, as well as with a kinase designated *CheA*. The latter transfers phosphate from the energy transporting molecule ATP to two other proteins, known as *CheY* and *CheB*. *CheY* is the effector of the bacterium's flagellar motor.

The binding of an attractant ligand to the receptor molecule's external domain decreases the activity of the kinase, and this in turn leads to a decrease in the amounts of *CheY–P* and *CheB–P* that are being made. *CheY–P*'s signalling action proceeds by its diffusing through the cytoplasm and by its binding to a component of the flagellum known as the switch (designated *FliM*). This binding increases the probability that the motor will spin in the clockwise direction. Because forward motion is produced by an *anti*-clockwise rotation of the flagellum, such clockwise rotation causes the cell to tumble and the ultimate result is that the creature then swims in a different direction. When the cell is swimming in the direction of increasing attractant concentration, therefore, more attractant becomes bound to the receptor molecule's external domain, less *Chey-P* is produced, less of this substance binds to the switch, and the bacterium continues its forward movement. There are a number of different categories of receptor molecule, and they activate the kinase to varying degrees. This differential mode of functioning imbues the system with the ability to integrate the incoming chemical information, the product of that integration determining the concentration of *CheY–P* in the cell's cytoplasm.

In addition to these primary mechanisms, there is an adaptation process which increases the range of sensitivity. A methytranferase, designated *CheR*, slowly methylates the inner domain of the receptor protein when the kinase activity has been reduced by the binding of attractant ligands. This tends to restore the activity of the kinase, even when the ligand remains bound. Cells exposed to a sudden change in ligand concentration will thus tend to recover, and continue with their current mode of swimming. *CheB* is a so-called methyesterase, which detaches methyl groups from certain molecules. It is more active when it is phosphorylated (that is to say, when a phosphate group is attached to it), and it appears to act on receptor molecules only when they are in a kinase-active state. In effect, this enables the receptor molecules to compare the occupancy of their ligand binding site (a measure of the current conditions in their local environment) with the occupancy of their methylation sites (a measure of past conditions) and activate the kinase in proportion to the measured difference. There is also a phosphatase, designated *CheZ*, which removes phosphate from *CheY–P*, thereby shortening its lifetime.

The upshot of these various condition-dependent chemical reactions is to imbue the cell with a short-term memory spanning about 4 s, with the concentration measured over the past 1 s being compared with the concentration measured over the previous 3 s, with positive and negative weighting, respectively. Compared with the mechanisms which produce permanent memory traces (as will be discussed in Chapter 13), these changes in chemical concentration are ephemeral. The bacterium and its close relations are always at the dictates of their local surroundings. The plots of their simple lives are written in their genes, and the only variations on the standard theme are those imposed by the environment.

10.3   Muscular Movement

The movement of large multicellular organisms, such as ourselves, is achieved through the contraction of muscles, which are themselves composed of numerous specialized cells. These are eucaryotes, of course, so the ultimate control is by ions, just as for the simpler eucaryotic cilia and flagella discussed earlier. However, in the case of the larger animals, simple diffusion would be too slow a process to fulfil the need of responding sufficiently rapidly to changes in the environment, so evolution has augmented that process by developing nervous systems. We thus see signals being swiftly dispatched to remote regions of the body, via the nerve cells (neurons), which will be discussed in detail in later chapters, with only the final (and very small) step in the signalling process being entrusted to diffusion (see Figure 10.9). Muscle action is triggered by the electrical signals received from a motor nerve. The nerve cell has one highly elongated extension, or process, which can have a length of many centimeters, and this terminates in intimate contact with the membranes of the muscle cells. It is known as an axon and it is surrounded, over most of its length, by myelin which acts as a sort of electrically insulating layer. The myelin is occasionally interrupted, at sites known as nodes of Ranvier, at which the electrical signal receives a boost. A nerve signal travels much faster along such a myelinated axon than it would if the sheath were absent. Nerve cells can also receive signals from other nerve cells by way of synaptic contacts to their dendrites, or minor extensions. It is worth emphasizing that muscles can only be made to contract, not expand, by stimulation from a nerve signal. Muscular expansion is thus achieved through the contraction of an opposing muscle. Skeletal muscles, which are those used to move the limbs, are thus arranged in pairs of flexors (used to bend a joint – as does the biceps muscle of a human arm) and extensors (used to straighten out the joint – as does the triceps muscle of a human arm).

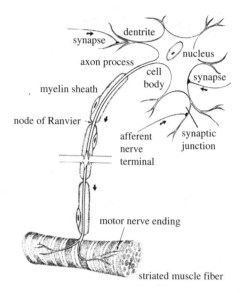

Figure 10.9   Muscle action triggered by the electrical signals received from a motor nerve

The advent of transmission electron microscopy, and its use particularly by Hugh Huxley in the 1950s, has permitted muscular microstructure to be investigated at high resolution, and muscles have been studied both in their expanded and contracted states. The data thereby acquired have shed decisive light on the issue of muscle function. Skeletal muscles (also known by the adjectives striped, striated and voluntary) consist of large elongated cells, the interiors of which display conspicuous striations at right angles to the long axis. Numerous elongated fibrils follow the direction of that axis, and their striations are seen to be the result of similar markings on the microfibrils of which they are composed.

Figure 10.10 shows the structure of a typical skeletal muscle at four different scales of magnification, and we first note certain (regular) features visible in the microfibrils, namely the *A-bands*, the *I-bands* and the *Z-lines*. The distance between two successive *Z*-lines is referred to as a sarcomere, and this distance is observed to decrease under muscle contraction, as does the width of the *I*-band. The width of the *A*-band, on the other hand, is not affected by contraction. These observations, taken collectively, led Huxley and Jean Hanson to put forward the sliding filament theory, in which there is mutual movement between certain of the filaments even though their lengths remain unaltered. (One could loosely compare the action with that of a cycle pump, in which there is an overall length change during use, though neither the barrel nor the piston change their lengths.)

As exciting as these observations were, a fuller account of muscle function had to await molecular-scale resolution of the various filaments. Much

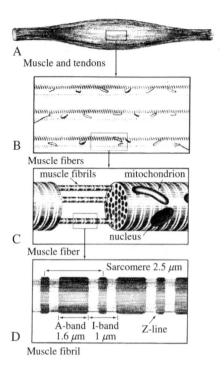

A
Muscle and tendons

B
Muscle fibers

muscle fibrils   mitochondrion

C

nucleus

Muscle fiber

Sarcomere 2.5 μm

A-band   I-band   Z-line
1.6 μm   1 μm

D

Muscle fibril

Figure 10.10 The structure of muscle was gradually elucidated as the power of microscopes increased

progress has been made in recent years, and we now know that there are two major types of filament, namely those consisting of myosin and those made of actin. The former are the thicker, with a diameter of 15 nm. They are composed of protein, the individual chains of which produce structures rather reminiscent of golf clubs, with long shanks and a head at one end. The shanks are sufficiently flexible to allow many of these to become entwined, so as to produce a complex elongated helical structure, with the head units sticking out in a periodic and systematic fashion (see the left-hand inset to Figure 10.11, which shows the overall arrangement from an end-on view). The distance between adjacent heads, in this arrangement, is 7 nm, but one must bear in mind that the heads are mutually offset by 60 degrees.

The actin filaments are also composite structures. Each actin polypeptide actually forms a globular protein, about 7 nm across, but it has adhesive spots at either end of a diameter, so the globules become linked up into a chain. Two such chains become mutually intertwined, so as to produce a double helical arrangement, the pitch of which is 36 nm (see the central inset to Figure 10.11). Down each of the two grooves in this double helix run the thinner stands of a third type of protein, namely tropomyosin, while the fourth and final component is troponin, the small (and heterogeneous) globules of which are attached to the

actin filaments, as shown in the right-hand inset to Figure 10.11. Rather interestingly, the spacing between the individual troponins is not 36 nm, as one might
have expected, but rather 40 nm. Viewed end-on, therefore, the troponins too are
seen to be mutually off-set. The distance between adjacent actin and myosin
filaments is 13 nm, so the myosin heads stick out into a space of this width.

When an arriving nerve impulse commands a muscle to contract, Ca^{2+} ions
diffuse to, and bind with, the troponin molecules, whereupon the latter change
their conformation. This has the effect of permitting the tropomyosin strands
to slip more deeply into the grooves between the actin filaments, thereby
exposing the active sites on those latter molecules. The upshot is that these
active sites can now interact with the myosin head groups, and because of the
above angular off-sets, the result is mutual sliding of the actin and myosin
filaments. It has been established, by Kanzuo Kitamura and his colleagues,
that a single myosin head moves along an actin filament with regular steps of
5.3 nm. When the Ca^{2+} ions retreat, the interatomic bonds established during
the contraction phase must be broken again, but *precisely* what happens during
the contraction-expansion cycle is still being investigated.

Given that muscles are features of eucaryotic systems, we should not be
surprised to find certain similarities with what was described earlier in connection with the dynein arms in eucaryotic cilia and flagella. Such similarities are
indeed observed, because there is an ATP-provoked change in shape of the
myosin head that closely parallels what occurs in dynein. Each of these proteins
possesses a groove, not unlike what is also seen in the β sub-units of the F_1
complex of ATP synthase (as was discussed in Chapter 9), the vital determination of the relevant molecular structure having been accomplished by Ivan
Rayment and his colleagues in 1993 (see Figure 10.12). The difference is, however, that ATP is *fabricated* in that other example whereas it is *broken down*
during the conformational changes of myosin heads and dynein arms. In

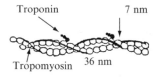

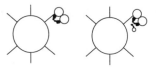

Troponin   7 nm

Tropomyosin   36 nm

Seen end-on, the arrangement of
the head groups of the individual
myosin molecules emerges as
having six-fold symmetry.

The participation of actin in the
sliding-filament mechanism is
controlled by tropomyosin and
troponin.

The actual triggering of the inter-
filamental sliding is achieved by
a subtle change of position of the
troponin molecules, these being
influenced by the local calcium
ion concentration.

Figure 10.11 Electron microscopial studies were crucial to the elucidation of muscle action
as they revealed both the existence and the arrangement of the filaments and indeed the
structure of the latter

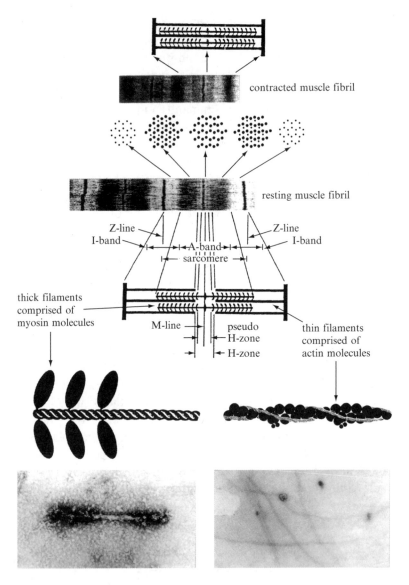

contracted muscle fibril

resting muscle fibril

Z-line
I-band
A-band
sarcomere
Z-line
I-band

thick filaments
comprised of
myosin molecules

M-line
pseudo
H-zone
H-zone

thin filaments
comprised of
actin molecules

Figure 10.11   (*continued*)

the case of the myosin head, it is that conformational change which enables the myosin to drag the actin filament along (see Figure 10.13), through a small distance, before becoming detached and making the step whose length Kitamura and his colleagues have measured.

The various stages of the process appear to be as follows. An ATP molecule binds to the groove in the myosin head, causing it to separate from the actin filament, and the ATP molecule then breaks down into ADP and inorganic

ATP binding site

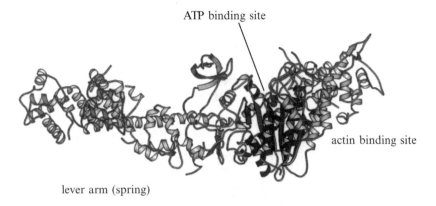

actin binding site

lever arm (spring)

Figure 10.12   Structure of the myosin S1 fragment (Protein Data Bank ID: 2MYS; Rayment *et al.*, 1993

phosphate, P_i. The conformation of the myosin head is simultaneously changed, such that its binding to the actin filament is now energetically favourable. As soon as this attachment has been achieved, further subtle modifications of the head's shape permit release of the P_i, and this leads to a grosser conformational change, which provides the power stroke that produces muscular contraction. Finally, because of the latter change in shape, the ADP molecule can now be released, and its replacement by an ATP molecule starts a new cycle, often referred to as a cross-bridge cycle (see Figure 10.13). The system will continue to cycle in this manner, with the myosin heads making their way gradually along the actin filaments, and progressing by what is known as the stepping distance, Δx, for each cross-bridge cycle, until the supply of Ca^{2+} ions is removed.

The duty ratio, r_{duty}, is defined as the fraction of cross-bridge cycling time during which a myosin head unit is attached to an actin molecule. If the durations of attachment and non-attachment are τ_{on} and τ_{off} respectively, we have

$$r_{duty} = \frac{\tau_{on}}{\tau_{on} + \tau_{off}} = \frac{\tau_{on}}{\tau_{cycle}} \qquad (10.4)$$

If movement is to be continuous, at least one head group will have to be attached to the actin molecule at any instant. This implies that there will have to be at least N_{min} heads, and that

$$N_{min} \cong \frac{1}{r_{duty}} \qquad (10.5)$$

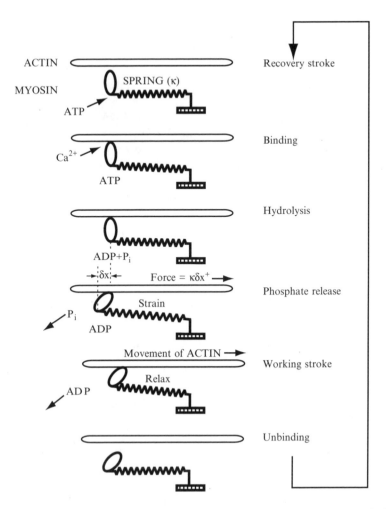

Figure 10.13 The molecular-level changes believed to underlie one complete cycle in the stepping mechanism that produces muscle contraction

Assuming that the system is ideally efficient in its utilization of ATP, that is to say that there is one cross-bridge cycle for each hydrolysis of an ATP molecule, the translation speed v of progression of the myosin along the actin will be given by

$$v = k_{ATPase} \Delta x \tag{10.6}$$

where k_{ATPase} is the rate constant for hydrolysis of an ATP molecule.

The power stroke would last the entire cross-bridge cycle only if τ_{off} were zero, which is very unlikely. In practice, therefore, the power stroke will advance the myosin head (relative to the actin filament) by a working distance

δx that is less than the stepping distance. If there is a sufficient number of myosin heads to give the smooth continuous motion referred to above, the velocity during the power stroke will not differ from that observed during the entire cross-bridge cycle (see Equation (10.6)), so we would have

$$v = \frac{\delta x}{\tau_{on}} \tag{10.7}$$

Moreover, for such smooth motion, we would expect that

$$\tau_{cycle} = \frac{1}{k_{ATPase}} \tag{10.8}$$

and taking Equations (10.4), (10.6), (10.7) and (10.8) together, we find that

$$r_{duty} = \frac{\tau_{on}}{\tau_{cycle}} = \frac{\delta x \cdot k_{ATPase}}{v} = \frac{\delta x}{\Delta x} \tag{10.9}$$

In practice, it is found that r_{duty} lies around 0.02, and that k_{ATPase} is approximately $20\,s^{-1}$ for each myosin head, while τ_{on} is about $10^{-3}\,s$. As was noted above, the experimentally observed working distance δx is 5.3 nm, so Equation (10.7) predicts a speed of about $5\,\mu m\,s^{-1}$ for the myosin heads relative to the actin filament. Because there are about 300 cross bridges in each half of a thick filament, and in view of the above-quoted duty ratio, the picture is thus one in which about six myosin heads are performing their working strokes at any one time, all the others being detached. Also, a given myosin head hydrolyses one molecule of ATP while making its 5.3 nm contribution to the overall 250 nm motion achieved during a cross-bridge cycle. When it has completed its task, other myosin heads are already starting to make their own contributions, thereby guaranteeing continuation of the smooth motion.

This picture would be reliable only if it were permissible to ignore the strain produced in the myosin molecule when ATP hydrolysis changes the conformation of the myosin head. It would then be possible to calculate the force generated in the muscle solely on the basis of that conformational change, together with information about the arrangement of the various molecules. As Andrew Huxley argued in 1957, part of the head-group conformational change must be supplanted into a more general strain throughout the entire myosin molecule, and this was the basis of his power-stroke model. As Huxley noted, both tensile and compressive strains will be possible, the latter arising from the possibility of a cross bridge detaching too late, as it were, and thus imposing a slight restraining force. This latter is appropriately referred to as the drag force.

Let us denote the tensile and compressive strains by the symbols δx^+ and δx^-, respectively, the former being referred to as the power-stroke distance while the latter is the drag-stroke distance. They are related to the above-defined working distance δx by

$$\delta x = \delta x^+ + \delta x^- \qquad (10.10)$$

Only in the case of a zero drag-stroke distance, therefore, will the power stroke equal the working stroke. Recalling the discussions of stress and strain in Chapters 3 and 5, we can now invoke the spring constant κ of the myosin molecule and write an expression for the force $\mathcal{F}_{on}$ generated by a cross bridge, while it is attached. This is

$$\mathcal{F}_{on} = \frac{\kappa(\delta x^+ - \delta x^-)}{2} \qquad (10.11)$$

The average macroscopic force generated by the muscle can then be found from a knowledge of the size of the duty ratio, that force naturally decreasing as the duty ratio becomes smaller. The two components on the right-hand side of Equation (10.11) are respectively an elastic driving force and a restraining drag force. As Huxley noted, the drag imposes a maximum speed, v_{max}, at which a muscle can contract, the effective value of τ_{on} in Equation (10.7) then becoming the time that the myosin head stays attached while it is being compressed. The upshot of this will be that the maintainable force will decrease as the filament sliding speed increases, and it is found experimentally that the force has dropped to zero when the speed reaches about $5\,\mu m\,s^{-1}$. The force is about $1.5\,pN$ (i.e. $1.5 \times 10^{-12}\,N$) when the speed is zero.

The classic study of maintainable force was carried out by Archibald Hill, in 1938, and it produced an empirical relationship between the force and the velocity, namely

$$(\mathcal{F} + a)(v + b) = (\mathcal{F}_0 + a)b \qquad (10.12)$$

$\mathcal{F}_0$ is the force at zero velocity and a and b are fitting parameters. Hill found that his experimental data could be optimally fitted when the following conditions were fulfilled

$$\frac{-a}{\mathcal{F}_0} = \frac{b}{v_{max}} = \frac{1}{4} \qquad (10.13)$$

Using these relationships, Equation (10.12) can be reduced to what has naturally come to be known as the Hill equation, namely

$$\mathcal{F}/\mathcal{F}_0 = \frac{(1 - v/v_{\max})}{(1 + 4v/v_{\max})} \tag{10.14}$$

One sees implicit acknowledgement of these trends in the sport of weightlift-ing. A heavier load can be raised if it is jerked suddenly from its starting position than if it is moved more smoothly. The lifter thereby exploits the higher performance of the muscles before the filaments have had time to pick up speed. There used to be three different lifting exercises in competitive events: the press (in which the weight had to be pushed up slowly, following rapid lifting to shoulder height), the snatch (in which the weight was lifted to maximum height directly from the floor) and the clean-and-jerk (which resembled the press, but with a sudden jerk replacing the final slow push). Not surprisingly, the greatest weights were lifted in that final discipline. The press was later done away with because it was too difficult for the judges to detect illicit jerking at the initial stage of the final movement.

There has been one glaring omission in our analysis thus far: no mention has been made of the thermal motion of the atoms in the various filamentary components. This could not be more serious because the atoms actually move faster than the components. The situation is loosely analogous to that in a hurricane, in which the wind speed is $30\,\mathrm{m\,s^{-1}}$ or higher, but in which the individual molecules can readily achieve speeds 100 times greater.

The presence of covalent bonds in the proteins that make up the filaments tells us that the atoms are not as free to move as they would be in a gas, but a considerable amount of movement is nevertheless possible. Those covalent bonds constrain the molecules to maintain their polymeric forms, but such forms are flexible, as we saw in Chapter 7. Paul Flory studied the statistical behaviour of such chains. The limiting case is that of the freely-jointed chain (also known as the worm-like chain, WLC), in which all values of the steric angles are equally allowed. This is clearly not strictly applicable to a real protein, because the Ramachandran diagram (see Chapter 7) would then be irrelevant.

The freely-jointed chain nevertheless displays some interesting physics, which we will find useful. Its individual links can be thought of as describing a random walk, of the type we considered in Chapter 5, when discussing diffusion. If the chain consists of n segments, each of standard length b, the fully outstretched chain would have a length nb. When the chain is in the random conformation, an analysis similar to the one originally carried out by Albert Einstein (see Chapter 5) shows that the mean-square end-to-end dis-tance will be nb^2. In real proteins, the sequences of amino-acid residues in-volved in secondary structure motifs are certainly not freely jointed, but this is a reasonable approximation for the random-coil segments that lie between such motifs. (It is such intermediate sections that are particularly vulnerable to attack by the peptide-cleaving enzymes known as proteases.) So the freely-jointed chain approximation is still useful.

Now the entropy of a chain will depend upon its conformational state, the fully outstretched form clearly being associated with a lower entropy. When the chain is in the random coil form, on the other hand, the individual links will be able to access a greater region of phase space, and the entropy will be larger. As the chain is gradually stretched out toward its maximum length, therefore, the entropy will gradually decrease. Given that the entropy contributes negatively to the free energy, this means that the totally random state is associated with a lowest free energy, and the free energy will thus increase as the chain is stretched out. In other words, there will be effectively a mechanical force resisting the stretching, and this is usually thought of in terms of an entropic elasticity (which is discussed in the article by John Marko and Eric Siggia, cited at the end of the present chapter). It is not difficult to show that the effective spring constant κ_{entropic} is then given by

$$\kappa_{\text{entropic}} = \frac{3k_B T}{nb^2} \qquad (10.15)$$

The denominator is simply the above-mentioned mean-square end-to-end distance, and we should not be surprised by the numerator because this is merely the total thermal energy associated with three kinetic-energy and three potential-energy degrees of freedom. A typical value for the spring constant is $5 \times 10^{-3}\,\text{N}\,\text{m}^{-1}$.

This brings us back to that issue of the influence of thermal motion on the mechanism being envisaged for the filamentary sliding. Let us recall the basic difference between the working of ATP-synthase and a typical ATPase (see Chapter 9). In the former, the ADP and inorganic phosphate units are forced together by the rotation of part of the enzyme with respect to certain other parts. The atoms in the various parts are thermally vibrating, of course, but the ATP-producing rotation acts over a super-atomic scale, and it thus prevails. In an ATPase, on the other hand, the individual thermal motions have to get their act together through their own devices, as it were. In effect, this means that certain molecular conformations will be more favourable than others because they lead to a more pronounced focussing of the thermal energy to the region where it is needed.

A myosin molecule is effectively an ATPase, because it hydrolyses that molecule when acquiring its energy of dissociation; and thermal motions must be important in the mechanism that permits incorporation of the ATP molecule into its binding pocket. However, it is not clear whether one is to regard the binding as then triggering a succession of local displacements that pass down the molecule or whether, contrariwise, the insertion of ATP into its pocket merely produces a bias in thermal motions that are at all times spread fairly uniformly throughout the myosin molecule.

Two of the techniques described in Chapter 6, namely atomic-force micro-scopy and optical tweezers, have been used to measure the elastic properties of relevant biopolymers. Such experiments obviate the need for approximations of the type embodied by Equation (10.15), and they make the assumption of a linear stress–strain relationship unnecessary. An example is seen in the several studies of the giant muscle protein known as titin reported by L. Tskhovrebova and his colleagues, and by Miklós Kellermayer and his colleagues, both in 1997, and by Piotr Marszalek and his colleagues 2 years later (as is illustrated in Figure 10.14). A titin molecule connects the end of each actin filament to the transverse structure that provides the anchoring points for the myosin super-chains, thereby preventing the actin and myosin filaments from being completely pulled apart. (An analogous job is done by the ferrule that stops the piston and tube of a bicycle pump from coming apart.) Bjørn Nielsen has used this data, and he has also replaced the two-state approximation invoked above by what he calls a tug-of-war model, in which the strain is permitted to vary continuously within the limit of the two end states. This produces an impressively close fit with Hill's experimental force–velocity data, as shown in Figure 10.15.

The work done by a muscle is found in the usual manner, as the product of the relevant force and distance. Its maximum value, w_{max}, will correspond to the condition of zero drag, and the distance in question will clearly be δx^+. Using Equation (10.11), we thus find that

$$w_{max} = \frac{\kappa(\delta x^+)^2}{2} \tag{10.16}$$

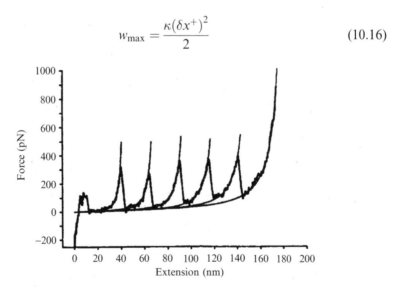

Figure 10.14   The force-extension curve for titin, determined by atomic force microscopy by Julio Fernandez and his colleagues (see Marszalek *et al.* in the reference list). The kinks are artefacts caused by the rupture of hydrogen bonds

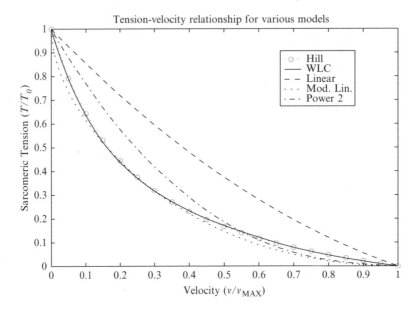

Figure 10.15 Velocity-tension curves produced by various models of muscle contraction. Bjørn Nielsen's tug-of-war version of the worm-like chain (WLC) model produces a particularly good fit to the experimental data reported in 1938 by Archibald Hill

Recalling that κ is typically $5 \times 10^{-3} \, N \, m^{-1}$, while δx^+ has been measured to be 5.3×10^{-9} m, we find that the maximum work done by the muscle will be about $6 \times 10^{-20} \, N \, m$. It is not difficult to show that this is about half the energy released by the hydrolysis of the relevant amount of ATP, so the process is rather efficient.

As a final point in this discussion of the sliding-filament model, we ought to take a closer look at the diffusion of Ca^{2+} ions invoked to explain the triggering of the molecular movements. The typical skeletal muscle cell has a diameter of about 200 μm (i.e. a radius of about 10^{-2} cm). Assuming that the relevant cytoplasmic diffusion coefficient is about $10^{-5} \, cm^2 \, s^{-1}$ (see Chapter 5), we see from Equation (5.1) that the time taken for the ions to move from the cell surface to the cell's axis would be about 10 s. This would make muscle function unacceptably sluggish.

Nature's solution to this problem has been to provide the muscle with an additional membranous structure that lies transverse to the cell's outer membrane (see Figure 10.16). This is the *sarcoplasmic reticulum*, which is directly activated by the arriving nerve impulse, and which is thereby provoked into releasing Ca^{2+} ions from its surface. The maximum diffusion length from the sarcoplasmic reticulum to the actin–myosin complex is just 2 μm (see Table 10.1), and this brings the diffusion time down to an acceptable 10^{-2} s. It should be noted, however, that the essentially one-dimensional diffusion of

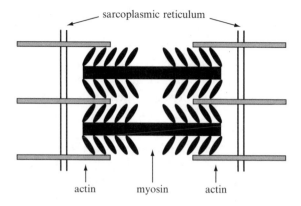

sarcoplasmic reticulum

actin   myosin   actin

Figure 10.16   The sarcoplasmic reticulum is the source of the calcium ions which cause the triggering of the sliding-filament mechanism

Table 10.1   The distance between successive slabs of sarcoplasmic reticulum, under various conditions

Length [µm]	State
3.65	Stretched
2.05	Normal
1.65	Bent

the Ca^{2+} ions, from the sarcoplasmic reticulum and in the direction parallel to the actin and myosin strands, would activate the troponin complexes in a *sequential* fashion. The exact significance of this still awaits elucidation.

10.4 Human Performance

Let us close this chapter with an example drawn from life – the present author's life, indeed. He somewhat rashly estimated that he would be able to run up the 50 m high Monument to the fire of London in less than 1 min. This resulted in a wager, which the 75 kg author won, by scaling the tower in just 53 s (and then almost passing out!). Let us see whether we can account for this feat at the molecular level.

To begin with, we recall the estimate of 800 kcal h^{-1} for climbing stairs, as given in Chapter 9. This is equivalent to a power output, P, of 960 W. Now when a mass m is raised a height h against gravity, the acceleration due to which would be g ($= 9.8\,\mathrm{ms}^{-2}$), and the time taken is t, the equivalent power P is given by

$$P = \frac{mgh}{t} \qquad (10.17)$$

Inserting the author's 75 kg, and his 53 s performance into Equation (10.17) gives 690 W, which agrees well with the above estimate, considering that heat loss from the body, as well as power consumed by the nervous system, have not been allowed for.

However, let us see whether such good agreement is obtainable at the molecular level. We start by estimating how many actin–myosin complexes there might be in the cross-sections of a calf muscle and a thigh muscle (assuming that these muscles do most of the work). The diameters of these muscles are each about 10^{-1} m, while the effective diameter d of a myosin filament and its surrounds – assuming close-packing, in which six actin filaments are deployed around each myosin filament, and each actin filament is surrounded by three myosin filaments – is about 20 nm. This gives the number N of myosin filaments in a cross-section of either of the above muscles

$$N \approx \frac{D^2}{d^2} = \frac{(10^{-1})^2}{(20 \times 10^{-9})^2} = 25 \times 10^{12} \qquad (10.18)$$

With six rows of side-chains per myosin filament, we have $6N \, (= 1.5 \times 10^{14})$ contact surfaces. Because the myosin heads take 5.3 nm steps along the actin filaments, while the total change of muscle length during contraction is about 3×10^{-2} m, we find that the total number, n_B, of bonds in the two muscles that are cyclically made and broken during the contraction is

$$n_B \approx \frac{2 \times 1.5 \times 10^{14} \times 3 \times 10^{-2}}{5.3 \times 10^{-9}} \approx 1.7 \times 10^{21} \qquad (10.19)$$

Finally, assuming that each step taken along an actin filament by a myosin head involves consumption of *all* the energy made available by an ATP $\rightarrow$ $ADP + P_i$ reaction (0.06 aJ – see Chapter 9), we see that the total energy, $\mathscr{E}_B$, involved in these bonds is

$$\mathscr{E}_B \approx 1.7 \times 10^{21} \times 0.06 \times 10^{-18} \approx 100 \, \text{J} \qquad (10.20)$$

Moreover, the author recalls having calculated that the steps were being taken at about four per second, so each contraction occupied just 0.25 s. This gives the final power, P, to be

$$P \approx \frac{100}{0.25} = 400 \, \text{W} \qquad (10.21)$$

which is a surprisingly good result, given all the approximations that we have been forced to make.

Exercises

10.1   Find the maximum work done by a muscle if it may be assumed that the drag at the microscopic level can be ignored. Assume that the working distance of a myosin head is 5 nm, and that the relevant spring constant is $5 \times 10^{-3}\,\mathrm{N\,m^{-1}}$.

10.2   Show that the maximum work calculated in Exercise 10.1 is about half the energy released by the hydrolysis of the relevant amount of ATP.

Further reading

Berg, H. C., (1993). *Random Walks in Biology*. Princeton University Press, Chichester.

Bray, D., (2001). *Cell Movements* 2nd edition. Garland, New York.

Carpenter, R. H. S., (1996). *Neurophysiology*. Arnold, London.

Corrie, J. E. T., *et al.*, (1999). Dynamic measurement of myosin light-chain-domain tilt and twist in muscle contraction. *Nature* **400**, 425–430.

Dickinson, M. H. *et al.*, (2000). How animals move: an integrative view. *Science* **288**, 100–106.

Duke, T. A. J., (1999). Molecular model of muscle contraction. *Proceedings of the National Academy of Science, USA.* **96**, 2770–2775.

Eisenbach, M., (1996). Control of bacterial chemotaxis. *Molecular Microbiology* **20**, 903–910.

Ghez, C., (1991). Muscles: effectors of the motor systems. In: *Principles of Neural Science*, E. R. Kandel *et al.*, eds. Elsevier, New York. pp 548–563.

Hill, A. V., (1938). The heat of shortening and the dynamic constants of muscle. *Proceedings of the Royal Society, Series B* **126**, 136–195.

Howard, J., (2001). *Mechanics of Motor Proteins and the Cytoskeleton*. Sinaur Associates, Sunderland, MA.

Huxley, A. F., (1973). A note suggesting that the cross-bridge attachment during muscle contraction may take place in two stages. *Proceedings of the Royal Society, Series B* **183**, 83–86.

Huxley, A. F., (1998). Support for the lever arm. *Nature* **396**, 317–318.

Huxley, H. E. and Hanson, E. J., (1954). Changes in the cross-striations of muscle during contraction and stretch and their structural interpretation. *Nature* **173**, 973–976.

Huxley, A. F. and Niedergerke, R., (1954). Structural changes in muscle during contraction; interference microscopy of living muscle fibres. *Nature* **173**, 971–973.

Katz, B., (1939). The relation between force and speed in muscular contraction. *Journal of Physiology* **96**, 45–64.

Katz, B., (1966). *Nerve, Muscle and Synapse*. McGraw-Hill, New York.

Kellermayer, M. S. Z. *et al.*, (1997). Folding-unfolding transitions in single titin molecules characterized with laser tweezers. *Science* **276**, 1112–1116.

Keynes, R. D. and Aidley, D. J., (1991). *Nerve and Muscle*. Cambridge University Press, Cambridge.

Kitamura, K. *et al.*, (1999). A single myosin head moves along an actin filament with regular steps of 5.3 nanometres. *Nature* **397**, 129–134.

Macnab, R. M., (1995). Flagellar switch. In: *Two-Component Signal Transduction*. Hoch, J. A. and Silhavy, T. J. (Eds.), ASM Press, Washington, DC, pp. 181–199.

Mahadevan, L. and Matsudaira, P., (2000). Motility powered by supramolecular springs and ratchets. *Science* **288**, 95–99.

Marko, J. F. and Siggia, E. D., (1995). Stretching DNA. *Macromolecules* **28**, 8759–8770.

Marszalek, P. E., (1999). Mechanical unfolding intermediates in titin molecules. *Nature* **402**, 100–103.

Nielsen, B. G., (2002). Entropic elasticity in the generation of muscle force—a theoretical model. *Journal of Theoretical Biology* **219**, 99–119.

Purcell, E. M., (1977). Life at low Reynolds number. *American Journal of Physics* **45**(1), 3–11.

Purcell, E. M., (1997). The efficiency of propulsion by a rotating flagellum. *Proceedings of the National Academy of Science, USA.* **94**, 11307–11311.

Rayment, I. *et al.*, (1993). Structure of the actin-myosin complex and its implications for muscle contraction. *Science* **261**, 58–65.

Rayment, I., (1996). Kinesin and myosin: molecular motors with similar engines. *Structure* **4**, 501–504.

Segall, J. E., Block, S. M. and Berg, H. C., (1986). Temporal comparisons in bacterial chemotaxis. *Proc. Natl. Acad. Sci. USA* **83**, 8987–8991.

Tskhovrebova, L. *et al.*, (1997). Elasticity and unfolding of single molecules of the giant muscle protein titin. *Nature* **387**, 308–312.

Vale, R. D. and Milligan, R. A., (2000). The way things move: looking under the hood of molecular motor proteins. *Science* **288**, 88–95.

Vogel, S., (1994). *Life in Moving Fluids*. Princeton University Press, Princeton, NJ.

Yonekura, K. *et al.*, (2000). The bacterial flagellar cap as the rotary promoter of flagellin self-assembly. *Science* **290**, 2148–2152.

11 Excitable Membranes

The signal-mediating cells of the nervous system are referred to by the name neurons (or sometimes, neurones). They are special, in that they are able to send signals to one another with velocities that greatly exceed those displayed by mere diffusion. Neurons have shapes that differ markedly from those of the body's other cells. Their membranes have long protuberances, known as processes, these coming in two general types. There are the numerous dendrites, which convey information toward the body of the cell (that is to say, the soma), and there is the single axon, which extends out from the soma, on the side remote from the dendrites. The latter display considerable ramification, reminiscent of the branches of a tree, and one refers to the dendritic arborization. The single axon often divides close to the soma, so as to produce several axon collaterals, and the extremities of the axon and the axon collaterals invariably display extensive branching as well. These gross neural features are illustrated in Figure 11.1. Electrochemical signals travel along the dendrites, impinge upon the centrally-located soma, which is the main body of the cell, and – if the threshold is thereby exceeded – a signal is dispatched out along the axon. Signals are passed from one neuron to another via the synapses, at which the transfer is purely chemical and involves molecules of a neurotransmitter.

Three properties of the neural membrane are of particular interest, namely the resting potential, V_{rest} the passive cable response, and the nerve impulse (also known as the action potential). The term excitability refers to the fact that such membranes are able to mediate the passage of these (electrochemical) impulses.

11.1 Diffusion and Mobility of Ions

The analytical development of an expression for the resting potential requires several preliminaries, of which the first is a relationship derived by Albert Einstein. We have earlier developed expressions which describe the motion of particles under the influence of concentration gradients, that is to say purely

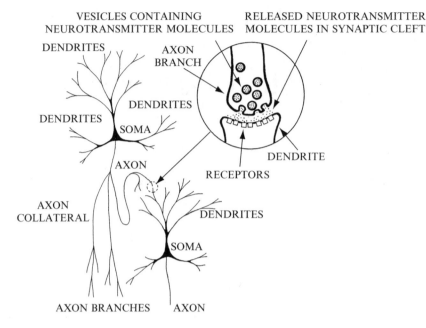

VESICLES CONTAINING   RELEASED NEUROTRANSMITTER
NEUROTRANSMITTER MOLECULES   MOLECULES IN SYNAPTIC CLEFT

DENDRITES

AXON
BRANCH

DENDRITES

DENDRITES

SOMA

DENDRITE

AXON

RECEPTORS

AXON
COLLATERAL

DENDRITES

SOMA

AXON BRANCHES   AXON

Figure 11.1 A highly schematic picture of a neuron showing the three major regions:
dendrites, soma and axon, together with a synapse (inset)

chemical diffusion (see Chapter 5). One notable result was that the flux, J_{chem}, is
given by

$$J_{chem} = -D\frac{dC}{dx} \qquad (11.1)$$

where D is the diffusion coefficient and C is the concentration of the chemical
species in question. In the case of the nerve cell membrane, there is the added
factor that an electric field is also present. Because the particles of interest bear
electrical charges, they will experience a force because of this field, and they will
thus acquire a drift velocity, v_{drift}. (Although the force actually causes acceleration, the particles experience numerous collisions with each other, and rapidly
attain a terminal velocity which is simply the drift velocity.) For a field strength
E, this is defined to be

$$v_{drift} = \mu E \qquad (11.2)$$

where the coefficient μ is known as the mobility. By considering particle
movement in a cylinder of unit cross-sectional area and unit length, we can
readily use this drift velocity to calculate the flux, because this quantity will

clearly also be linearly related to the concentration of particles. The expression for the flux is

$$J_{elec} = v_{drift}C = \mu EC \tag{11.3}$$

In a situation in which the driving forces due to the electric field and the concentration gradient are mutually opposed, there will be equilibrium if the fluxes caused by these two forces are equal in magnitude. Using Equations (11.1) and (11.3), this equilibrium is thus described by

$$-D\frac{dC}{dx} = -\mu EC \tag{11.4}$$

Equilibrium also implies that there will be no net force on any arbitrarily-positioned small region. Let us consider an imaginary cylinder of cross-sectional area A and length dx, and note that the concentration gradient (of perpetually moving particles) will give rise to a pressure gradient. This is to be opposed by E, as before, so the situation will be as in Figure 11.2.

If the electrical charge on each particle is q, it will experience a force Eq due to the field. Because the imaginary cylinder contains $CAdx$ such particles, the electrical force on the cylinder is

$$\mathcal{F}_{elec} = EqCAdx \tag{11.5}$$

This is to be opposed by the force due to the difference in pressure at the two ends, and recalling that pressure is defined as force per unit area, we have

$$\mathcal{F}_{chem} = A[(P + dP) - P] = AdP \tag{11.6}$$

Equating (11.5) and (11.6), for equilibrium, we get

$$EqC = \frac{dP}{dx} \tag{11.7}$$

We also recall that the perfect gas equation (see Appendix C) relates pressure, volume and temperature, by the equation

$$PV = nk_BT \tag{11.8}$$

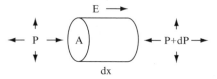

Figure 11.2 The Einstein equation is derived by considering the equilibrium between electrical and mechanical factors

where n is the number of particles and k_B is Boltzmann's constant; and because n/V is simply the concentration C, this may be written

$$P = Ck_BT \qquad (11.9)$$

Using this to rewrite Equation (11.7), we have

$$EqC = k_BT\frac{dC}{dx} \qquad (11.10)$$

Combining Equations (11.10) and (11.4), which arose from equilibrium between forces and equilibrium between fluxes, respectively, we have

$$\mu = \frac{qD}{k_BT} \qquad (11.11)$$

which is the desired Einstein equation relating mobility to the diffusion coefficient.

Until now, the moving entities have merely been referred to as *particles*, and we have invoked the perfect gas equation, which is not strictly valid for molecules in a real gas, let alone molecules in the sort of liquid that surrounds a biological membrane. Moreover, mobility is not strictly a constant, independent of the strength of the field E. It is nevertheless found that the Einstein equation continues to be a good approximation, even for electrolytes, in which the particles are *ions*.

11.2 Resting Potential

We now proceed toward deriving an expression for the resting potential, V_{rest}, and begin by considering the situation in which two different species of ions, positive and negative, denoted by the subscripts $+$ and $-$, are present. As before, we consider the general situation in which motion is induced by both an electric field and a concentration gradient. We begin by confining our attention to the case in which the concentrations of the two species are equal (as would be the case if the ions were the positive and negative species in an electrolyte that was globally neutral). The electrical current, due to the two influences identified above, is given by

$$i = -q(D_+ - D_-)\left(\frac{dC}{dx}\right) + qEC(\mu_+ + \mu_-) \qquad (11.12)$$

This is the Nernst-Planck equation, (named after Walther Nernst and Max Planck), and it is to be noted that the positive sign in the final parenthesis arises from the fact that the two species of ions, which carry oppositely-signed charges, move in opposite directions under the influence of the electrical field, so the contributions to the current are additive. Noting that

$$\left(\frac{dC}{dx}\right)\bigg/ C = \frac{d(\ln C)}{dx}$$

we can rewrite Equation (11.12) to produce

$$i = qC(\mu_+ + \mu_-)\left\{E - \frac{(D_+ - D_-)}{(\mu_+ + \mu_-)}\frac{d(\ln C)}{dx}\right\} \qquad (11.13)$$

The first term inside the brace is, of course, simply the electrical field. By analogy, the second term is the chemical (non-electrical) field, and it naturally arises from the concentration gradient. Equation (11.13) is, in fact, a more general form of Ohm's Law, since it may be expressed in the simpler form

$$i = \sigma(E + E_{\text{chem}}) \qquad (11.14)$$

The coefficient σ that appears before the parentheses is a concise way of expressing the combined product of the terms appearing before the brace in Equation (11.13), and it is known as the conductivity. For convenience, the relationship will be stated by an equation

$$\sigma = qC(\mu_+ + \mu_-) \qquad (11.15)$$

In electrical equilibrium, $E = -E_{\text{chem}}$, and there is no flow of current.

We will soon be in a position to consider the situation in which electrolytes having different compositions are separated by a membrane. As a final preliminary, we note that if electrodes are placed at positions a and b, and the concentrations at these two points are C_a and C_b, the potential difference between the two electrodes will be

$$\Delta V = V_a - V_b = -\int_a^b E_{\text{chem}} dx = \frac{(D_+ - D_-)}{(\mu_+ + \mu_-)}\ln\left(\frac{C_b}{C_a}\right) \qquad (11.16)$$

We note that no potential difference will exist if either the Ds are equal or the Cs are equal. It is at this point that the Einstein equation (Equation (11.11)) becomes useful, for we can use it to eliminate the diffusion coefficients in Equation (11.16). This produces

$$\Delta V = \frac{k_B T}{q} \frac{(\mu_+ - \mu_-)}{(\mu_+ + \mu_-)} \ln \left(\frac{C_b}{C_a} \right) \qquad (11.17)$$

To obtain an idea of the size of potential difference, we note that $k_B T/q$ equals 0.026 V at 23°C, this applying to the situation when $q = q_e$, q_e being the single electronic charge. In a sodium chloride electrolyte, the mobility of Cl^- ions is approximately 1.5 times that of the Na^+ ions, so $(\mu_+ - \mu_-)/(\mu_+ + \mu_-)$ is roughly -0.2. Unless some form of barrier intervenes, the potential difference cannot be maintained indefinitely, and it will gradually decay to zero because of the movement of the ions.

The biological membrane acts as just such a barrier, and one that is selectively permeable to certain types of ion. Suppose that its permeability to negative ions is much greater than it is for positive ions (i.e. $\mu_+ \ll \mu_-$), and replace the subscripts a and b in Equation 11.17 by o and i, to denote outside and inside, respectively. Equation (11.17) then becomes

$$\Delta V = V_i - V_o = -\frac{k_B T}{q} \ln \left(\frac{C_i}{C_o} \right) \qquad (11.18)$$

and it gives the potential of the *inside surface with respect to the outside*. This is the Nernst equation, and it is of paramount importance in the study of nerve membranes. The quantity C_i/C_o is known as the Donnan ratio (after F. G. Donnan).

We have now reached the point at which the resting potential, V_{rest}, can be calculated. For the well-known case of the squid giant axon, the concentrations of Na^+ ions inside and outside are, respectively, 50 nmol/l and 460 nmol/l, while for K^+ ions the corresponding concentrations are 400 nmol/l (inside) and 10 nmol/l (outside). For Cl^- ions, the values are 70 nmol/l (inside) and 540 nmol/l (outside). These values can be inserted into the Nernst Equation (11.18) to give the potential differences that arise from the concentration differences for each of the ionic species. These potential differences can be looked upon as being equivalent to batteries that are applied across the membrane. Two very important points must be made regarding the Nernst equation, and both dictate whether the potential difference is positive or negative. First, it is absolutely vital that one does not confuse inside with outside. (From the concentration values given above, this immediately shows that the Na^+ and K^+ 'batteries' must lie in opposing directions.) The second point is that the right-hand side of the Nernst equation loses its negative sign in the case of *negative* ions, because of the change of sign of q.

In reality, the mobilities of the various ions are merely very small rather than zero, but the approximations are still acceptable and the small but finite mobilities produce conductivities (see Equations (11.13) and (11.14)). The equivalent circuit for the membrane is shown in Figure 11.3, and the polarities and voltages of the imaginary batteries are such as to oppose, and just balance, the tendency of the relevant ions to diffuse in the direction of the concentration gradient. (Thus the Na^+ battery is *positive inside*, and it repels the Na^+ ions that are predominantly on the *outside*, keeping them on that side of the membrane.)

The total current flowing across the membrane (which, by our convention, will be in the direction from inside to outside) will be equal to the individual currents due to the motions of the individual ionic species. Thus

$$I_{total} = \Sigma_j I_j \qquad (11.19)$$

where j stands for Na^+, K^+, Cl^-, and indeed for any other ionic type that is present, such as Ca^{2+}. For each of these we have

$$I_j = g_j V_j \qquad (11.20)$$

where V_j can be calculated from the Nernst equation, provided that the relevant ionic concentrations are known. The total membrane conductance is simply the sum of the individual conductances (which are, of course, inverse resistances):

$$g_{total} = \Sigma_j g_j \qquad (11.21)$$

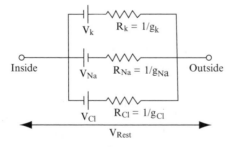

Figure 11.3   The chemical gradients across the nerve membre of the various ions can be thought of as being equivalent to batteries, while the small leakages through the channels can be thought of as passing through electrical resistors; these token components can be regarded as constituting an electrical circuit from which the resting potential can be calculated

and this is related to the resting potential by

$$I_{total} = g_{total} V_{rest} \qquad (11.22)$$

Equations (11.19)–(11.22), taken together, permit one to calculate the resting potential in terms of the individual conductances and Nernst potentials. The expression is

$$V_{rest} = \frac{\Sigma_j g_j V_j}{\Sigma_j g_j} \qquad (11.23)$$

This is a particularly simple form of what is known as the Goldman equation (after D. E. Goldman).

It is frequently justifiable to consider only the axon, and that part of the soma adjacent to the axon (i.e. the axon hillock), as having (protein) ion channels in its membrane. It is these channels which determine the magnitude of the resting potential that was discussed in the preceding paragraph. It is a reasonable approximation to assume that the resting potential is nevertheless uniform over the entire nerve cell, that is to say in the axon, the soma and the dendrites. The dendrites lack such channels and are thus incapable of support-ing the passage of an action potential (at least, in the simplified view that will be adequate for our purposes here). Their membranes therefore respond in a passive manner to changes in the voltage across them. These changes may be either a depolarization or a hyperpolarization, both with respect to the resting potential. Also, because the resting potential is itself a negative voltage, the changes will respectively produce a less negative and a more negative mem-brane voltage. As will be shown in the next chapter, the effect of such an instantaneous change varies in both time and space, in an exponentially decaying manner. It is particularly interesting to note that the time constant for the temporal decay is about 4 ms.

Much is now known about the protein molecules which mediate the passage of ions across the membrane. It turns out that a single molecule is responsible for maintaining the concentration gradients of sodium and potassium, and because it has to carry out its task by working against those gradients it is said to be an active transporter, or ion pump. It is this enzyme which is responsible for maintaining the resting potential, and it was discovered by Jens Skou in 1957. Not surprisingly, it was given the name $Na^+, K^+ ATPase$, because it breaks down ATP molecules into ADP and P_i, while translocating sodium and potassium ions. Other ion pumps have since been discovered, and they collectively use about one third of all the ATP molecules produced by the body; the maintenance of the nervous system in its ready-to-signal state is clearly of vital importance.

In the early 1950s, Richard Keynes and Alan Hodgkin had demonstrated that Na$^+$ ions flow into a nerve cell when it conducts a signal, and that ATP is consumed when such ions are translocated in the opposite direction. They also showed that the latter process is inhibited when ATP synthesis is disrupted. Following Skou's discovery of the enzyme, it was found that three Na$^+$ ions are transported out of the cell for each pair of K$^+$ ions redeployed in the opposite direction (see Figure 11.4). The actual atomic arrangement in Na$^+$,K$^+$-ATPase has still not been determined.

If the degree of depolarization of the axonal membrane exceeds the threshold value, at any instant, a nerve impulse (or action potential) is generated. Impulses will continue to be emitted out along the axon for as long as the threshold is exceeded, the rate of emission being related to the degree of the excess voltage over the threshold. The threshold typically lies at around $-$ 50 mV. Because the resting potential is typically of the order of $-$ 100 mV (inside the membrane, with respect to the outside), a *depolarization* of about 50 mV will be required.

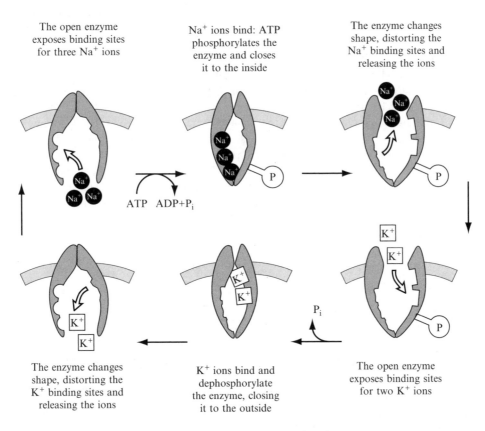

Figure 11.4 The active (molecular) pump known as Na$^+$, K$^+$-ATPase cycles between two states – open to the inside or open to the outside of the cell – and it thereby transports potassium ions into the cell and sodium ions out to the extracellular space

(If the depolarization does not reach this value, the axon's subsequent electrical behaviour is merely that which is dictated by the passive cable properties mentioned earlier.)

If the threshold is exceeded, the (protein) ion channels undergo a radical change in conformation and their conductances are suddenly increased. This leads to the rapid events first investigated by Alan Hodgkin and Andrew Huxley, as will be discussed in detail in the following chapter.

Exercises

11.1   The concentrations (in units of $10^{-3}\,mol^{-1}$) of sodium, potassium and chlorine ions inside a squid axon have been determined to be 50, 400 and 70, respectively. The corresponding concentrations outside the axon's membrane were found to be 460 (Na^+), 10 (K^+) and 540 (Cl^-). The membrane conductance for sodium ions, g_{Na}, was found to be essentially zero, while the conductances for potassium and chlorine have been measured to be $g_K = 2.7 \times 10^{-6}\,mho\,cm^{-2}$ and $g_{Cl} = 1.0 \times 10^{-6}\,mho\,cm^{-2}$. What is the resting potential for this squid axon (at 37°C)?

11.2   In 1983, 4 year old Jimmy Tontlewicz fell through a hole in the ice on a lake in Minnesota, and when the rescue team pulled him out of the water 20 minutes later, his body temperature had fallen from the normal 37°C to 29°C. Calculate the sign and magnitude of the change in the resting potential of Jimmy's neurons, if the relevant data are as follows. The concentrations (in units of $10^{-3}\,mol\,l^{-1}$) of sodium-, potassium- and chloride ions inside the axon are found to be, respectively, 60, 390 and 65. The corresponding concentrations outside the axon were 440 (Na^+), 15 (K^+) and 510 (Cl^-). The membrane conductance for sodium ions, g_{Na}, was $0.12 \times 10^{-6}\,mho\,cm^{-2}$, while the conductances for potassium and chloride were $g_K = 2.6 \times 10^{-6}\,mho\,cm^{-2}$ and $g_{Cl} = 1.3 \times 10^{-6}\,mho\,cm^{-2}$.

Further reading

Carpenter, R. H. S., (1996). *Neurophysiology*. Arnold, London.

Deutsch, S. and Micheli-Tzanakou, E., (1987). *Neuroelectric Systems*. NYU Press, New York.

Hille, B., (1984). *Ionic Channels of Excitable Membranes*. Sinauer, Sunderland, MA.

Hodgkin, A. L. and Huxley, A. F., (1952). A quantitative description of membrane current and its application to conduction and excitation in nerve. *Journal of Physiology (London)* **117**, 500–544.

Hodgkin, A. L. and Katz, B., (1949). The effect of sodium ions on the electric activity in the giant axon of the squid. *Journal of Physiology (London)* **108**, 37–77.

Hodgkin, A. L. and Keynes, R. D., (1955). Active transport of cations in giant axons from *Sepia* and *Loligo*. *Journal of Physiology (London)* **128**, 28–60.

Hodgkin, A. L. and Rushton, W. A. H., (1946). The electrical constants of a crustacian nerve fibre. *Proceedings of the Royal Society (London)* **B 133**, 444–479.

Hodgkin, A. L., Huxley, A. F. and Katz, B., (1952). Measurement of current-voltage relations in the membrane of the giant axon of *Loligo*. *Journal of Physiology (London)* **116**, 424–448.

Katz, B., (1966). *Nerve, Muscle and Synapse*. McGraw-Hill, New York.

Keynes, R. D. and Aidley, D. J., (1991). *Nerve and Muscle*. Cambridge University Press, Cambridge.

Koch, C., (1998). *Biophysics of Computation: Information Processing in Single Neurons*. Oxford University Press, Oxford.

Koch, C. and Segev, I. (eds.), (1989). *Methods in Neural Modeling: From Synapses to Networks*. MIT Press, Cambridge, MA.

Skou, J. C., (1957). The influence of some cations on an adenosine triphosphatase from peripheral nerves. *Biochimica et Biophysica Acta* **23**, 394–401.

Sten-Knudsen, O., (2002). *Biological Membranes: Theory of Mass Transport, Membrane Potential and Electrical Impulses*. Cambridge University Press, Cambridge.

12 Nerve Signals

It is frequently justifiable to consider only the axon of a neuron, and perhaps that part of the soma adjacent to the axon (i.e. the axon hillock), as having (protein) ion channels in its membrane. It is these channels which give rise to the resting potential that was discussed in the preceding chapter. It is a reasonable approximation to assume that the resting potential is nevertheless uniform over the entire nerve cell, that is to say the axon, the soma and the dendrites. The latter, lacking such channels, are thus incapable of supporting the passage of nerve impulses. Their membranes therefore respond in a passive manner to changes in the voltage across them. These changes may be either a depolarization or a hyperpolarization, both with respect to the resting potential. Because the resting potential is itself a negative voltage, those changes will respectively produce a less negative and a more negative membrane voltage. The effect of such an instantaneous change varies in both space and time in an exponentially varying manner. It is of obvious importance that we consider the physics underlying the spatial and temporal functionalities.

12.1 Passive Response

The general situation regarding a nerve process will be as shown in Figure 12.1, its length being imagined as being divided up into segments, each having unit magnitude. The conducting medium located inside the process, that is to say the axoplasm, offers a resistance R_i per unit length, while the corresponding value outside is R_o. As we saw in connection with the discussion related to Figure 11.3, the membrane offers a resistance to the passage of (ion-mediated) current. For our initial purposes here, we need not think of this resistance as comprising separate components for the various types of ion, and we can likewise ignore the equivalent batteries shown in Figure 11.3. The net membrane resistance per unit length is taken to be R_m. Finally, because the bulk of the membrane (that is, the lipid bilayer) is an insulator, it will function as an electrical capacitor, with a capacity per unit length of C_m.

Although the neurons in the brain are quite densely packed, the spacing between them is nevertheless such that the individual processes can be regarded

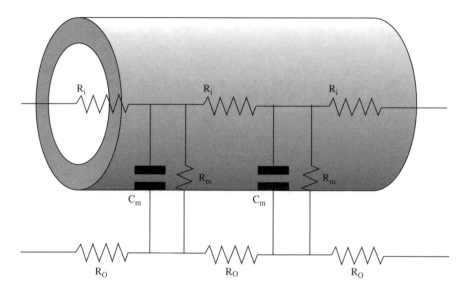

Figure 12.1 Idealized representation of a nerve process (axon or dendrite) showing the various associated resistances and capacitances

as being immersed in a large volume of conducting fluid, and this means that R_o is essentially zero. For the neurons lying in the more peripheral parts of the nervous system, this is an even safer assumption. Consequently, the electrical potential, V, may be regarded as being constant throughout the extra-neuronal space. That will be one of the underlying assumptions in the balance of this chapter.

Even though Figure 12.1 becomes somewhat simpler when we can ignore R_o, the general situation will still be complicated when things are changing with time. A full account would, under such circumstances, demand that we are able to derive the spatiotemporal potential $V_{x,t}$. Later in this chapter, we will indeed be confronted with precisely that challenge. Our first task is rather simpler, however, because we merely wish to examine the spatial distribution of the electrical potential along the inside of the process, V_x, when a steady state prevails. Figure 12.1 can then be simplified to the circuit shown in Figure 12.2, in which we see that the capacitance effects can be ignored.

It is clear that V_x will gradually diminish as x increases, partly because of the losses incurred in traversing the resistance of the axoplasm and partly because of leakage through the membrane. For the first of these components, using Ohm's Law, we see that

$$I_i = -\frac{1}{R_i}\frac{dV_x}{dx} \tag{12.1}$$

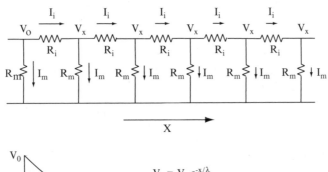

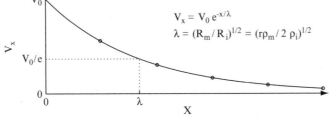

Figure 12.2   The equivalent circuit for a nerve cell process when the extracellular resistance can be ignored; the lower part of the figure shows the steady-state variation of voltage with distance

while the membrane-leakage effect causes further diminution of the longitudinal current, I_i, according to the relationship

$$-\frac{dI_i}{dx} = I_m = \frac{V_x}{R_m} \tag{12.2}$$

Combining Equations (12.1) and (12.2), we find that

$$\frac{R_m}{R_i}\frac{d^2 V_x}{dx^2} = V_x \tag{12.3}$$

The general solution of the differential Equation (12.3) is

$$V_x = A\exp\left(\frac{-x}{\sqrt{R_m/R_i}}\right) + B\exp\left(\frac{x}{\sqrt{R_m/R_i}}\right) \tag{12.4}$$

where A and B are constants. We can eliminate the second term because it is inconsistent with the fact that V_x tends to zero (and not infinity) as x becomes large. If a potential V_0 is applied at one end of the process, that is $V_x = V_0$ at $x = 0$, the particular solution becomes

$$V_x = V_0 \exp\left(\frac{-x}{\sqrt{R_m/R_i}}\right) \tag{12.5}$$

We therefore find, as anticipated, that the voltage signal falls away exponentially, V_x decreasing to $1/e$ of its value over the length constant $\sqrt{(R_m/R_i)}$. This indicates that the spatial propagation of the applied signal will be favoured by a low R_i and a high R_m.

A word of warning will not be out of place here, regarding the dimensions of the axial and membrane resistances. The exponent in Equation (12.5) has to be dimensionless, of course, and this indicates that $\sqrt{(R_m/R_i)}$ must have the dimension of length. In other words, R_m and R_i cannot have the same dimensions, as we might otherwise have assumed. We must thus take a closer look at these parameters. Now it is always the case that the resistance offered by a conductor of length L in the direction of the applied potential gradient, and with cross-sectional area A is

$$R = \rho L/A \tag{12.6}$$

where ρ is the relevant resistivity. Applying this to the axial component, we have

$$R_i = \rho_i L_1/\pi r^2 \tag{12.7}$$

where r is the radius of the process, and where L_1 denotes unit length, because we divided the axial direction up into segments having that size. For the membrane component, we find that

$$R_m = \rho_m L_m/2\pi r L_1 \tag{12.8}$$

L_m being the membrane thickness. The dimensions of the right-hand sides of Equations (12.7) and (12.8) are the same, as indeed they ought to be. Dividing (12.8) by (12.7), we find that

$$R_m/R_i = r\rho_m L_m/2\rho_i L_1^2 \tag{12.9}$$

from which it is clear that L_m need not be specified, as long as we define ρ_m in units of resistance times *area* and ρ_i in units of resistance times *length*.

As noted earlier, the denominator in Equation (12.5) is referred to as the length constant, λ, and adopting the above convention for the resistivities, we may write

$$\lambda = \sqrt{R_m/R_i} = \sqrt{r\rho_m/2\rho_i} \tag{12.10}$$

this indicating that the length constant increases only as the square root of the radius of the process. A bare crustacean axon with a 30 μm diameter typically has resistivities of $\rho_m = 5000$ ohm-cm^2 and $\rho_i = 50$ ohm-cm, which gives a length constant of 2.7 mm. Such a rapid dying away with distance would put long processes at a considerable disadvantage. The longest axons in the human body, for example, are about 1 m in length. In order to obtain a length constant of around this magnitude, assuming that these same resistivities apply, we would have to have an axon diameter in the region of 5 m! This is clearly not a practical proposition. It is important to note, however, that the transmission of nerve impulses would simply not be possible if there was *no* passive cable response, that is to say if λ was zero, because in that case the voltage change associated with a nerve signal would not be able to supplant itself from one part of the process to another.

Moreover, for signalling in dendrites, the passive cable response is sufficiently effective. A typical dendritic arborization in the brain extends over a distance of a few hundred micrometres, and it is worth bearing in mind that the cerebral cortex itself is a sheet of neuronal tissue that is only 3 mm thick. So a length constant of 2.7 mm is quite adequate for handling the signals that pass along the dendrites and impinge upon the soma. Indeed, passive cable response underlies what could be called the arithmetic processing that takes place in that region of every neuron.

We must now move on to consider what happens when things are changing as a function of time. In such a case, and despite the fact that the response is still a passive one, we can no longer ignore the influence of the membrane capacitance, so the appropriate circuit diagram is that shown in Figure 12.1 (naturally with R_o put at zero). Equation (12.1) remains valid, except that we must now replace I_i and V_x by $I_{i,t}$ and $V_{x,t}$ respectively. Equation (12.2), on the other hand, acquires an extra term on the right-hand side, for we must add $C_m(dV_{x,t}/dt)$ to $I_{m,t}$. The upshot of this is to produce the time-dependent counterpart of Equation (12.3), namely

$$\frac{R_m}{R_i}\frac{d^2 V_{x,t}}{dx^2} = V_{x,t} + R_m C_m \frac{dV_{x,t}}{dt} \qquad (12.11)$$

the general solution of which is (see Appendix E)

$$V_{x,t} = \frac{V_0}{2}\left\{ e^{-x/\lambda}\mathrm{erfc}\left(\frac{x/\lambda}{2\sqrt{t/\tau}} - \sqrt{t/\tau}\right) + e^{x/\lambda}\mathrm{erfc}\left(\frac{x/\lambda}{2\sqrt{t/\tau}} + \sqrt{t/\tau}\right)\right\}$$

where

$$\lambda = \sqrt{R_\mathrm{m}/R_\mathrm{i}}$$
$$\tau = R_\mathrm{m} C_\mathrm{m} \tag{12.12a}$$

The symbol *erfc* denotes the complementary error function.

Again we find that the system displays exponential variation, but now with respect to both distance and time. Indeed, the solution is essentially that given by Equation (12.5), but with the additional time-dependent terms (see Figures 12.3(a) and 12.3(b), which display the reaction of the system to a *sudden* voltage change V_0, at $t = 0$ and $x = 0$. The first term within the brace of Equation (12.12a) is appropriate for a gradual (and asymptotic) decrease in voltage, while the second describes an analogous increase. The important word here is *gradual*, because we see that the addition of that capacitative term makes the system's response sluggish, the accommodation of the voltage to any change in circumstances requiring a certain amount of time. Also, analogously to the case for the spatial counterpart, we have now acquired another constant, namely the time constant, τ.

Alan Hodgkin and William Rushton have derived the counterpart of Equation (12.12a) for the case of sudden current injection rather than sudden voltage change. If the current injected is $2I_0$ (at $t = 0$ and $x = 0$), so that I_0 flows in either direction, parallel with the axis of the dendrite or the axon, the resulting time- and distance-dependent response is described by

$$V_{x,t} = \frac{\lambda R_\mathrm{i} I_0}{2} \left\{ e^{-x/\lambda} \, \mathrm{erfc}\left(\frac{x/\lambda}{2\sqrt{t/\tau}} - \sqrt{t/\tau} \right) - e^{x/\lambda} \mathrm{erfc}\left(\frac{x/\lambda}{2\sqrt{t/\tau}} + \sqrt{t/\tau} \right) \right\} \tag{12.12b}$$

and one should note the change of sign between the two terms inside the brace, compared with Equation (12.12a). Figure 12.3(c) shows the resulting temporal variation of the voltage at different distances from the point of current injection. Finally, examples of the temporal response of the system, to current impulses of varying strength and polarity, but of a standard duration, are shown qualitatively in Figure 12.3(d).

The capacitance of the above-mentioned crustacian nerve membrane is about $1\,\mu\mathrm{F}\,\mathrm{cm}^{-2}$, and multiplying this by the corresponding value of membrane resistance, quoted earlier, we see that Equation (12.12a) produces a time

---

Figure 12.3   (a) The solution of Equation (12.12a) as a function of the normalized distance for the various times after imposition of the step voltage, V_0; the numbers for the various curves are in units of the membrane time constant $\tau = R_\mathrm{m} C_\mathrm{m}$

(b) The solution of Equation (12.12a) as a function of the normalized time after imposition of the step V_0 for various distances from the point of imposition; the numbers for the various curves are in units of the membrane distance constant $\lambda = \sqrt{(R_\mathrm{m}/R_\mathrm{i})}$

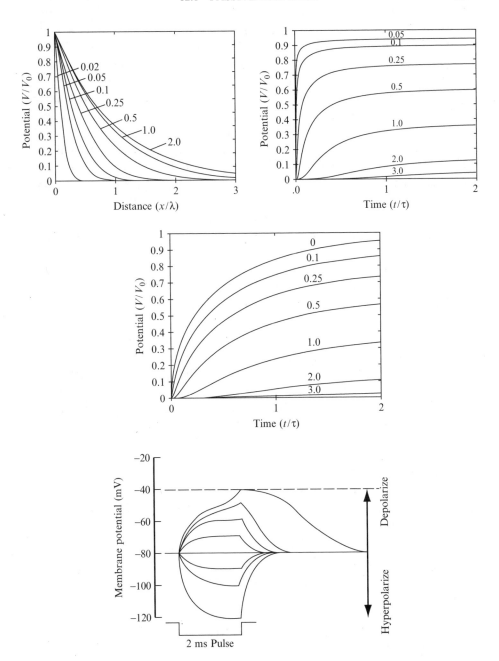

(c) The solution of Equation (12.12b) as a function of the normalized time after injection of a current pulse I_0 for various distances from the point of imposition; the numbers for the various curves are in units of the membrane distance constant $\lambda = \sqrt{(R_m/R_i)}$

(d) The temporal variation of the membrane potential following 2 ms injections of current of various strengths for the cases of both depolarization and hyperpolarization

constant of about 5 ms. Any attempt to change the voltage over a time period much shorter than this will be thwarted by what could loosely be called the system's electrical inertia. It should be noted that in contrast to what was the case with the membrane's length constant, the time constant does not vary with changing diameter of the process. This is because the linear *increase* of capacity with increasing membrane area is precisely counterbalanced by the linear *decrease* of resistance with increasing membrane area.

Together, the length and time constants determine the speed with which an electrical signal can be *passively* propagated along the nerve membrane. That speed is given by the quotient λ/τ, which has the correct dimension of length divided by time. And from what was discussed in the preceding paragraph, we see that its variation with the diameter of the process will be the sole responsibility of the length-constant contribution. The speed will thus increase as the square root of the diameter. If there is sufficient space available, it is clearly advantageous for a process to be wide rather than narrow. Inserting the values quoted above for a crustacian nerve fibre, we find a speed of about $0.5\,\mathrm{m\,s^{-1}}$. During the 19th Century, Hermann von Helmholtz measured the speed of conduction of nerve impulses, and found a value that exceeded this passive speed by a factor of around 50. The difference between these speeds is due to what we must now focus on, namely the additional effects that arise because of the passage of ions across the nerve membrane. The appearance of such effects is conditional upon the presence, in the membrane, of ion channels and, as mentioned earlier, it is a reasonable approximation to consider these channels as being present only in the axonal membrane.

12.2 Nerve Impulses (Action Potentials)

Figure 12.4 shows what happens if the depolarization exceeds a certain threshold value. We see that there is a gradual and systematic departure from the asymptotically exponential behaviour observed in Figure 12.3(d), the reaction ultimately being so pronounced that it actually reverses the membrane's polarity. (Indeed, we see that the departure from exponential behaviour is observable even *before* the stimulus reaches the threshold level.) This super-exponential response is the all-or-nothing nerve impulse, also known as the action potential. It has a standard amplitude, irrespective of the actual degree of above-threshold depolarization. (However, as we will see later, the *rate* of emission of impulses does increase as the degree of above-threshold depolarization increases.) The threshold usually lies at around $-50\,\mathrm{mV}$. Because the resting potential is typically about $-100\,\mathrm{mV}$ (inside the membrane, with respect to the outside), this means that a *depolarization* of around 50 mV is required for action-potential generation. If the depolarization does not reach this critical

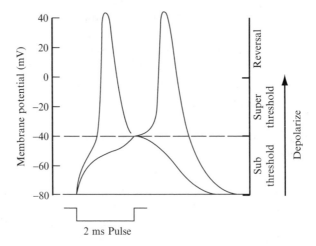

Figure 12.4   The temporal variation of the membrane potential following 2 ms injections of current of various strengths for depolarizations in excess of the threshold value

level, the axon's subsequent electrical behaviour is merely that dictated by the passive cable properties discussed above.

Let us follow the analysis published by Alan Hodgkin and Andrew Huxley in 1952, which provided the first quantitative explanation of action-potential generation and propagation. In the preceding chapter, particularly in connection with Figure 11.3, we learned that the concentration imbalance of the various ionic species, between the axoplasm and external fluid, can be regarded as stemming from equivalent batteries, one for each type of ion. We also saw that the need for active pumping, by the Na^+, K^+-ATPase molecules, arises because of the leakage of the various ions, in the direction of their respective concentration gradients. In Figures 12.1 and 12.2, as well as in Figure 11.3, the corresponding leakage paths were likened to electrical resistances, R. In what follows, it will be more convenient to think in terms of conductances, g, their relationship to the resistances being simply $g = 1/R$, as indicated in Figure 11.3.

Hodgkin and Huxley exploited a technique known as the voltage clamp, which permitted them to vary the membrane voltage in a controlled manner. We need not go into the technical details here, so suffice it to say that the control is exercised by a feedback amplifier, connected to two electrodes, one in the axoplasm and one outside the axon. This amplifier automatically supplies just the right amount of current required to maintain the membrane potential at the desired level. They were thereby able to prevent the runaway fluctuation of potential seen in Figure 12.4, and could thus determine the various ionic conductances under a variety of conditions. As a result, they were able to demonstrate that g_{Na} and g_K vary with both time and membrane potential, though *not* with membrane current, during the impulse, whereas g_{Cl} remains constant, as do V_{Na}, V_K, V_{Cl}, R_i and C_m. (In practice, Hodgkin and Huxley included in the Cl parameters the effects of various other ionic species, which also make a small contribution to the leakage current.)

It was found that the influence of membrane potential on the conductances could be summarized as follows: first, depolarization causes a transient increase in g_{Na}, and a slower but maintained increase in g_K; secondly, these changes are graded, and they can be reversed by repolarizing the membrane. To this, we may add that the above-mentioned runaway (i.e. positive feedback) behaviour arises because a depolarization-induced increase in sodium conductance increases the inward flow of sodium ions (in excess of the leakage flow, that is), which causes a further depolarization, this leading to a further increase in sodium conductance, and so on. In other words, the sodium conductance increases in a manner which could loosely be described as being explosive. We will later consider speculations as to the underlying molecular physics of this behaviour, but for the time being let us note that these changes are opposite to what would be expected on the basis of Ohm's law; a depolarization diminishes the voltage across the membrane, and yet the ion-mediated current through the membrane *increases*.

As Hodgkin and Huxley noted, the total membrane current comprises both capacitative and ionic components, so

$$I_{total} = C_m \frac{\partial V'_{x,t}}{\partial t} + I_{ionic} \qquad (12.13)$$

the sign of I_{total} being such that an *inward* current is positive. $V'_{x,t}$ has been given its prime to denote the fact that the membrane voltage is now being defined with respect to the resting potential, a *depolarization* therefore being regarded as positive. The total ionic current is naturally given by

$$I_{ionic} = I_{Na} + I_K + I_{Cl} \qquad (12.14)$$

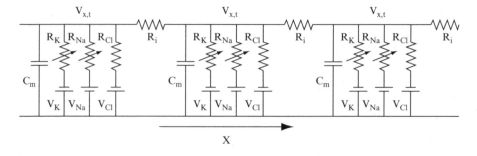

Figure 12.5 The equivalent circuit for a nerve cell process when the extracellular resistance can be ignored, but when the voltage dependences of the various ionic conductivities must be allowed for

while its individual components will have magnitudes determined by the following relationships

$$I_{Na} = g_{Na}\left(V'_{x,t} - V'_{Na}\right)$$

$$I_{K} = g_{K}\left(V'_{x,t} - V'_{K}\right) \qquad (12.15)$$

$$I_{Cl} = g_{Cl}\left(V'_{x,t} - V'_{Cl}\right)$$

Here too, the prime on the second term within the parentheses indicates, in each case, that the equivalent battery voltage is being defined with respect to the resting potential (i.e. $V'_{Na} = V'_{Na} - V'_{rest}$, and so on).

It was at this point in the analysis that Hodgkin and Huxley introduced the data on conductance obtained from their voltage clamp experiments. Examples from their comprehensive compilations are shown in Figures (12.6(a) and 12.6(b), for sodium and potassium respectively. From these data, it was possible

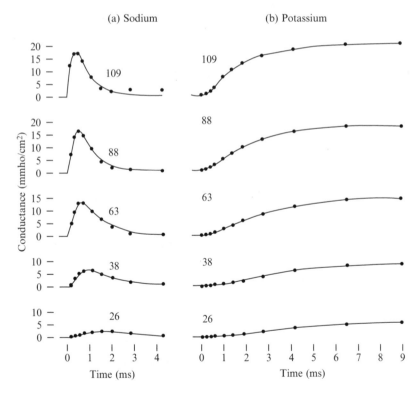

Figure 12.6   The increase of sodium and potassium conductances associated with five different degrees of depolarization (indicated by the numbers – in millivolts – indicated on each curve); the data were obtained by Hodgkin and Huxley using the voltage clamp technique

to demonstrate that the time and voltage dependencies of the sodium conductance can be described by the following empirical relationships.

$$g_{Na} = m^3 h g_{Na, \, max}$$

$$\frac{dm}{dt} = \alpha_m(1 - m) - \beta_m m \qquad (12.16)$$

$$\frac{dh}{dt} = \alpha_h(1 - h) - \beta_h h$$

in which the parameters α and β are functions of voltage, but not of
time. $g_{Na, \, max}$ is the maximum value attained by g_{Na}, during the time-course
of its variation. The voltage-clamp data provided the following expressions for
the α and β parameters.

$$\alpha_m = 0.1(V' + 25) \Big/ \left(\exp\left(\frac{V' + 25}{10}\right) - 1 \right)$$

$$\beta_m = 4 \exp(V'/18)$$

$$\alpha_h = 0.07 \exp(V'/20) \qquad (12.17)$$

$$\beta_h = 1 \Big/ \left(\exp\left(\frac{V' + 30}{10}\right) + 1 \right)$$

the values of V' being given in mV, while the units for α and β are ms^{-1}. For the
potassium conductance, similarly, it was found that

$$g_K = n^4 g_{K, \, max}$$

$$\frac{dn}{dt} = \alpha_n(1 - n) - \beta_n n$$

$$\alpha_n = 0.01(V' + 10) \Big/ \left(\exp\left(\frac{V' + 10}{10}\right) - 1 \right) \qquad (12.18)$$

$$\beta_n = 0.125 \exp\left(\frac{V'}{80}\right)$$

During the propagation of an action potential, and assuming that the external resistance R_o can indeed be ignored, the net membrane current will be given
by

$$I_m = \frac{1}{R_i} \frac{\partial^2 V_{x,t}}{\partial x^2} \qquad (12.19)$$

this following directly from Equation (12.3). Alternatively, we may write

$$i_m = \frac{r}{2\rho_i} \frac{\partial^2 V_{x,t}}{\partial x^2} \tag{12.20}$$

where i_m is the membrane current density, r is the axon radius and ρ_i is the specific resistance of the axoplasm. Combining Equations (12.13), (12.14), (12.15) and (12.20), we obtain

$$\frac{r}{2\rho_i} \frac{\partial^2 V_{x,t}}{\partial x^2} = C_m \frac{\partial V_{x,t}}{\partial t} + g_{Na,max} m^3 h\left(V' - V'_{Na}\right)$$

$$+ g_{K,max} n^4 \left(V' - V'_K\right) + g_{Cl}\left(V' - V'_{Cl}\right) \tag{12.21}$$

This equation contains derivatives with respect to both distance and time, and it is thus not easy to solve. However, we can make use of the experimentally-observed fact that the shape of the action potential remains unchanged as it propagates. In other words, the variation of $V_{x,t}$ with time, at a given value of x, is identical to the variation of $V_{x,t}$ with position, at a given time. This automatically implies that

$$\frac{\partial^2 V_{x,t}}{\partial x^2} = \frac{1}{v^2} \frac{\partial^2 V_{x,t}}{\partial t^2} \tag{12.22}$$

v being the velocity of conduction of an action potential.

As a final step, Hodgkin and Huxley inserted (12.22) in (12.21), to obtain the ordinary differential equation

$$\frac{r}{2\rho_i v^2} \frac{d^2 V_{x,t}}{dt^2} = C_m \frac{dV_{x,t}}{dt} + g_{Na,max} m^3 h(V' - V'_{Na})$$

$$+ g_{K,max} n^4 (V' - V'_K) + g_{Cl}(V' - V'_{Cl}) \tag{12.23}$$

which can be solved numerically, if v is known. When Hodgkin and Huxley carried out their pioneering investigation, a value of v could *not* be assumed, of course, because one of the goals of their analysis was to account for the experimental result. Instead, therefore, they had to insert various values of v into Equation (12.23), and then solve the equation to see if it produced the correct variation of $V_{x,t}$ with time. This is not a particularly difficult task for a modern computer, but such machines were not available in 1952, when the original study was made, and the trial-and-error process was a laborious one. Hodgkin and Huxley found that $V_{x,t}$ tended toward $+\infty$ or $-\infty$, depending on whether the assumed value of v was too small or too large. The correct value of v, on the other hand, brings $V_{x,t}$ precisely back to zero (that is to say, the resting condition) when the action potential has run its course.

Hodgkin and Huxley found that this situation was obtained for an assumed conduction velocity of $18.8\,\mathrm{ms}^{-1}$, this comparing favourably with the experimentally-observed velocity of $21.2\,\mathrm{ms}^{-1}$. (The values of the various other experimental parameters were as shown in Table 12.1.) These pioneers also noted that a conveniently simple expression for the conduction velocity can be extracted from the above equations, namely

$$v = \sqrt{\xi r / 2\rho_i C_m} \qquad (12.24)$$

the value of the constant ξ depending only on the physical properties of the membrane. Here again, we note that the conduction velocity increases as the square root of the axon radius, just as was the case for the passive cable response. For the squid axon used by Hodgkin and Huxley, it was found that $\xi = 10.47\,\mathrm{ms}^{-1}$.

When Hodgkin and Huxley carried out this important study, knowledge about membrane structure was still only rudimentary. It was known that the membrane comprised both lipid and protein, but the mutual arrangement of these two components had not been determined. In fact, the widely-held assumption was that the membrane consisted of an inner lipid layer with a thin film of protein on either side (see Chapter 8). It was thus quite bold of Hodgkin and Huxley to postulate that there were nevertheless pores spanning the width of the membrane, through which the sodium and potassium ions passed during the action potential. Subsequent electrochemical investigations, particularly by Hodgkin and Richard Keynes, gradually established the essential correctness of this picture.

In fact, the experimental data permit one to make a rough estimate of the number of sodium and potassium ions which pass through each square centimetre of the membrane during a single action potential. We know that the change in membrane voltage during the sodium influx is about 0.1 V, and multiplying this by the membrane capacity per square centimetre, namely $1\,\mu\mathrm{F}$, gives the total electrical charge transferred per square centimetre (i.e. we invoke the well-known relationship $Q = VC$). This produces the result 10^{-7} Coulombs cm^{-2}. Now the charge on a single sodium ion is 1.6×10^{-19} Coulombs, so the number of ions passing through each square centimetre of the membrane, during a single action potential, must be 0.6×10^{12}. This is equivalent to the transfer of one ion at each of favoured sites spaced about 1.2×10^{-6} cm apart. The number of potassium ions transported in the other direction, per square centimetre per action potential, will naturally be the same, because the two types of ion bear the same charge, and because the magnitude

Table 12.1 The cable constants for various nerve and muscle fibres (at around 20°C). The lower part of the diagram shows the electrolyte concentrations (in mM/l) and the resting potentials across the cell membranes in the cases of the frog muscle and the squid giant axon

Fibre	Fibre diameter, μm	Length constant (in large outside vol.) (mm)	Time constant (msec)	Membrane resistivity ρ_m, (ohm-cm^2)	Membrane capacity C_m, $\mu f/cm^2$	Resistivity of cell interior ρ_i, ohm-cm	Resistivity of medium ρ_o, ohm-cm
Squid nerve	500	5	0.7	700	1	30	22
Lobster nerve	75	2.5	2	2000	1	60	22
Crab nerve	30	2.5	5	5000	1	60	22
Frog muscle	75	2	24	4000	6	200	87

Frog muscle

External		Internal	
Na$^+$	120	Na$^+$	9.2
K$^+$	2.5	K$^+$	140
Cl$^-$	120	Cl$^-$	(3–4)
			−90mV

Squid axon

External		Internal	
Na$^+$	460	Na$^+$	50
K$^+$	10	K$^+$	400
Cl$^-$	540	Cl$^-$	40–100
			−60mV

of the change in membrane voltage is also the same (though oppositely-directed).

As biophysicists, we would naturally like to know whether these numbers can be accounted for on the basis of the density of channels in the membrane, the density of ions in the surrounding aqueous medium, and finally the mobility of those ions. It has been demonstrated experimentally that the density of sodium channels in the axon membrane is 1.2×10^{12} cm^{-2}, while for the potassium channels the figure is 1.8×10^{12} cm^{-2}. These numbers immediately reveal that not every channel needs to be active during a given action potential, because there are actually more channels than there are ions moving across the membrane.

The given densities are equivalent to a mean distance between adjacent channels of about 0.6×10^{-6} cm for either type. Now from Table 12.1, we recall that the concentration of sodium ions in the extra-cellular fluid around the squid axon is 460 mmol l^{-1} while the number for the intracellular potassium is 400 mmol l^{-1}. These numbers, in turn, indicate a mean distance between neighbouring ions of about 1.6 nm for the extra-cellular sodium and about 1.1 nm for intracellular potassium. Table 12.2 gives the mean neighbouring distances for all the ions, both inside and outside the axonal membrane. So we see that the ions in solution are more closely spaced than are the channels in the membrane.

It remains for us to convince ourselves that an ion will, under the prevailing circumstances, be able to reach the entrance to the nearest channel in the time available, this being an acceptably small fraction of the total duration of an action potential, which is roughly 1 ms. Of course, some ions will happen to be located in just the right place when needed, but let us consider a worst-case scenario in which, using the above numbers, the channel-ion distance is half of that 1.6 nm, namely 0.8 nm. We have to demonstrate that this distance could be covered in the time available, given the prevailing electric field and ionic mobility. The relevant mobility data are listed in Table 12.3, and we see that the number for sodium is $5.2 \, \mu$m s^{-1} per V cm^{-1}. Calculation of the field strength is straightforward, because we know that there is a voltage difference of about 0.1 V across the roughly 5 nm thick membrane; the field is thus 2×10^{5} V cm^{-1}. An ion could therefore move at about 1 ms^{-1}, under the

Table 12.2 The mean neighbouring distances (in units of nm) for the various ions inside and outside a squid nerve cell membrane

Inside	Ion	Outside
3.2	Na$^+$	1.6
1.7	K$^+$	5.5
2.9	Cl$^-$	1.4

Table 12.3 The mobilities and other characteristics of ions found in association with nerve membranes

Ion	Ionic crystal radius, Å	Limiting conductance in H_2O	Absolute mobility in H_2O, $(\mu m/sec)/$ $(volt/cm)$	Calculated hydration numbers
Na^+	0.95	50.11	5.2	4.5
K^+	1.33	73.52	7.64	2.9
Cl^-	1.81	76.34	7.91	2.9

prevailing electric field, which is equivalent to a millimetre during the entire action potential. We see, in other words, that there is certainly adequate time for the required ionic movements.

In fact, even simple diffusion, in the absence of an electric field, would be sufficient to transport an ion over the required distance. We recall from Chapter 5 the relationship between the mean square distance diffused, the time available, and the diffusion coefficient, namely $r_{RMS} = \sqrt{(Dt)}$. Using the typical value $D = 10^{-5}\,cm^2\,s^{-1}$ and taking $t = 10^{-3}\,s$ for the duration of the action potential, we arrive at the value $r_{RMS} = 10^{-4}\,cm$. This is well in excess of the spacing between the favoured sites discussed above. We can conclude that evolution has led to the typical axon being adequately provided with channels.

The Hodgkin–Huxley model was also bold in its assumption of membrane-spanning pores, because the interior of the lipid bilayer is hydrophobic whereas the sodium and potassium ions clearly have a great affinity for water. When it became clear that there are proteins which stretch from one side of the membrane to the other, it was natural to conclude that some of these might have tube-like holes through which ions could pass. The problem had then, in effect, been turned around, because it was now necessary to explain why such a hole would not always be open, allowing the ionic concentration gradient to gradually collapse. This gave rise to models in which the channel possessed one or more charged regions, at certain positions along its length, the idea being that the ions would be held up at such points because of Coulombic attraction.

There was also the need to explain the voltage sensitivity of the channels, this permitting the holes to be open only if the threshold depolarization had been achieved. Here the theoreticians were able to draw upon the admirable efforts of those who had undertaken the difficult task of determining the structures of membrane-bound proteins, because this work revealed that a common feature of such molecules is alpha helices lying at right angles to the plane of the membrane, and spanning its 5 nm width. We recall from Chapter 7 that the alpha helix owes its stability to the numerous hydrogen bonds, the axes of which are roughly parallel to that of the helix itself. Now each of these bonds is equivalent to an electric dipole, and because the bonds all point in the same direction, the overall result will be what could be called a giant dipole, lying roughly at right angles to the plane of the membrane. Such a dipole will experience a torque, just as does a magnetic dipole when it is under the influence of a magnetic field. A possible channel-opening mechanism might thus be as follows. The torque acting on the alpha helices when the voltage across the membrane is at its resting value is sufficient to keep the holes closed, by causing each helix to be pressed up against its neighbour, but when the membrane is depolarized to a sufficient degree, the diminished torque permits the holes to open, and the ions are then able to make the journey from one side of the membrane to the other.

A further question remains, however: do the ions pass through the holes as isolated entities, or are they accompanied by some of the water molecules? The point is that the surface charge density on one of these ions is so high that a number of water molecules will become bound to it. In fact, because sodium is a smaller ion than potassium, while both have the same charge, the surface charge density will be higher in the case of sodium, so it will attract the larger retinue of water molecules (or hydration shell as it is more correctly called), namely 4.5 on average, compared with potassium's 2.9 on average.

The publication of the three-dimensional atomic structure of a potassium channel, by Roderick MacKinnon and his colleagues in the spring of 1998, was clearly a major milestone in this branch of science, and it supplied answers to these questions. The various parts of Figure 12.7 show the structure from different viewing angles, and at various degrees of resolution. The position of the potassium ion as it passes through the pore is indicated schematically by the large central circle. The lower view indicates the two co-operating mechanisms by which the channel stabilizes an ion in the middle of the membrane: a large aqueous cavity 'protects' the ion from the hydrophobic interior of the membrane, and the alpha helices, slanted with respect to the plane of the membrane, orient their (negative) carboxyl termini towards that cavity. From the dimensions of the mole, it seems that some of the water molecules would indeed accompany each ion, as it passes through the constriction.

It is common for the axons of nerve cells to be surrounded by what is known as a myelin sheath, the myelin being provided by nearby Schwann cells (named after Theodor Schwann). These have the ability to extend their membranes so as to wind these around the axon, spiral fashion, thereby providing the axon with an insulating layer. The length of the axon thus insulated is, of course, determined by the size of the Schwann cell, and it is usually of the order of a few

Figure 12.7   Plan (above) and side (below) views of the potassium channel of *Streptomyces lividans* published by Roderick MacKinnon and colleagues in 1998

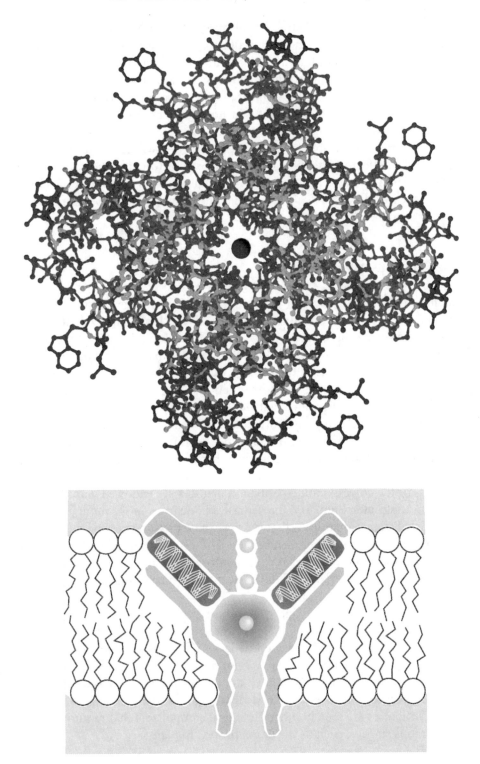

millimetres. The individual stretches of myelination do not precisely butt up against each other, leaving instead small gaps of the order of 1 μm. Such gaps are referred to as nodes of Ranvier. This arrangement of interrupted stretches of insulation is very important to the functioning of the nervous system. In effect, the insulation increases the local membrane resistance by a very large factor, and it similarly makes the constant ξ in Equation (12.24) very large. This, in turn, greatly increases the velocity of action potential conduction. When the pulse reaches the next node of Ranvier, however, it will revert to its normal velocity. The effect of the myelination is therefore to cause the propagating action potential to move forward in a series of jumps, and the overall velocity is thereby greatly increased. This common mode of propagation is referred to as saltatory conduction.

12.3   The Nervous System

The job of the nervous system is, of course, to mediate the passing of nerve signals from one part of the body to another. The task of an individual neuron is to receive signals via its dendrites and, if the resulting depolarization at the axon hillock is sufficient to exceed the threshold, to pass signals out along its axon, and axon collaterals. If the position of the neuron, in the overall scheme, is such that it receives signals from other neurons, those signals will be chemical in nature, and this will require that the neuron's dendrites be equipped with protein molecules known as chemoreceptors. The same will be true if the neuron lies at the input surface of either the gustatory (taste) or olfactory (smell) organs, because those two senses also function in a chemical manner. The tactile (touch) and auditory (hearing) senses are essentially mechanical, and in their cases the dendrites of the corresponding peripheral neurons are endowed with mechanoreceptor molecules. Similarly, the dendrites of neurons at the input periphery of the visual (sight) apparatus, that is to say the retinas, possess photoreceptor molecules in their bounding membranes. The list of receptor molecules also includes the thermoreceptors that pass on information about the body's surface temperature, at a given location, and nociceptors, which mediate the experience of pain.

By way of examining a receptor neuron in more detail, let us focus on the light-sensitive rod cell, of which there are about a million in each human retina. This cell is elongated in the direction lying at right angles to the retinal plane, and its dimensions are approximately 1 μm by 40 μm (see Figure 12.8(a)). The light-sensitive end contains a series of plate-like membranes, stacked one above the other, such that their planes lie at right angles to the long axis of the cell. Similar structures are seen in other biological components that require a large membranous surface, such as the grana present in plants.

These membranes contain the photo-active molecules of rhodopsin (see Figure 12.8(b)), which consist of the protein opsin and an attached prosthetic group known as 11-*cis*-retinal (see Figure 12.8(c)) which functions as a chromophore. The body itself cannot fabricate this latter substance, so we ingest it in the related form all-*trans*-retinol (see Figure 12.8(d)). This important substance is better known under the name vitamin A, and if we consume insufficient amounts of it, we are likely to suffer from night blindness. Opsin consists of 348 amino-acid residues, arranged as seven membrane-spanning helices, designated A-G. The 11-*cis*-retinal is bound to the opsin molecule via a lysine side-chain on helix G, this bond being known as a Schiff base (see Figure 12.8(e)). This Schiff base is stabilized by the carboxylate of a glutamate residue on helix C, this counterion profoundly influencing the properties of the rhodopsin complex.

The membrane itself is unusually rich in polyunsaturated fatty acid, particularly docosahexaenoic acid, the 22-carbon chain of which has no less than six double bonds. (That fatty acid is also prominent in cerebral grey matter, which suggests that it plays a vital role in excitable tissue in general.) When a light photon impinges upon 11-*cis*-retinal, the absorbed energy is utilized to modify that molecule's configuration, thereby producing all-*trans*-retinal (see Figure 12.8(f)). This involves a rotation of 180° at the eleventh carbon–carbon bond (counting from the closed-ring end of the molecule).

We considered the energy barrier opposing such rotation in Chapter 4, but this present case involves a carbon–carbon *double* bond. An experimental investigation has revealed that the rotational potential energy function, $\mathscr{E}(\theta)$, has the following form:

$$\mathscr{E}(\theta) = 0.15 \sin^2 \theta \qquad (12.25)$$

the units of energy being aJ. The energy available from a single light photon is, of course, $h\nu$, where ν is the frequency of the incident electromagnetic radiation. For visible light, we have frequencies of roughly $10^{15}\,\mathrm{s}^{-1}$, and with Planck's constant h being 6.63×10^{-34} Js, this implies photon energies of about 0.6 aJ. The maximum value that $\sin^2 \theta$ can attain is unity, so we see that there is indeed sufficient energy in a single photon to provoke the change from 11-*cis*-retinal to all-*trans*-retinal.

This isomerization, as it is called, leads within a few picoseconds to the production of the bathorhodopsin form of the chromophore, the protein thereby being put into a relatively high energy configuration. There then follows a series of further (thermally-activated) conformational changes, which take place at ever-decreasing rates. Thus bathorhodopsin is transformed into lumi-rhodopsin within a few nanoseconds, and this transforms to metarhodopsin I within a few microseconds. There then follows a transformation to metarhodopsin II within a few milliseconds. The final goal is reached when the latter,

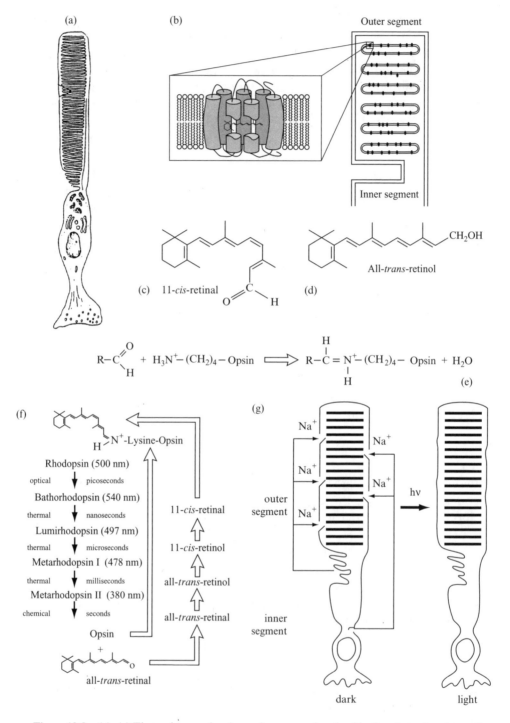

Figure 12.8 (a)–(g) The various molecules and processes involved in the photochemistry of vision

within a few seconds, is hydrolysed (that is to say broken apart by the action of water), so as to produce what is collectively known as bleached rhodopsin – separated molecules of opsin and all-*trans*-retinal. Although this chain of events is still the object of much research, the current opinion is that the key triggering event in the visual process is the transformation between the two forms of metarhodopsin.

The final act of this fascinating scenario is rather surprising, given what was discussed earlier in this chapter in connection with the action potential. The overall functioning of the photoreceptor cell involves the *closing* of ion channels! In the absence of incident light, the activity of Na^+, K^+ATPase molecules located at the other end of the cell, remote from the above-discussed membranous discs, actively extrudes Na^+ ions, which diffuse toward the cell's photoactive end and re-enter through sodium channels. When light falls on that photoactive terminus, the ensuing chain of molecular events, discussed above, leads to the closing of those sodium channels, thus permitting the membrane voltage to hyperpolarize (see Figure 12.8(g)). This could not lead to the generation of an action potential, of course, and in any event receptor neurons do not emit such nerve impulses. Instead, the passive cable response provokes a voltage change at the junction between the receptor cell and the next cell in the visual network. (As described in Chapter 14, that next cell is a bipolar cell.) Many of the details of these mechanisms were elucidated through the individual efforts of Ragnar Granit, Haldan Hartline and George Wald.

As is indicated in Figure 12.9, which is a highly schematic depiction of a neuron, the output from the extremity of a neuron's axon is always chemical. In this figure, the entire dendritic arborization has been collapsed down to a single token dendrite. For neurons that lie on the peripheral surfaces of the sensory organs, this is actually a good representation of reality, whereas neurons lying deeper in the nervous system often have hundreds of dendritic branches. Similarly, the somas of peripheral sensory neurons often *do* lie to one side, as

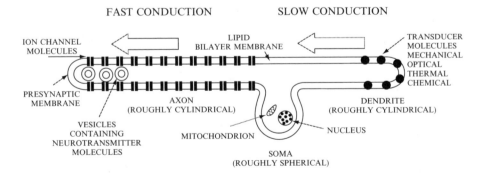

Figure 12.9   A highly schematic representation of a neuron

shown in Figure 12.9, rather than *between* the dendrites and the axon, as shown in Figure 11.1.

The chemical output from the axon's extremity is in the form of molecules known as neurotransmitters. The destination of these emitted molecules usually lies very close at hand, namely the dendritic surface of a nearby neuron. Indeed, the separation distance is usually only about 20 nm. Such a close juxtaposition of axon terminus and dendrite is referred to as a synapse, and the region that must be traversed by the diffusing neurotransmitter molecules is known as the synaptic cleft. At such an arrangement, the extremity of the axon terminal is referred to as the pre-synaptic membrane, while that part of the dendrite that lies at the other side of the synaptic cleft is the post-synaptic membrane. As mentioned earlier, the post-synaptic membrane possesses chemoreceptors. The highly schematic arrangement shown in Figure 12.10 is composed of four copies of the neuron from Figure 12.9, one of these now being the receiving neuron while the remaining three copies are positioned so as to be able to dispatch neurotransmitters toward it. In other words, this little arrangement of four neurons involves three synapses.

There are two situations that are not covered by this simple picture. One involves the synaptic contacts that certain neurons make with muscles, at the so-called neuromuscular junction, rather than other neurons. Such innervation of a muscle fibre by a neuronal axon was illustrated in Figure 10.9. In the other situation, the synapse itself is replaced by an arrangement in which the terminal region of the axon directly injects molecules into the blood stream. This is a common feature of the endocrine system, by which the nervous system exercises internal control over certain involuntary bodily functions. We thus see that these two exceptional situations are in fact indispensable, because it is they alone which handle the nervous system's only two outputs, namely that which controls the glands and that which controls the muscles.

When a neurotransmitter molecule has diffused across the synaptic cleft, it can dock with the appropriate region of the surface of one of the chemoreceptor molecules in the post-synaptic membrane, and this leads to events which ultimately change that membrane's potential. The nervous system employs a variety

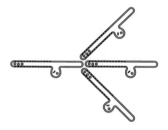

Figure 12.10 One can depict the synaptic interactions of several neurons with a single receiving neuron, with several duplicates of the schematic neuron shown in Figure 12.9

of different types of neurotransmitter, and the molecules of a given type can dock only with one specific type of chemoreceptor molecule. Depending upon the type of neurotransmitter the pre-synaptic neuron has dispatched, the change of potential of the post-synaptic membrane will either be a depolarization or a hyperpolarization. Neurotransmitters which cause a depolarization are said to be excitatory, while those that provoke a hyperpolarization are referred to as being inhibitory. It is important to bear in mind that a given type of neuron can manufacture and dispatch *only one type* of neurotransmitter, so neurons themselves are said to be either excitatory or inhibitory. However, the dendrites of a given neuron can, and usually do, have several different types of chemoreceptor in their membranes. This means, amongst other things, that a given neuron can receive excitation from some of its neighbours and inhibition from others.

To sum up, then, neurons come in two varieties: those that excite other neurons, via the mediating agency of their synaptic contacts with the latter, and those which exert an inhibiting influence. The neurons that feed signals to the endocrine system or to the muscles are *always* excitatory.

Let us briefly consider the activation of muscles in a little more detail. In this case, the changes of voltage of the post-synaptic membrane are referred to as end-plate potentials (usually abbreviated to e.p.p.'s). It is observed that even in the absence of a nerve impulse, there is intermittent spontaneous secretory activity from the pre-synaptic membranes, at an average rate of about one pulse/s for the entire set of axon branches innervating a single muscle fibre. Such an event corresponds to the discharge, into one of the synaptic clefts, of the contents of a single vesicle, and it gives rise to an elemental e.p.p. with an average amplitude of 0.4 mV (the distribution about that value being Gaussian). The simultaneous arrival of a nerve impulse, at the extremities of all the axon branches, greatly magnifies the secretory activity, such that several hundred of these elemental events occur within about a millisecond.

By analysing the experimentally observed e.p.p.'s provoked by the arrival of a nerve impulse at a *single* pre-synaptic membrane, Bernard Katz was able to demonstrate that vesicle discharges follow a Poisson distribution. The data for 198 observed e.p.p.'s showed well-defined peaks at 0.4, 0.8, 1.2 and 1.6 mV, so there was a strong indication that these are simple multiples of the spontaneous no-impulse e.p.p. If the mean number of discharged vesicles provoked by a single nerve impulse is m, the probability p_x of observing an e.p.p. comprising x elemental units is given by

$$p_x = \frac{m^x}{x!} e^{-m} \qquad\qquad (12.26)$$

For a sufficiently large number N of observations, such as the above-mentioned 198, the numerical value of Np_x should be in good agreement with the experimental data. Figure 12.11 shows that this was indeed the case. This work had therefore demonstrated that neurotransmitter release is quantized, and it greatly strengthened the view that a single quantum corresponds to the contents of a single pre-synaptic vesicle.

It would be appropriate, at this point, to calculate the time it takes for a neurotransmitter molecule to diffuse across the synaptic cleft. This a straight-forward and simple matter, because we can use Equation (5.1), and insert the width of the synaptic cleft as the distance r_{rms}, while taking the value $10^{-5}\,\mathrm{cm^2\,s^{-1}}$ as the diffusion coefficient, since we are dealing with diffusion through a liquid. This produces the result

$$t = r_{rms}^2 D^{-1} = \left(2 \times 10^{-6}\right)^2 \times 10^5 = 4 \times 10^{-7}\ \mathrm{s} \qquad (12.27)$$

and we see that the time taken for this diffusion process is small, even compared with the approximately $10^{-3}\,\mathrm{s}$ duration of an action potential.

Another little calculation will reveal the degree of co-operativity required of neurons in the brain's networks. With the possible exception of the Purkinje cells in the cerebellum (named in honour of Jan Purkinje), it never happens that a neuron is provoked into activity as the result of signals from just one other neuron. In any case, one would guess that the degree of co-operativity is far

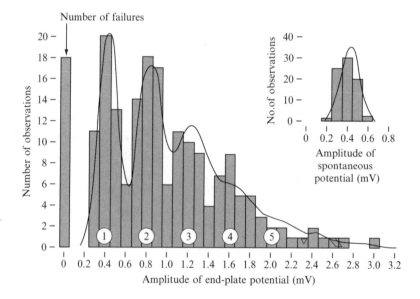

Figure 12.11 Histograms of end-plate potential and (inset) spontaneous potential ampli-tudes from a mammalian end plate; the latter is seen to be well fitted by a Gaussian distribution, and the former has peaks which occur at integer multiples of the maximum for the spontaneous case (i.e. 0.4 mV) (Based on an original by Bernard Katz)

greater, because each neuron makes synaptic contacts with between 10 000 and
100 000 other neurons.

An estimate of the actual degree of co-operativity can be obtained from
the voltage changes involved in the generation of action potentials (also
referred to as nerve impulses or spikes). Taking the resting potential as being
roughly $-100\,mV$, and the threshold potential as being roughly $-50\,mV$, we
see that a depolarization of about $50\,mV$ is required for action potential
generation.

Now a typical excitatory post-synaptic potential lies at around 0.5–$1.0\,mV$, so
if the effect of about 100 of these arrived at the soma at precisely the same time,
and if there was no attenuation during transit along the dendrites, between 100
and 50 other neurons could co-operate to 'fire' the receiving neuron. However,
there *is* attenuation in the dendrites, and the signals from the other neurons
arrive at *staggered times*. In practice, therefore, *several hundred other neurons
will have to co-operate, in order to 'fire' the receiving neuron.*

Exercises

12.1 It is generally the case that a characteristic velocity can be approximated
through the expedient of dividing a length constant by a time constant. Discuss
the extent to which this approach permits one to compare passive and active
voltage changes in a nerve-cell membrane.

12.2 Write a computer program (using the language of your choice) to simulate the
temporal variation of membrane potential described by the Hodgkin–Huxley
analysis, and investigate the changes caused by varying the relevant parameters.
Employ the transformation used by Hodgkin and Huxley in order to keep the
wave-form stationary on your screen.

Further reading

Aidley, D. J., (1989). *The Physiology of Excitable Cells.* Cambridge University Press,
Cambridge.

Carpenter, R. H. S., (1996). *Neurophysiology.* Arnold, London.

Deutsch, S. and Micheli-Tzanakou, E., (1987). *Neuroelectric Systems.* NYU Press, New
York.

Dowling, J. E., (1987). *The Retina: An Approachable Part of the Brain.* Harvard University
Press, Cambridge, MA.

Doyle, D. A. *et al.*, (1998). The structure of the potassium channel: molecular basis of K+
conduction and selectivity. *Science* **280**, 69–77.

Eccles, J. C., (1964). *The Physiology of Synapses.* Springer-Verlag, Berlin.

Fatt, P. and Katz, B., (1951). An analysis of the end-plate potential recorded with an
intercellular electrode. *Journal of Physiology (London)* **115**, 320–370.

Hille, B., (1984). *Ionic Channels of Excitable Membranes*. Sinauer, Sunderland, MA.

Hodgkin, A. L. and Katz, B., (1949). The effect of sodium ions on the electric activity in the giant axon of the squid. *Journal of Physiology (London)* **108**, 37–77.

Hodgkin, A. L. and Huxley, A. F., (1952) A quantitative description of membrane current and its application to conduction and excitation in nerve. *Journal of Physiology (London)* **117**, 500–544.

Hodgkin, A. L., Huxley, A. F. and Katz, B., (1952). Measurement of current–voltage relations in the membrane of the giant axon of *Loligo*. *Journal of Physiology (London)* **116**, 424–448.

Katz, B., (1966). *Nerve, Muscle and Synapse*. McGraw-Hill, New York.

Katz, B., (1969). *The Release of Neural Transmitter Substances* Liverpool University Press, Liverpool.

Katz, B. and Miledi, R., (1963). A study of the spontaneous miniature potentials in spinal motoneurons. *Journal of Physiology (London)* **168**, 389–422.

Keynes, R. D. and Aidley, D. J., (1991). *Nerve and Muscle*. Cambridge University Press, Cambridge.

Koch, C., (1998). *Biophysics of Computation: Information Processing in Single Neurons*. Oxford University Press, Oxford.

Koch, C. and Segev, I., eds., (1989). *Methods in Neural Modeling: From Synapses to Networks*. MIT Press, Cambridge, MA.

MacKinnon, R. *et al.*, (1998). Structural conservation in prokaryotic and eukaryotic potassium channels. *Science* **280**, 106–109.

Sten-Knudsen, O., (2002). *Biological Membranes: Theory of Mass Transport, Membrane Potential and Electrical Impulses*. Cambridge University Press, Cambridge.

Wald, G., (1968). The molecular basis of visual excitation. *Science* **162**, 230–239.

Zeki, S., (1993). *A Vision of the Brain*. Blackwell, Oxford.

Zurer, P. S., (1983). The chemistry of vision. *Chemical Engineering News* **61**, 24–35.

13 Memory

There is still something missing from the picture of the nervous system given in the previous chapter; there is no obvious way in which it could store memories. One could, in principle at least, measure the electrochemical response at any position, due to the imposition of a given set of electrochemical stimuli elsewhere, but this response would always be the same; for the same set of stimuli, applied to the same place, the same set of responses would be observed. Memory requires something else in addition. It must involve some sort of change in the system. The response to a given stimulus must, somehow, depend upon whether the system has been exposed to that stimulus previously.

13.1 Hebbian Learning

In 1949, Donald Hebb published a book entitled *The Organization of Behaviour*. It has become one of the pillars of neuroscience, because of the seminal ideas that it contained. One of these was a surprisingly simple theory for the changes that provide the nervous system with the possibility of storing memories. Hebb suggested that the transmission efficiency of a synapse, what is now simply referred to as the synapse's strength, is modifiable by use. The exact wording of the relevant passage in his volume reads as follows.

When an axon of cell *A* is near enough to excite a cell *B* and repeatedly or persistently takes part in firing it, some growth process or metabolic change takes place in one or both cells such that *A*'s efficiency, as one of the cells firing *B*, is increased. The most obvious and I believe much the most probable suggestion concerning the way in which one cell could become more capable of firing another is that synaptic knobs develop and increase the area of contact between the afferent axon and efferent soma.

Hebb thus suggested that the synaptic contact between two cells would be strengthened if the one was persistently helping the other to emit nerve impulses. Implicit in his conjecture was the further suggestion that if the firing of one cell was only rarely (or even never) accompanied by the almost simultaneous firing of another cell with which it made a synaptic contact, then that contact would be weakened. This type of synaptic modification has come to be known as Hebbian learning, and it has been invoked in countless theories of brain function.

The idea is beautifully simple and straightforward. Just as water running down the side of a mountain will gradually cut a channel, which will subsequently become the preferred route, so will experience tend to increase the transfer efficiency of certain synapses, and weaken others, thereby creating favoured pathways for later nerve signals to follow. Also, just as the drying up of a river during a drought does not change its course, so too will the adopted neural pathway survive the periods when that part of the brain is inactive.

Some care is required with the river analogy, however, for it is valid only in so far as the distribution of the incident rain is reasonably uniform. And by the same token, the above discussion of Hebbian learning assumes a standard injection of nerve impulses – the sort of thing that might arise from a fixed sensory input, for example. It was the obvious need to be able to account for a wide variety of *different* sensory inputs that Hebb put forward another hypothesis in his book, one that is every bit as important as the learning mechanism that now bears his name. He suggested that for a given set of sensory stimuli only a very small fraction of all possible neural pathways would be in use, and he further proposed that these active routes would lie adjacent to one another, or very nearly so, because of their mutual electrochemical influences. Such a tight group of interrelated neurons, firing in or close to unison, has come to be known as a neuronal assembly, and the existence of these clusters relieves us of the problem of keeping track of what is going on in the bulk of the brain's roughly hundred thousand million nerve cells; at any one time, Hebb was postulating, the great majority of them would simply be inactive.

Hebb's hypothesis of modifiable synaptic strengths is so attractive that it seems almost obviously correct. At one fell swoop, he had presented neuroscience with a beautifully simple mechanism that was able to explain the main features of memory. The way in which new memories are laid down in the appropriate regions of the brain, and the manner in which they can be recalled, were suddenly revealed to be transparently simple, and it required little of the imagination to see that his conjectured mechanism could easily be extended so as to permit individual memories to be superimposed upon one another.

A number of years were destined to pass, however, before Hebb's brilliant insight was to receive experimental support. When this came, it was supplied by observations on a rather lowly creature whose taxonomic name is *Aplysia californica* (see Figure 13.1). This is a rather large aquatic snail, which has acquired the nickname sea hare not because of its speed across the seabed but because of the large ear-like structures that stick out from its head. *Aplysia* has been the subject of prolonged study by Eric Kandel and his colleagues, and their work required the preliminary demonstration that this creature is capable of displaying the basic forms of classical conditioning that had been established through Ivan Pavlov's work on dogs, half a century earlier.

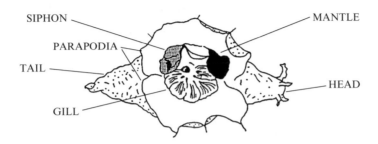

Figure 13.1   The aquatic mollusc *Aplysia* has a nervous system comprising only about 20 000 relatively large neurons but it displays some of the types of conditioning seen in more sophisticated creatures and is thus a convenient subject for investigations of memory mechanisms

It is not *Aplysia's* large ear-like protuberances that are of interest in this respect; four of its other externally visible attributes are far more important, namely the tail, the siphon, the gill, and a centrally-located structure called the mantle, which is bordered by the gill. The gill is naturally a vital organ in that any threat to its integrity will imperil the creature's oxygen supply. Both the siphon and the mantle serve as detectors of impending disturbance, and they induce the gill to promptly contract.

To appreciate the work of Kandel and his colleagues, we ought to briefly recall what Pavlov's research had revealed. Using a dog's tendency to salivate at the prospect of food, he first identified what is known as an unconditioned reflex. He defined this as a reaction in which response and stimulus are physiologically related. Such properties of food as dryness and texture are good examples of stimuli which fulfil this criterion. In a conditioned reflex, on the other hand, the stimulus has no obvious physiological role, a good example of this category being colour. One can appreciate the fact that Pavlov's definitions were not trivial from contemplation of the fact that a dog salivates more profusely at the sight of dry bread than fresh meat. He then demonstrated that the dog's capacity for salivation diminishes if it is repeatedly shown food to which it is denied access. The dog's salivation response is said to have become inhibited under such circumstances.

Let us now return to the beautiful work of Kandel and his collaborators. They first demonstrated the possibility of something that is akin to inhibition, namely habituation. If the siphon, instead of being strongly provoked, is subjected to a gentle flow of sea water, using a fine jet for example, the gill-withdrawal reflex gradually gets weaker. The change is due to a gradual diminution of efficiency in the synapse between the siphon's sensory neurons and the gill's motor neurons. The researchers then went on to demonstrate that *Aplysia's* repertoire of responses also includes the opposite of inhibition, namely sensitization. When the creature is given a stimulus that it does not like, such as a sharp pinch on the tail, the gill-withdrawal reflex becomes enhanced. Finally, Kandel and his colleagues were able to show that the sea hare is even capable of displaying a conditioned reflex. If the tail and the siphon are irritated at about the same time, subsequent stimulation of the siphon alone produces a particularly vigorous gill-withdrawal reflex.

This set of observations was interesting in its own right, but the real value of the research lay in what it was able to reveal about the underlying biochemical changes at the synaptic level. The neural circuitry that imbues the sea hare with its various types of response is quite simple, as nervous systems go (see Figure 13.2). It is not the simplest imaginable, however, because a single neuron with its dendrites embedded in the siphon (or mantle) and its axon activating muscles in the gill would, in principle at least, be sufficient to facilitate the gill-withdrawal reflex. A circuit that primitive, on the other hand, would not permit the differentiated response that had been revealed in the creature.

In the somewhat more complicated arrangement actually found in *Aplysia*, the neuron that has its dendrites embedded in the siphon sends its axon not directly to the gill, but rather to a synaptic contact with another neuron which, in turn, does indeed have its axon in contact with the gill. The former neuron is known as the sensory neuron, while the latter is called the motor neuron. (In

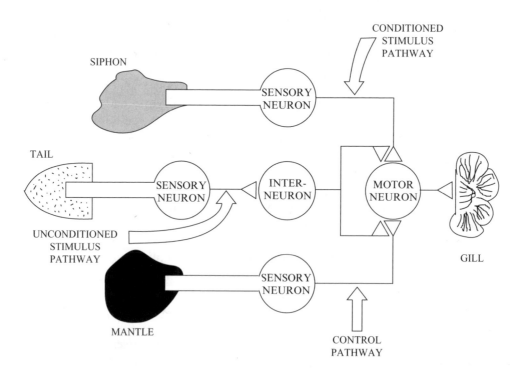

Figure 13.2 In this highly schematic diagram groups of parallel neurons are represented by a single neuron

practice, the route is not served by single neurons of these two types; that would leave the nervous pathway much too vulnerable. There are actually several pairs of sensory and motor neurons, and they act in parallel. For the sake of simplicity, however, our discussion here will invoke only one neuron of each type.)

It is this slightly more complicated circuitry which makes the various types of conditioning possible and Kandel and his colleagues were able to establish that biochemical changes occur at the various synapses. When the gill-withdrawal reflex becomes habituated, for example, this is due to a change in the synapse between the sensory and motor neurons. It had already been shown that when a nerve signal reaches that region, calcium ions enter the immediate presynaptic region through its bounding membrane. Habituation, it was discovered, develops because there is a gradual diminution in the membrane's ability to pass these calcium ions.

The opposite happens in the case of sensitization, namely an enhancement in the membrane's capacity for allowing the passage of calcium. An additional type of neuron is required, however, to make this feasible. It is called an interneuron. The dendrites of a second sensory neuron are located in the tail, while its axon makes a synaptic contact with such an interneuron (which in this case is referred to as a facilitating interneuron). The latter, in turn, makes a synaptic junction with that part of the siphon sensory neuron's axon that is immediately adjacent to the latter's synapse onto the gill-withdrawal motor neuron. This involves, therefore, *a synapse onto a synapse*.

Provoking the tail causes sensitization because a signal is transmitted from that region's sensory neuron to the facilitating interneuron, whereupon the latter releases the neurotransmitter serotonin. This, in turn, influences the synapse between the siphon sensory neuron and the gill-withdrawal motor neuron, increasing its ability to take in calcium ions. The biochemical changes that occur in the case of the conditioned reflex are rather more complicated, and we need not go into them here. Suffice it to say, then, that Hebb's hypothesis has been fully endorsed by this beautiful set of experiments on the sea hare.

13.2   Neural Networks

The feasibility of synaptic modification having thus been established, let us turn to the question of how memories are stored in a network of mutually interacting neurons. In 1961, Karl Steinbuch put forward the idea for a neural network that would display association. His main interest was the possibility of capturing the essence of this brain faculty in an electrical device. Such circuits that take their inspiration from the brain without attempting to faithfully copy its actual anatomy are now referred to as artificial neural networks.

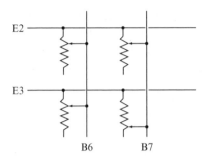

Figure 13.3   A small part of the Steinbuch network

Figure 13.3 shows a small part of Steinbuch's circuit, the horizontal and vertical sets of lines representing the axons and dendrites of neurons. Joining the various horizontal–vertical pairs, one sees inter-linking resistors, the strengths of which can be varied. These represented the synapses, and the general idea was to vary the latter so as to capture associations between matched activity patterns in the horizontal and vertical conductors (that is, artificial neural axons and dendrites). Steinbuch likened this to biological conditioning, as symbolized by his use of the words *Bedeutungen* (German for *meaning*) and *Eigenschaften* (German for *property*).

Steinbuch's procedure was quite straightforward. He worked out the set of synaptic strengths that would be required to capture the correlation between a given activity pattern in the horizontal conductors and the corresponding desired activity pattern in the vertical conductors. Thereafter, and when the correct adjustment had been made to each synapse, presentation of the same input pattern to the horizontal set would produce the desired output pattern in the vertical set. In other words, the network had learned to associate the input and output patterns.

This was interesting in its own right, but the really impressive thing is that such a network can actually store several sets of correlations simultaneously, in a superimposed manner. When Steinbuch worked out the adjustments that would be required to capture a second correlation between a different pair of input and output patterns, he simply *added* these to the previously-adjusted synaptic values, instead of starting over from scratch. Thereafter, presentation of the first input pattern still produced the first output pattern, but presentation of the second input pattern now produced the second output pattern. This was the first demonstration of what has come to be called parallel distributed processing (or PDP, for short). The twin hallmarks of this mode of memory storage are that each individual item in memory is distributed over many synapses, and that each synapse participates in the storage of many different items,

Christopher Longuet-Higgins, Peter Buneman and David Willshaw subsequently demonstrated association in a network that was simplified right down to the bare bones, as it were. The arrangement of their axons and dendrites was similar to that of Steinbuch, and the junctions between all pairs of these were again handled by variable synapses, the initial strengths of which were taken to be zero. Their sweeping simplification was that once a synapse has been activated, through its capturing the correlation of simultaneous activity in the corresponding axon and dendrite, its strength immediately adopts its maximum possible value, and remains unaltered thereafter.

As in the case of the Steinbuch network, the individual input–output associations were superimposed upon one another, and Figure 13.4 shows the situation following storage of the following four associations: $B3$, $B4$ and $B8$ with $E1$, $E3$ and $E7$; $B1$, $B5$ and $B7$ with $E2$, $E5$ and $E8$; $B2$, $B3$ and $B6$ with $E2$, $E4$ and $E6$; and finally $B4$, $B6$ and $B7$ with $E1$, $E2$ and $E3$. If the threshold is suitably adjusted, all the original associations can be recovered, though there is admittedly a single error in one of them. The reader should verify that by assuming a threshold of 3, and by letting all input patterns consist of sets of 1s and 0s (e.g. pattern $B1$, $B2$ and $B5$ is represented by 11001000) , the input $E1$, $E2$ and $E3$ actually gives a *four*-neuron response consisting of $B4$, $B6$, $B7$ and (erroneously) $B3$. All the other original associations are reproduced *without* error. Perhaps the greatest merit of this primitive neural network is its transparent demonstration of capacity limitation, for if *all* the synapses have been involved in the capture of one or more correlations, they will *all* have adopted their maximum possible strength and the network will have lost its discriminative powers.

Let us consider in more detail how the trained network is reacting to a stimulus. The prescription for calculating the activity in each and every neuron, at any instant, is that one is to sum the products of the activities of the other neurons and the strengths of the synapses by which they are connected to the neuron in question. This is done simultaneously for each neuron in turn. Putting this formally, we have:

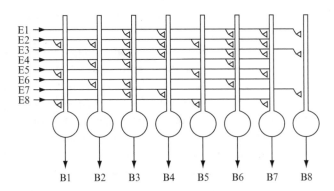

Figure 13.4   The associative circuit proposed by Longuet-Higgins, Buneman and Willshaw

$$V_i \to 1 \quad \text{if} \quad \sum_{j \neq i} V_j S_{ji} \geq V_{\text{threshold}} \tag{13.1}$$

and

$$V_i \to 0 \quad \text{if} \quad \sum_{j \neq i} V_j S_{ji} < V_{\text{threshold}} \tag{13.2}$$

where V_i is the voltage on neuron i, at any instant, and where S_{ji} is the strength of the synapse which joins neuron j to neuron i. The totality of synapses in the network will have strengths which can be written into a synaptic connection matrix $[S_{ji}]$.

When considering the nerve action potential, in the preceding chapter, we saw that the threshold usually lies around $-50\,\text{mV}$, which means that it lies $50\,\text{mV}$ above the resting potential of $-100\,\text{mV}$. In the scheme employed by Longuet-Higgins, Buneman and Willshaw, the threshold voltage, $V_{\text{threshold}}$, has been assumed to be a negotiable variable, whose value is related to the fraction of the input neurons which are active for the input patterns being considered. A neuron is defined as being active if, and only if, the above sum of products is greater than or equal to this threshold. If that sum is less than the threshold value, the neuron in question is deemed to remain inactive. In the example we considered, just three of the eight input neurons was active in any of the input paterns, and $V_{\text{threshold}}$ was accordingly set at the value 3. One could speculate about the biological justification for introducing such a normalizing factor into the calculations. Let us turn to the issue immediately.

By forcing the model to store memories in which almost half the neurons are active while the others are inactive, we have produced a situation that is quite unlike the one envisaged by Hebb, when he put forward the idea of the neuronal assembly (see above). The latter was conjectured to involve only a *small* fraction of the brain's vast number of neurons, at any one time. Although the actual fraction of neurons participating in one of these ephemeral assemblies has yet to be measured, it would certainly lie well below the level invoked in the above model.

Another criticism that could be levelled against this type of model is that it invokes only binary neurons; the neuronal units can adopt either of only two alternatives, namely active or inactive, the latter being of a standard intensity. Real neurons are capable of a more graduated response (see Figure 13.5). However, it would be unfair to mark this up on the debit side for the model, since many previous theories had made the same assumption and it can be shown that the introduction of such graded responses does not invalidate the mechanism. A neuron emits action potentials out along its axon only if the depolarization of its bounding membrane, in the vicinity of the axon hillock, exceeds the threshold value. When this is the case, the pulses are dispatched at a

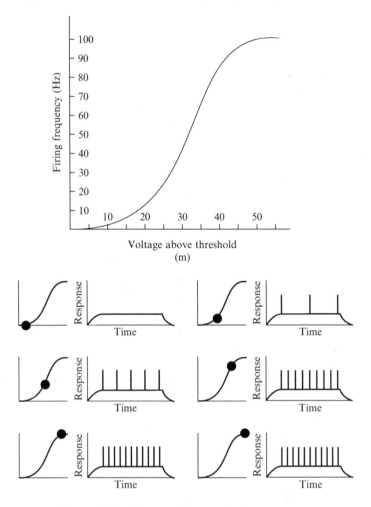

Figure 13.5   The graded responses from a neuron

frequency that depends upon the amount by which the threshold has been exceeded. This frequency does not increase indefinitely, however, because a limit is imposed by the speed with which the atoms in the protein ion channel molecules can relax back to their original (pre-pulse) positions. This fact implies that the voltage-frequency characteristic of the neuron has a sigmoidal shape, as shown in the upper part of Figure 13.5. Typical emissions from a neuron are shown on the right of the lower part of Figure 13.5, the operating voltage in each case being indicated by the small black circle's position on the sigmoidal curve shown on the left. The sigmoidal curve is often assumed to represent the input–output characteristic of each unit in an artificial neural network. In both real and artificial systems, the elements currently exerting the greatest influence will be those in which the summed input exceeds the

threshold by the greatest amount, the advantage being particularly large if the
emission frequency is at the saturation value.

Another deficiency is that the network is always being forced to make a
choice between alternative stable memories. It cannot merely relax to a condition in which all activity dies away to zero, as is frequently the case in a real
neural network. The insights into the basic manner in which a neural network
functions nevertheless make this model a major milestone in the history of the
subject. If the model's neurons are looked upon as conceptual (and perhaps
multi-cellular) entities, rather than as individual neurons, the discord with
certain anatomical facts might prove to be unimportant.

Returning to the issue of threshold, let us now consider networks which are
closer to those encountered in biology. In Figure 13.6 the symbols shown
permit ready discrimination between the various versions of neurons, despite
their crudity. This convention will be adhered to throughout this book, with
additional special cases being described as they arise. The neurons coloured
white are excitatory, while the black ones are inhibitory. This distinction is
not made in the case of an artificial neurons, the excitatory and inhibitory
effects being embodied at the synaptic level. If it were not for the ubiquitous
presence of inhibitory neurons, the activity level in the typical neural network
would rapidly reach its saturation value. The influence that a pre-synaptic
neuron has on a post-synaptic neuron depends upon the type of neurotransmitter molecules that the former releases into the synaptic cleft, and it

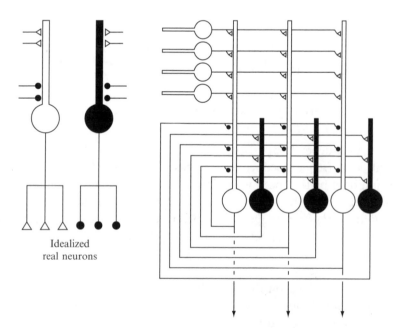

Idealized
real neurons

Figure 13.6   Idealized real neurons showing the excitatory and inhibitory effects

is found that a given neuron can release only one type of neurotransmitter. These signalling molecules fall into two classes. The ones that tend to depolarize the post-synaptic membrane are termed excitatory, while those that hyper-polarize the post-synaptic membrane are said to be inhibitory. Glutamate and acetylcholine are common examples of the former type, and the most prominent member of the inhibitory class is GABA (gamma-amino-butyric acid). In the circuits shown in Figure 13.6, the excitatory synapses stem only from excitatory neurons, and the inhibitory synapses only from inhibitory neurons. The inhibitory neurons in such circuits are frequently referred to as inhibitory interneurons, because their synaptic connections are always made locally. (Their job, in the scheme of things, is to inhibit excitatory neurons, and because these target neurons lie locally, the inhibitory neurons are often said to be providing lateral inhibition.) This contrasts with the synapses made by excitatory neurons, which often lie at considerable distances from the somatic regions of those neurons. In the case shown here, the inhibitory neurons are of the so-called feedback type, which is to say that they are feeding inhibition back to the neurons which activate them.

Another type of arrangement (see Figure 13.7) is also frequently found in the brain, namely that involving feed-forward inhibitory neurons, which receive

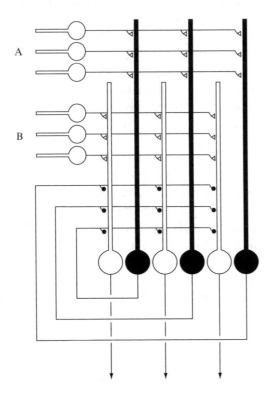

Figure 13.7   This type of early arriving inhibition is able to set up a sort of opposing cloud, through which only the strongest excitatory signals can penetrate

their activation from other neurons lying at greater distances. Such feed-forward inhibition is achieved by adding activation of the inhibitory neuron's dendrites via synapses from excitatory neurons lying more distant (as is already shown for the excitatory neurons). In either case, the role of the inhibitory interneurons is to dampen down the activity level in the excitatory group. Such inhibition is usually strong enough to prevent activity in all but the most active excitatory neurons, a process which is usually referred to as winner-take-all although the term *winners-take-all* might be more appropriate, because there are usually more than one strongly activated excitatory neuron (another commonly encountered term is *feed-the-fattest*).

13.3   Auto-association

A feature of Figure 11.1 which has not been properly touched upon, so far, is the axon collateral. A relatively short distance from the place where the axon leaves the soma (that is to say, the axon hillock), the single axon is frequently observed to branch into two or more strands. When this happens, the main route is still referred to as the axon, while the subsidiary conduits are known as axon collaterals. There is an important sub-class of the latter, namely those which arch back and form synaptic contacts with the dendrites of the very neurons from which they originated. These are known as recurrent collaterals, important examples of which are seen in connection with the so-called CA3 pyramidal neurons of the region of the brain known as the hippocampus.

By invoking such recurrent collaterals, Teuvo Kohonen demonstrated that a group of neurons can auto-associate. In other words, the network will be able to identify input patterns to which it has previously been exposed. Such an auto-associative network is shown in Figure 13.8. (Using a convention favoured by the present author, the strength of a given synapse is indicated by the size of the small triangular symbol used to denote an excitatory synapse. All the synapses have initially been drawn the same size, because no memories have yet been imposed on the network.) In this network too, individual memories can be superimposed, and the fascinating thing is that complete input patterns can be recovered even if the network is fed only *partial* versions of these. This is analogous to recovering an item of one's belongings from a lost-property office even though one supplies the attendant with only a partial description of it. A network capable of performing in this fashion is said to be content-addressable.

The interest in this auto-associative network is attributable to the fact that although it displays those basic properties that any believable network must possess, its dynamical behaviour is nevertheless very easy to understand. During the years following their appearance, the Kohonen papers became

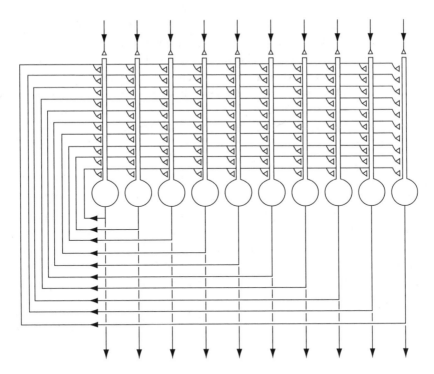

Figure 13.8 An auto-associative neural network; the key feature is the presence of recurrent collaterals

some of the most cited neural network publications in the scientific literature, and this work will remain a classic.

Kohonen took as his starting points something which already enjoyed virtually universal support, namely the concept of Hebbian learning, and he invoked this in order to adjust the synaptic strengths. In the network shown in Figures 13.8 and 13.9, every neuron is connected to every other neuron, not just once but twice, because for every synaptic contact that provides a signalling pathway between any two neurons, there is a similar contact running in the counter direction. All the synapses are, in other words, reciprocal, and because all pairs of counter-running synapses will have the same strength (see below), the overall structure has a remarkably high degree of symmetry.

Kohonen made another rather bold hypothesis. He assumed that the short-term dynamical behaviour of such a network would have no influence on the laying down of memories. The latter were thus envisaged as being determined only by the longer-term statistical distribution of neural firing patterns, a conjecture that does not sound unreasonable. This enabled him to make a remarkably simple implementation of the Hebbian learning principle; he considered each pair of neurons in turn, made the two counter-running synapses between them stronger (in the direction of increasing excitation, that is) if both neurons were active, and weaker if one was active while the other was inactive. All the synapses were treated in this way, during the storage of a given memory, and the entire procedure was repeated for each new memory.

To make the application of this strategy quite clear, let us contemplate an illustration involving just 10 neurons, as shown in Figures 13.8 and 13.9. We can again conveniently denote the firing state of any neuron by a **1** or a **0**, depending upon whether the neuron is active or inactive. And let us also assume that we are initially to impose just one memory on the system, which we may call *memory alpha*. Let us assume that the corresponding firing pattern is: **1010101010**. Because, as was noted earlier, such a network has no directionality, and also because every neuron is connected to every other neuron, the placement of the individual neurons in a diagram of the network is without significance, but for the sake of illustration we will imagine them as being located as shown in the figure. Because the reciprocal pairs of synapses are forced to have the same strengths, by dint of their linking the same two neurons, the diagram will be symmetrical.

The Hebbian prescription directs us to strengthen the two synapses lying between a **1** and a **1**, while those between a **1** and a **0** are to be weakened. In the

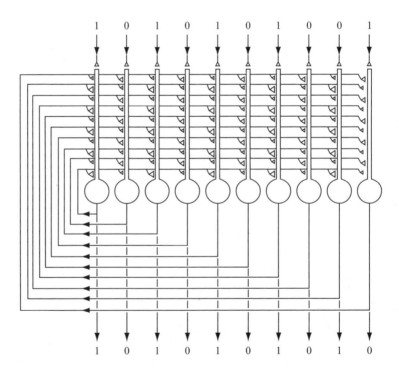

Figure 13.9   An auto-associative neural network in which learning has taken place; the size of each (excitatory) synapse denotes its strength

Kohonen implementation of Hebb's rule, the increment of strengthening or weakening is standardized to unity, although this is an arbitrary feature of the model; any other standard increment would have served the same purpose. In fact, it is more likely that synaptic changes, as a function of captured correlations between neuronal activities, will vary in a sigmoidal manner, consequently displaying saturation. Such saturation seems biologically plausible, given the obvious limitation on the underlying chemical resources that will be responsible for the synaptic changes.

Our illustration has now become as shown in Figure 13.9 and we might note in passing that there are equal numbers of strengthened (larger triangles) and weakened (smaller triangles), the remaining synapses not having been changed because they were linking pairs of neurons neither of which had been active. (Careful inspection of Figure 13.9 allows one to draw conclusions as to what *input* vector the circuit has been trained on.

Armed with the rules embodied in Equations (13.1) and (13.2), we are now in a position to follow the excitation levels in the various neurons, as a function of time. We will consider what happens if the network is presented with a pattern which *merely resembles* the one used during the learning phase. Let us assume that this test pattern is **1010101001**. It differs from the original only in the final two digits, which have been interchanged. When the test pattern is presented to the 10 signal-injection synapses (shown entering from above, in Figure 13.9), five of the neurons will be activated, while the others will remain quiescent. Five of the recurrent collaterals will be activated, therefore, and they will in principle be in a position to impart new activation to all 10 of the neurons. We can assume that this new activity arrives sufficiently *after* the original activity that the latter will have died away, and that it is thus not able to contribute to the new states of the various neurons.

The activity induced in each neuron will thus be the sum of five contributions, and each of these will itself be the product of the signal strength in one of the axon collaterals and the strength of the corresponding synapse. The signal strengths will be equal, of course, because the signals in the active neurons of the initial pattern were all equal. The synapses, on the other hand, are not all equal. This is the whole point of the training; the network's memory lies in the *non-uniformity* of the synapses.

Moreover, if we suitably adjust the value of $V_{threshold}$ (assuming this to have been achieved biologically, through the action of inhibitory interneurons not shown in the diagrams), and then use the prescriptions given in Equations (13.1) and (13.2), we readily find that the activity pattern at the next temporal epoch will have become **1010101010**. The ninth neuron will have become activated despite its not receiving any input from the initial pattern. The reason for this is not difficult to find. It arises from the fact that this neuron is nevertheless being fed by four active collaterals, through four strengthened synapses, whereas it is only getting a single 'dose' of weakened activity, through a weakened synapse. That the 10th neuron, conversely, is switched off is attributable to the fact that four of its activity contributions are arriving through weakened synapses.

Let us now put the behaviour of the auto-associative network on a more mathematical footing, beginning by considering the situation in which the recurrent collaterals are absent. In this case, each incoming axon merely activates a neuron which then sends the same activity onwards. If V_{in} is the input 'vector', and V_{out} is the output 'vector', this situation is described by the equation

$$V_{out}(t + \Delta t) = V_{in}^*(t) \tag{13.3}$$

where the Δt is the time that elapses as the signals move (slowly) down the dendrites. (We ignore the times taken for the signals to move – quickly – through the axons.) The asterisk is a reminder that a component of a vector can exercise an influence only if it is above the threshold. If we then add the recurrent collaterals, the expression for the output vector acquires an extra term

$$V_{out}(t + \Delta t) = V_{in}^*(t)$$
$$V_{out}(t + 2\Delta t) = [S] \cdot V_{out}^*(t + \Delta t) \tag{13.4}$$

As noted earlier, the merit of such an auto-associative circuit is that it can respond with the 'correct' output vector even if it is presented with an input vector $V_{in, partial}$ which differs from V_{in} because some of the components are incorrect. These erroneous components are compensated for by the matrix of connections $[S]$.

Our network has shown us that it can store a single memory, and that it can auto-associate even when it is fed with a slightly incorrect input pattern. If we now tried to impose a second memory, we would find that this is possible, using the additivity principle discussed earlier. There would, however, still be the limitation of memory capacity that applies to all associative networks. If that capacity was exceeded, we would discover that the network is no longer capable of properly recalling. It will have started to show saturation effects, like the ones that were mentioned in connection with the primitive network studied by Longuet-Higgins, Buneman and Willshaw.

If the capacity has not been exceeded, we will be able to superimpose the stored memories upon one another; a given synapse participating in the recording of all the memories, and each memory being distributed over all the synapses, as noted earlier. Destruction of a certain fraction of the synapses can be tolerated by the system, the latter retaining its ability to supply complete responses to incomplete clues. The network is thus said to be robust against partial synaptic loss, and to degrade gracefully. There is much evidence that the real brain possesses the same qualities, and many have taken this to indicate the essential validity of the auto-associative model.

Exercises

13.1   Study carefully the neural network shown below, in which *all* the delay times for
the passage of signals from synapses to the relevant soma are 10 ms (there being

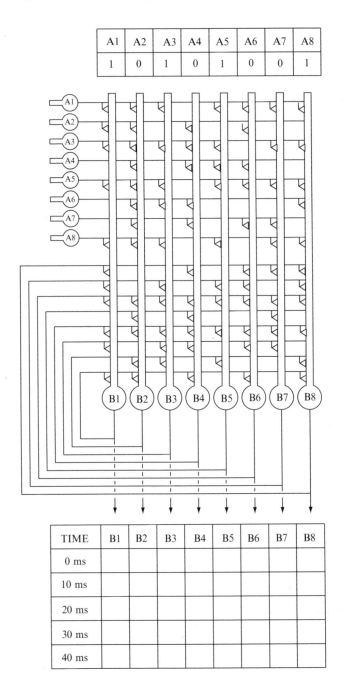

A1	A2	A3	A4	A5	A6	A7	A8
1	0	1	0	1	0	0	1

TIME	B1	B2	B3	B4	B5	B6	B7	B8
0 ms								
10 ms								
20 ms								
30 ms								
40 ms								

no decrease of signal amplitude during that passage), and in which there is a
standard threshold of 3 units. The synapses *all* have the standard strength of
unity, and the output pattern from the set of neurons designated *A* comprises
four *1*s and four *0*s. A *single* instantaneous example of such a pattern is
transmitted *from* the neurons *A* at time zero, and it is *not repeated*. Indicate in
the appropriate boxes the corresponding output from the neurons *B*, at the
various times shown, for the case in which the output pattern from neurons *A* is
as shown in the diagram.

13.2   Write a brief essay discussing the importance to neuroscience of (a) the *Aplysia*
studies of Eric Kandel and his colleagues, and (b) the conjectures of Donald
Hebb.

Further reading

Aleksander, I. and Morton, H., (1990). *An Introduction to Neural Computing.* Chapman and
Hall, London.
Amit, D. J., (1989). *Modeling Brain Function: The World of Attractor Neural Networks.*
Cambridge University Press, Cambridge.
Beale, R. and Jackson, T., (1990). *Neural Computing: An Introduction.* Adam Hilger, Bristol.
Byrne, J. H. and Berry, W. O., eds., (1989). *Neural Models of Plasticity: Experimental and
Theoretical Approaches.* Academic Press, San Diego.
Churchland, P. S. and Sejnowski, T. J., (1992). *The Computational Brain.* MIT Press,
Cambridge, MA.
Cotterill, R. M. J., ed., (1988). *Computer Simulation in Brain Science.* Cambridge University
Press, Cambridge.
Cotterill, R. M. J., ed., (1989). *Models of Brain Function.* Cambridge University Press,
Cambridge.
Cotterill, R. M. J., (1998). *Enchanted Looms: Conscious Networks in Brains and Computers.*
Cambridge University Press, Cambridge.
Durbin, R., Miall, C. and Mitchison, G., eds., (1989). *The Computing Neuron.* Addison-
Wesley, Wokingham.
Hebb, D. O., (1949). *The Organization of Behavior.* Wiley, New York.
Hertz, J., Krogh, A. and Palmer, R. G., (1991). *Introduction to the Theory of Neural
Computation.* Addison-Wesley, Redwood City, CA.
Kandel, E. R. and Hawkins, R. D., (1992). The biological basis of learning. *Scientific
American* 267(3), 52–60.
Koch, C., (1998). *Biophysics of Computation: Information Processing in Single Neurons.*
Oxford University Press, Oxford.
Koch, C. and Segev, I., eds., (1989). *Methods in Neural Modeling: From Synapses to
Networks.* MIT Press, Cambridge, MA.
Kohonen, T., (1987). *Content-Addressable Memories.* Springer-Verlag, Berlin.
Kohonen, T., (1989). *Self-Organization and Associative Memory.* Springer-Verlag, Berlin.
Minsky, M. L. and Papert, S. A., (1988). *Perceptrons.* MIT Press, Cambridge, MA.
Nauta, W. J. H. and Feirtag, M., (1986). *Fundamental Neuroanatomy.* Freeman, New York.
Peretto, R., (1992). *An Introduction to the Modeling of Neural Networks.* Cambridge Uni-
versity Press, Cambridge.
Ritter, H., Martinetz, T. and Schulten, K., (1992). *Neural Computation and Self-Organizing
Maps: An Introduction.* Addison-Wesley, Reading, MA.

Shepherd, G. M., (1990). *The Synaptic Organization of the Brain*. Oxford University Press, New York.

Squire, L. R., (1987). *Memory and Brain*. Oxford University Press, New York.

Steinbuch, K., (1961). Die Lernmatrix. *Kybernetik* **1**, 36–45.

Taylor, J. G. *et al*., eds., (1992). *Neural Network Dynamics*. Springer-Verlag, Berlin.

Wasserman, P. D., (1989). *Neural Computing: Theory and Practice*. Van Nostrand Reinhold, New York.

White, E. L., (1989). *Cortical Circuits: Synaptic Organization of the Cerebral Cortex, Structure, Function and Theory*. Birkhäuser, Boston.

Willshaw, D., Buneman, O. P. and Longuet-Higgins, C., (1969). Non-holographic associative memory. *Nature* **222**, 960–962.

14 Control of Movement

One of the present author's favourite cartoons shows two people looking at the elephants in a zoo. One asks the other, rhetorically: 'Did you know that elephants never forget?' After contemplating this for a few moments, the other person retorts: 'Ah, but what would an elephant actually have to remember?' The implication is that because an elephant cannot comprehend more than a minor fraction of the rich flow of input which is readily understandable to the average human being, it would have to remember rather little.

The point might seem to be well taken, given that elephants have not been able to match our culture and our technology. However, evolution is only indirectly concerned with such things; they play a role only if they influence our chances of survival until the age at which we reproduce. Human life is indeed sophisticated, but that fact does not in itself tell us much about the underlying biology. Shakespeare's plays and sonnets, Mozart's operas and symphonies and Michaelangelo's paintings and sculptures are deeply impressive, but they shed very little light on the subjects to which this final chapter will ultimately be aiming: consciousness, intelligence and creativity. Although this might seem to have nothing to do with those high-flown issues, the very important fact is that the answer to the second person's question is simply this: 'An elephant has to remember no more and no less than any other creature, great or small, namely how to *move* under the prevailing conditions in its environment and within its own body.'

14.1 The Primacy of Movement

In Chapter 10, we learned that the smallest creatures, that is to say unicellular animals, move about through the agencies of their cilia and flagella, whereas the larger examples, such as ourselves, move through the flexing of muscles. Indeed, as Charles Sherrington noted: *To move is all mankind can do, and for such, the sole executant is muscle, whether in whispering a syllable or in felling a forest.* A good case could be made for calling all movements reflexes, if we include those generated as a consequence of the body's monitoring its own internal state, as when hunger initiates foraging for food. It is true that this definition would seem to exclude movements resulting from prior thought, but until we have a proper understanding of what thought actually is, that conclusion might be rather hasty. Amongst other things, therefore, this chapter must seek to shed light on the nature of mental processes.

Meanwhile, we should recall that *E. coli* can discriminate between about 30 different varieties of chemattractants and chemrepellents (see Chapter 10), and that a relatively lowly creature such as *Aplysia* displays classical conditioning (see Chapter 13). This underscores the fact that the repertoire of movements does not have to be impoverished, merely because these are all automatic. There is nevertheless the important issue of whether a given reflex can be modified by experience. If it cannot, the memory of it is, in effect, innate and its possessor acquires it through its genetic inheritance. A creature having *only* this type of reflex is totally at the mercy of its environment, which then dictates its every response. The plot of such a creature's life could be said to be written in its genes. A more subtle situation prevails if the reflex can be modified by chemical events arising from chance correlations between environmental factors. The creature will still be under the exclusive dominion of its surroundings, but it will *seem* to be exercising volition in remembering its experiences. The very important possibility then arises that *all* such behavioural modifications are actually deterministic, even in sophisticated creatures such as ourselves. That too is an issue which will have to be considered later in this final chapter.

Figure 10.9 was to be interpreted as illustrating the final stage in the activation of a muscle by the nervous system. The neuron shown in that figure would itself receive signals from other neurons. Indeed, we saw that this is the case in *Aplysia*; its nervous system includes neurons which lie intermediate between the sensory receptors and the final effectors. In the simplest type of control, the nerve impulses travel only in one direction; from the receptor

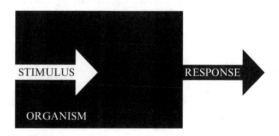

Figure 14.1 The mystery surrounding the functioning of the nervous system is often compared to that of a black box whose structure and workings are unknown; the stimulus may be detected by any one of the senses, or by a combination of these, and the externally-directed response is always just movement

neurons which detect signals from the environment, right through to the effector neurons which finally activate the muscles, the progress of impulses is always in what could be called the forward direction. In Figure 14.1, the neurons in question would be located between the words *stimulus* and *response*, and they would form what is known as a feed-forward neural network. The resulting control is said to be of the ballistic type. The incoming environmental signals, or their counterparts generated internally by the immediate needs of the body, lead to movement which is not subject to further adjustment, once it has been initiated. A good analogy would be the firing of an artillery shell from a gun, in contrast with the more controlled situation that is obtainable with a guided missile.

14.2 Ballistic Control in a Simplified Visual System

By way of illustrating ballistic control, let us consider a grossly simplified situation in which a visually observed obstacle is to be avoided. This example is shown in Figure 14.2, and although the rudimentary visual system depicted does not faithfully reproduce all the biological details, we will find that the figure does have a certain didactic value. In order to simplify things as much as possible, we will consider a one-dimensional eye, with its retina being composed of an orderly array of just one type of receptor cell. The obstacle occupies the left-hand seven units of the 10 units encompassed in the eye's visual field, and its avoidance will entail the organism ballistically veering to the right, so as to pass through the right-hand block of three unoccupied units. The obstacle and the navigable gap at the right extremity collectively constitute the visually observed object.

After passing through the pupil, the inverted image of the object falls on the receptor cells, whose large number is here represented by a mere 10 cells. These, in turn, make synaptic contacts with the ganglion cells, which are the retina's output. (The actual configuration of cells in the retina is more faithfully reproduced in Figure 14.4, but we do not need to consider such detail in this initial simplified treatment of the situation.) It is not practicable to draw in all these synapses, but they should be imagined as being present at all the points where the receptor cell axons cross the ganglion cell dendrites.

This synaptic array is reminiscent of the ones we encountered when considering the seminal studies of Steinbuch, and of Longuet-Higgins, Buneman and Willshaw, in Chapter 13, and it is particularly relevant to recall a complication that arose in the latter. We saw that the threshold for producing (onward travelling) nerve signals had to be adjusted, to allow for different degrees of activity in the incoming set of axons. In the example we considered (see Figure 13.4), just three of those eight axons carried activity, for any given input pattern. For the object shown in Figure 14.2, and if we say that the

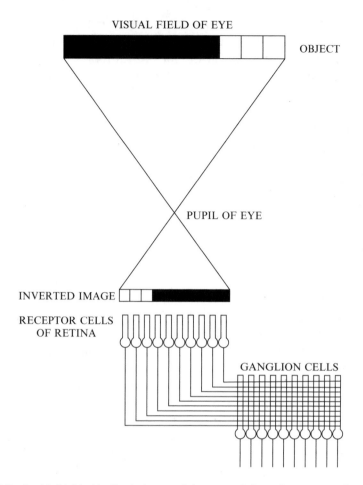

Figure 14.2   In this highly idealised picture of the eye and the retina, many cell types have been omitted for the sake of simplicity. The output of the receptor cells is thus shown to feed directly to the retina's output neurons, namely the ganglion cells

(black) obstacle gives off (i.e. reflects) no light, we again have a situation in which just three neurons are activated, this time out of a possible 10. However, we see that the system is ill-equipped to handle the more general situation in which the obstacle occupies an arbitrary number of the 10 units in our one-dimensional object; the action-potential threshold of the ganglion cells would have to be readjusted every time the eye is confronted with a new situation. This primitive eye is therefore a rather ineffective device.

There is a further deficiency. This concerns the strengths of the synapses between the receptor cell axons and the ganglion cell dendrites. By what agency are they to be adjusted to their ideal operational values? We recall that the overall aim is to match what is seen to the desired motor response, so that the system can behave in a ballistic manner. In general, it has to be admitted that there will be later stages in the system at which synaptic strength adjustments could be made; the responsibility for making the correct (obstacle-avoiding) movements will not rest exclusively on the ganglion cells. Nevertheless, we would hope that the pattern of synaptic strengths between the receptor axons and the ganglion dendrites is such as to optimally serve those later movement-mediating stages. This is indeed the case, but it is achieved at the expense of introducing other types of cell, which function as intermediaries. We will come to them shortly, but meanwhile let us note that their effect is to produce the situation illustrated by Figure 14.3.

Apart from the doubling up of the ganglion cells, which will be explained in a moment, we see that there has been a profound change. Whereas every ganglion cell could in principle be influenced by the entire visual field, in Figure 14.2, we now have a situation in which the receptive field of any given ganglion cell is only a small fraction of the eye's visual field. This has been achieved by the simple expedient of removing the majority of the receptor-ganglion synapses. In our highly simplified one-dimensional system, each ganglion cell (except those

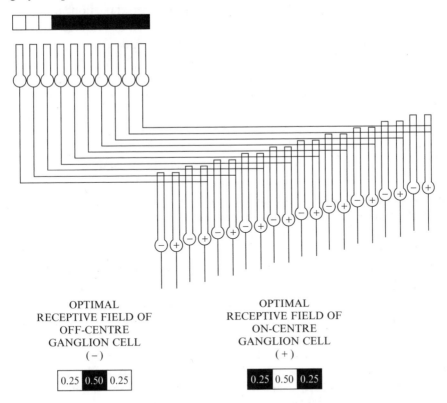

OPTIMAL
RECEPTIVE FIELD OF
OFF-CENTRE
GANGLION CELL
(−)

| 0.25 | 0.50 | 0.25 |

OPTIMAL
RECEPTIVE FIELD OF
ON-CENTRE
GANGLION CELL
(+)

| 0.25 | 0.50 | 0.25 |

Figure 14.3 A more realistic version of Figure 14.2 would take account of the fact that each ganglion cell receives signals from only a limited number of receptor cells. This means that the receptive field of a given ganglion cell is no longer the entire visual field seen collectively by the eye

located at either end of the line) is responsive to the inputs of just three receptor cells, and thus to just three of the ten units of the idealised object. Moreover, the maximum response of a given ganglion cell is obtained not when its receptive field is uniformly illuminated, but rather when there is a marked contrast between what is being received by its central portion and what is falling on its outer regions. In fact, two sub-classes are found to be present, and this is the reason for the doubling up of the ganglion cells in Figure 14.3.

The optimal receptive field (that is to say, the illumination which elicits the greatest response) of the so-called on-centre ganglion cell has light in the central portion and dark in the two side portions. Conversely, the off-centre ganglion cell responds maximally to a dark central portion flanked by two light portions. In the case of the real retina, these optimal receptive fields are two-dimensional, of course, and they have circular symmetry. There are two types, just as we have been discussing in our one-dimensional model, and they too display preferences for light and dark central portions, respectively. So they too display what are referred to as centre-surround preferences.

For the sake of argument, let us assume that the central portion accounts for 50% of the total response of a ganglion cell, while each of the edge portions is responsible for 25%. Moreover, let us assume that this contribution to the response will be produced if the portion in question is receiving the illumination it prefers, whereas the contribution will be reversed in sign if the illumination is of the non-preferred variety. We find immediately that uniform light or uniform dark, spanning all three portions, will produce zero response because there will be a cancelling out of a 50% contribution by two opposing 25% contributions.

Now let us enquire as to the responses of the various ganglion cells shown in Figure 14.3. We can start by deciding to ignore the two cells at either end of the row, because they do not have the full complement of inputs. In the real retina, with its hundreds of thousands of cells, such edge-located cells would be a negligible minority, and in any case, their missing inputs would probably be compensated for. Working from the left, for the object present in Figure 14.3, the next pair of cells, one with an off-centre preference and the other with an on-centre preference, would show zero response because they are receiving uniform illumination. There is a response from the on-centre member of the third pair, on the other hand, because it gets a 50% contribution from its central portion, whereas the contributions from the two edge portions mutually cancel. Conversely, the off-centre member of this third pair will not respond because it is getting, in effect, a net minus 50%. Moving on to the fourth pair from the left, we find that the situation is reversed: it is now the off-centre ganglion which is receiving input that comes closest to its optimum, whereas the on-centre cell has a net minus 50% contribution, and thus remains inactive.

Both members of the fifth pair, and indeed all the remaining pairs, receive uniform illumination, and they therefore display zero response. By giving the individual ganglion cells limited receptive fields, we thus find that the number of them which can respond to the type of object we have been considering is strictly limited. Indeed, and ignoring those end effects, we have just two active ganglion cells out of a possible 20, and it is easy to see that this will be the situation irrespective of where in the visual field the edge of the obstacle is located. Moreover, although things are usually more complicated than this, it would not be difficult in principle to arrange for activity in these two particular cells to cause the animal to move in the appropriate direction, and thereby avoid the obstacle.

The cellular arrangement in the actual human retina is indicated schematically in Figure 14.4, and we see that the centre-surround receptive fields of the ganglion cells are actually produced by the interplay of the receptors and the ganglions with the neurons of three other types, namely the horizontal cells, the bipolar cells and the amacrine cells. Oddly enough, and contrasting with what happens in the eye of a frog, for example, the image impinging on the light-sensitive receptors has to pass cells of the other four types. The rod receptors (R), which are sensitive to brightness, and the cone receptors (C),

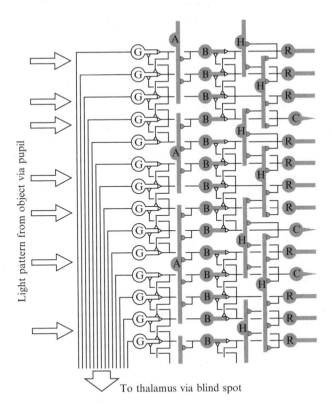

Figure 14.4   The mammalian retina has a complicated structure that comprises neurons of six different types as indicated in this highly schematic diagram

which monitor contrast, motion, size and wavelength, feed signals to the bipolar cells (B), and the latter send signals to the ganglion output cells (G). Signals from each neuron are passed on to several neurons of the next type, so the system involves much convergence and divergence. Moreover, the neurons of the remaining two classes, the horizontal cells (H) and the amacrine cells (A), mediate the lateral spread of signals, helping to eliminate the effect of (optical) noise and enhance the contrast at discontinuities in illumination. Only the ganglion cells emit action potentials, the responses of all the other neurons being of the more graduated kind normally associated with dendrites. Horizontal and amacrine cells do not have axons. Many of the synapses in the retinal circuit permit the two-way passage of signals, and those that do so have been drawn as semi-circles rather than triangles. The receptive field of a given ganglion cell is determined by the number of receptors which can influence its output. Given all this processing that takes place at the back of the eye, it is not surprising that the retina is usually said to be an extension of the actual brain.

14.3 More Sophisticated Modes of Control

The strengths of the synapses which connect neurons in the later parts of the animal's nervous system can be modified as the result of experience. In the situation we have been considering here, for example, the occasional bumping into the obstacle by the moving animal would be expected to cause at least mild pain, and the generation of such pain could be coupled to a mechanism which weakened certain synapses, or strengthened others, or even led to both of those types of adjustment. The animal's subsequent ballistic behaviour would thereby be modified, and the efficiency with which it moved around in its environment would be increased. A nervous system having the possibility of such modification is said to possess parametric feedback control. In mammals, there is evidence that such feedback, and the associated behavioural modification, invokes the dopamine pathways in the midbrain (dopamine being one type of neurotransmitter). Some systems are more sophisticated than this, in that the synaptic changes can be brought about not only by actual occurrences (such as collisions with the obstacle, in the above example) but even by occurrences that are merely *anticipated*. Of course, it is rather likely that the power of anticipation, in such systems, would itself have to be gradually acquired by learning from prior experience. Systems having this additional capacity for modification are said to possess parametric feed-forward control and in our own species there is evidence which suggests that the power of anticipation is provided, at least in part, by the cerebellum.

This latter mode of control is reminiscent of what we humans can readily manage. While performing a particular motor act, we may be reminded of something disagreeable that happened the last time we did such a thing, and this may cause us to terminate the execution of the act before it has been completed. One could say that the act has been vetoed. There is an obvious connection, here, with the imagination, but we should not overlook the possibility that there could be animals which possess parametric feed-forward control even though they are incapable of thought.

If the senses remain on duty, after the ballistic command has led to initiation of the motor act – which would usually be the case – one might guess that this could lead to the issuing of further ballistic commands, and thus to on-going correction of the animal's motor trajectory. That trajectory would then become the outcome of a series of ballistic sub-trajectories, and the situation would resemble more closely that of the guided missile mentioned earlier. However, we must bear in mind that this would involve continuous updating of what the muscles were being commanded to perform, and we cannot ignore the inertial factors in those muscles. Some time will be occupied in implementing the desired corrections.

14.4 The Heterogeneous Structure of Muscle Fibres

Nature's way of allowing for this need has been to produce muscle structures which are more complicated than one might have guessed. Figure 10.9 shows the activation of a muscle by a nerve cell, but it is not always the case that the effector cell is directly activated by what could be called a higher centre. In sufficiently advanced species, the skeletal muscles have a heterogeneous structure. The job of limb movement is carried out by what are known as the extrafusal fibres, which make up the bulk of the muscle; but interspersed amongst these are the smaller intrafusal fibres, which form either end of what are known as the spindles, these latter functioning as static and dynamic strain gauges. They measure the difference between the current and desired lengths of the overall muscle, and also the rate of change of that difference. These structures are depicted schematically in Figure 14.5(a).

Control of skeletal muscle function requires (proprioceptive) information about the instantaneous length of a muscle and about the force it is currently generating. This is supplied by two types of receptor which act in mutually complementary fashion. The muscle spindles, which are actually composites of the intrafusal fibres and the nuclear fibres (chain and bag types), lie parallel to the extrafusal fibres of the skeletal muscles, while the Golgi tendon organs are in series with the latter. Both types of receptor are innervated by efferent nerve fibres, but the reactions of these to a change in the length of the muscle differ radically. Those that serve the spindles increase their signalling rate when the muscle is stretched, while contraction causes a decrease. The fibres that innervate the tendon organ act oppositely. In humans, precision of control is aided by

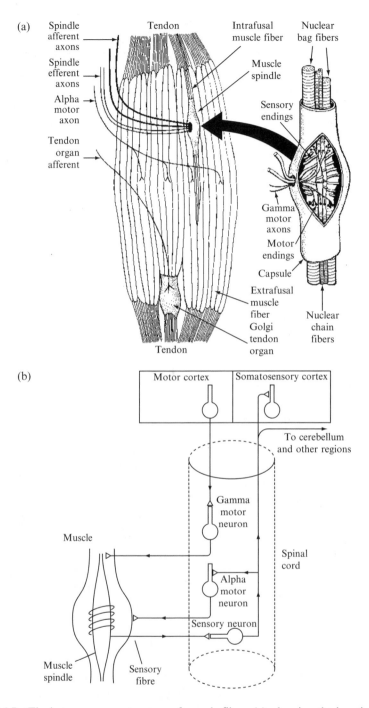

Figure 14.5 The heterogeneous structure of muscle fibres (a), showing the location of the spindles and tendon organs, and (b) the way in which the spindles are incorporated in the overall control circuit

separation of the innervating fibres, the skeletal muscles being activated by the alpha motor neurons and the spindles by the gamma motor neurons (the latter being further divided into static and dynamic groups).

One of the most striking features of the central nervous system is the profusion of looped routes in which pairs of regions are served by signalling paths in both directions. Good examples are seen in the interplay between the cortex and muscles. In the simplified case shown in Figure 14.5(b) a pyramidal cell in the motor cortex contracts a muscle by first activating the intrafusal fibres of a muscle spindle via a gamma motor neuron located in the spinal cord. This causes the spindle to contract and afferent signals are thereby passed to a sensory neuron which, acting via an alpha motor neuron, causes the extrafusal muscle fibres to contract. The gamma motor neuron continues to be activated, and this permits the spindle to monitor the state of the muscle, thereby allowing the muscle to adopt the required length. Signals from the muscle spindle are also passed to the somatosensory cortex, the cerebellum and other brain structures. The author of this book believes that muscle spindles, together with their simulated versions – in efference copy loops within the brain – play a key role in consciousness.

The upshot is a type of servo control, which has obvious advantages over the ballistic variety described above. An even more sophisticated variant of this type of control uses a prediction (based on prior experience) of the outcome of a sequence of motor acts rather than the actual result. A duplicate, known as an efference copy, of the commands dispatched from the higher control centres to the gamma neurons, and hence ultimately to the muscles, is used to predict the manner in which the system should behave. This prediction is automatically compared with the actual outcome, and any discrepancy is used to correct the muscular trajectory. This type of control, designated internal feedback, therefore represents a felicitous combination of the servo and parametric feedback strategies.

14.5   Central Pattern Generators

When a sequence of muscular movements is routinely required to be brought into action, as in the case of the rhythmic motion of the legs when a mammal walks, or the rhythmic undulation of the fins when a fish swims, such functions can be delegated to self-contained units known as central pattern generators. Nerve signals are dispatched to such units from the higher control centres, thereby bringing them into play, and the unit then assumes a considerable degree of autonomy in controlling the required sequence of muscular events.

A good example has been documented by Sten Grillner and his colleagues in the case of the lamprey, which originated from a group of animals that diverged from the main vertebrate line about 450 million years ago, during evolution, before the modern version of the fish had appeared. It may thus be regarded as a prototype vertebrate, and it offers the experimental advantage of a nervous system which contains orders of magnitude fewer neurons than are present in vertebrates such as ourselves. The creature's swimming action is produced by alternating contractions of muscles lying on the left and right sides of its body, these being activated by a local neural network which is reminiscent of the flip-flop components seen in some electronic circuits. The layout of this neural network is shown schematically in Figure 14.6, only one cell of each type being drawn so as to facilitate interpretation of its mode of functioning. The flip-flop units are located in the spinal cord, at intervals along the length of the body, and they are under the immediate control of what are known as reticulospinal neurons in the brainstem. The latter come in two varieties, namely the phasic neurons and the tonic neurons, and they both use glutamate as a neurotransmitter to activate their targets.

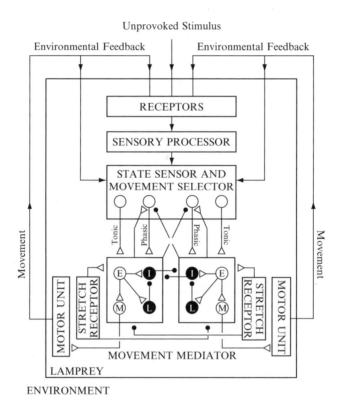

Figure 14.6 The central pattern generator circuit of the lamprey. Only one of the roughly one hundred such units is shown. Excitatory neurons and their synapses are indicated by the open circles and triangles while their inhibitory counterparts are shown as large and small black circles; the synapses shown as contacting the boxes, rather than individual neurons, exert more widespread influence

Those targets are the groups of neurons schematically drawn in the boxes in Figure 14.6, each group comprising excitatory neurons (E), motoneurons (M), midline-crossing inhibitory neurons (I) and lateral inhibitory interneurons (L). The latter two groups of neurons exert their influence through the neurotransmitter glycine, while the Es employ glutamate. The Ms use acetylcholine in order to activate their target muscles. The flip-flop function is clearly served by the midline-crossing of the axons of the I cells, but it is augmented by the presence of stretch-receptor neurons of both the excitatory and inhibitory types (let us call these SRE and SRI, respectively). These also use glutamate and glycine, respectively.

The mode of operation is as follows. Activity of the neurons in one of the groups, on the left say, will tend to suppress the activity in the group on the other side (known as the contralateral side), and it will also cause muscle contraction on its own side (the ipsilateral side). The stretch receptors on the contralateral side (the right-hand side in the present example) will therefore be activated, and some of these (the SRIs) will send inhibition to the left-hand side, while others (the SREs) will excite the neurons in the group (shown in the box) on the right-hand side. The overall activity will thus be transferred to the right-hand group, and muscular contraction will soon follow on the right-hand side. What was previously true of the left-hand side will now apply to the right-hand side, and the process will be repeated, one whole cycle of activity having occurred when the focus of activity has again returned to the left-hand side. An additional factor which reinforces the flip-flop pattern is the disinhibition that results from the E–L–I pathway, because the L cells inhibit the I cells.

The creature's propulsion is achieved by the alternating activation of the motor neurons, on the left and right sides, these being located in about 100 segments uniformly deployed along the length of its body. The higher the level of the stimulating activity they receive from the reticulospinal neurons, the faster will the lamprey swim. Those activating neurons receive *their* stimulation from neurons located further toward the head, and from a variety of sensory receptors. The undulatory waves that continuously pass down the body during locomotion (see Figure 14.7) involve a roughly 1% phase difference between adjacent segments, and because there are about 100 segments this means that approximately one wavelength of displacement is present along the creature's body, at any instant. This inter-segmental phase lag is independent of the cycle duration, and thus of the speed through the water.

The lamprey is capable of other types of movement, such as burrowing, and its full behavioural repertoire requires the involvement of neurotransmitters other than those which serve the basic locomotion just described. Serotonin (also known as 5-HT, which stands for 5-hydroxy-triptamine) and dopamine are both present in the vicinity of the spinal neurons, and they exert complementary influences on them, through subtle modifications of the ion channel mechanisms. (Even greater variability of behaviour is achieved through the

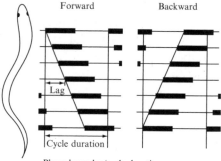

Phase lag = lag/cycle duration

Figure 14.7 In the lamprey, the paired units of the type shown in Figure 14.6 are activated at mutually staggered times, and this provokes forward or backward swimming depending on the relative phase lags

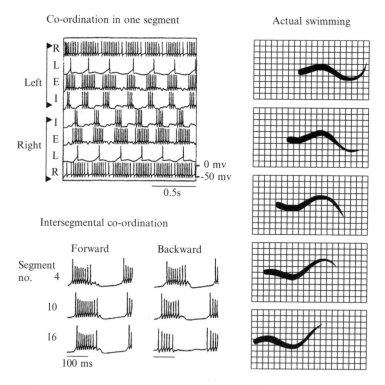

Figure 14.8 The phase lags indicated in Figure 14.7 are influenced by the excitability of the each type of neuron, and this in turn is determined by the interplay between different neurotransmitters. The plots refer to the units shown in Figure 14.6, and the data were obtained in the computer simulations reported by Sten Grillner and his colleagues

presence of peptides such as neurotensin, somatostatin and tachykinins.) It is found that an increased excitability in the rostral (head-end) part of the spinal cord increases the phase lag along the entire backbone, whereas an increase of excitability at the caudal (tail-end) region can actually reverse the phase pattern, such that the creature will swim backwards (see Figure 14.8). Grillner and his colleagues found that they could vary the inter-segment phase lag continuously over the range $+2\%$ to -1%. The lamprey clearly benefits from such variability, amongst other things because this permits it to swim through narrow channels, there then being considerably more than one wavelength of displacement along the length of its body at any instant. Finally, it should be mentioned that a brief injection of activation from the tonic reticulospinal neurons (see Figure 14.6) on one side of the body will cause the lamprey to turn in the appropriate direction. This function is of obvious utility for goal-directed locomotion and it involves the type of mechanism that would ultimately use the products of the general type of visual discrimination process discussed in connection with Figures 14.2 and 14.3.

It is fascinating to note that the neural mechanisms just decsribed are merely sophisticated counterparts of the chemical-concentration-controlled mechanism that produces analogous behaviour in the bacterium, as described in Chapter 10. A less obvious connection is seen when one compares the stretch receptors in the lamprey with the spindles in the muscles of more advanced creatures. It is not unlikely that spindles evolved through incorporation into the muscles of formerly independent stretch receptors.

14.6 Conditioned Reflexes

So far in this chapter, we have invoked only circuits of the feed-forward type. Moreover, nothing has been said about the manner in which the neural networks that control movement become adapted as a result of experience. An animal has to learn which environmental stimuli, and which of its movements, lead to significant consequences, appropriately detecting whether these are beneficial or detrimental to its goals. Because natural environments tend to be complex, this discriminatory task will unavoidably include correlating inputs to more than one sensory modality. The foraging bee, for example, has to learn to associate such visual clues as the shape and colour of a flower with the olfactory clue of its smell. Indeed, and as is well known, these pieces of sensory information can even be associated with the direction and distance of a flower's location, because the bee can inform its hive-mates of those parameters.

The learning of a meaningful association is not a purely intellectual event, even in advanced species such as our own. There must always be an underlying evaluation of its significance for the body, though the connection in the case of the human being may be tenuous and subtle. In the bee, it is tangible and direct. The driving force for much of this insect's behaviour is nectar, of course, and sucrose stimulation of its antennae and proboscis elicits what is known as appetitive behaviour, which is to say the development of a desire to satisfy hunger. As has been demonstrated by Martin Hammer, this manifests itself in an enhanced proboscis-extension response to an odour (the conditioned stimulus, CS), and the achievement of a sucrose reward represents reinforcement of the unconditioned stimulus (US) in olfactory conditioning. As was first established by Ivan Pavlov's famous studies with dogs, a temporal requirement must be fulfilled if conditioning is to occur, this being that the CS must precede the US by a brief interval (typically 0.5 s). Bees have to distinguish between odours that lead to a reward and those which do not, and it is found that learning of an unrewarded odour is indeed retarded in situations that involve several different olfactory inputs.

The adaptation which underlies foraging behaviour can be described by what is known as predictive error-correcting learning. The dynamics of such learning were studied by Robert Rescorla and Allan Wagner, their rule being captured in the following equation

$$\Delta V_i = \alpha_i \beta (\lambda - \sum V_i) \tag{14.1}$$

where ΔV_i is the change in the associative strength (V_i) of stimulus i. This latter quantity is a measure of the degree to which the stimulus predicts reinforcement. The coefficients α_i and β are learning-rate parameters related to CS and US, respectively. The parameter λ is the maximum associative strength that can be achieved for a given reinforcer, and ΣV_i is the combined associative strength of all stimuli present in a given situation (it can be interpreted as being the expectancy of reinforcement).

If the particular CS labelled i occurs in the absence of US, λ is zero and the term $\lambda - \Sigma V_i$ is negative; this means that Equation (14.1) will forecast inhibitory learning for stimulus i. Another important case occurs when one of several simultaneously presented CSs already predicts the US with 100% reliability, because ΣV_i will then already equal λ; with $\lambda - \Sigma V_i$ being therefore zero, there will be blocking of learning for all the other stimuli.

Because the reinforcing result of a captured correlation would be expected to be widespread in the animal's nervous system, and to make a particular impact on those regions which immediately serve sensory input, there is a strong hint that the adaptation should involve feedback routes from structures lying late in the system, where the reward would be expected to make its impact. Studies of the nervous system of the bee have tended to endorse this view. Neuromodulatory effects have been detected which influence cellular excitability of (and synaptic transmission between) neurons serving the sensory inputs. These effects involve sequences of biochemical modifications that are not unlike those we encountered when considering *Aplysia* in Chapter 13. Moreover, there are clear indications that these modifications ultimately stem from reward detection processes occurring late in the insect's neuronal pathways.

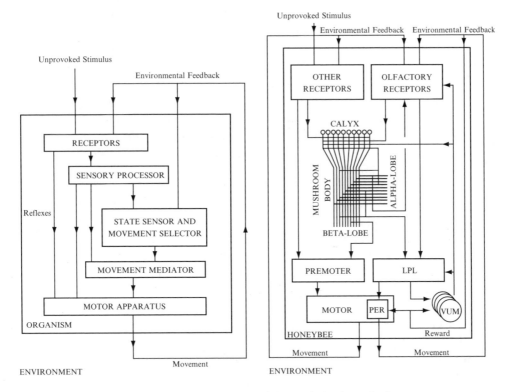

Figure 14.9 Highly schematic illustration of part of the honeybee's nervous system (PER is the proboscis extension response) and (left) a rationalization of its main components

A particularly well-studied example involves neurons in the so-called VUM region (see Figure 14.9) – VUM standing for ventral unpaired medial. One type of these is the VUMmx1 neuron, which responds to the amine known as octopamine. This substance has been found to enhance olfactory reward conditioning and memory retrieval. The VUM neurons' primary (*US*) response is to sucrose, the octopamine apparently being implicated in the associative conditioning. The result of this adaptation is the above-mentioned modification on the sensory input side, via the feedback routes indicated in the figure. The upshot of these changes is the generation of predictive behaviour. Here too, one sees more subtle influences on the functioning of the nervous system, and hence of the animal, which are mediated by dopamine. Indeed, in another commonly studied insect, the fruit fly *Drosophila*, one finds an interplay of dopamine and serotonin, just as is found in the lamprey discussed earlier in the present chapter.

The roles of some of the components shown in Figure 14.9 are still not fully understood. The unit marked 'motor' is equivalent to the 'movement mediator' shown in the inset. It may be regarded as a forerunner of the motor cortex found in more advanced species. Particularly noteworthy features are the

premotor unit, which permits the selection of *sequences* of muscular movements, and the behaviour-modifying reward system. The lateral protocerebral
lobe (LPL) is known to control appetitive behaviour through its ability to
combine olfactory signals with those emanating from other sources of input,
the mushroom body (MB) pathway possibly supplying experience-dependent
contextual information. As indicated in the figure, signals from the mushroom
body also impinge upon other so-called premotor units (of which the LPL is
also an example), and some of these could control such foraging-related
functions as flight, landing and tactile exploration. In the bee, a single neuron
in the so-called alpha-lobe appears to connect with all the neuronal groups
(known as glomeruli) of the antennal lobe. This indicates that the information
processed by the mushroom body could mediate remembering of odours
having behavioural significance. It could thereby help to facilitate the retrieval
of context-dependent memories. Moreover, the routes which implicate the
mushroom body could, in addition, mediate context-dependent learning in
premotor units, this ultimately producing acquisition of experience-dependent
behavioural routines.

A feature of Figure 14.9 that is of more general interest is those premotor
units. One might wonder why the final products of the sensory processing are
not fed directly to the motor output unit, or at least to central pattern generators, of the type we discussed in connection with the lamprey. The reason is
quite straightforward. If the creature's motor repertoire is going to routinely
include *non*-rhythmic sequences of motor acts, these will have to be encoded in
a suitably-connected set of neurons lying earlier in the nervous system than the
motor output units. This is precisely the job carried out by the premotor units.
The differential timings underlying such sequences of movements may ultimately be provided by the slow passive transport of signals along dendrites, as
was discussed in Chapter 12; one could imagine the selective strengthening of
synaptic contacts lying at different distances from the somas of the various
neurons, thereby producing a variety of different dendritic delay times. The
temporal coding potential inherent in the dendrites has been studied by Bartlett
Mel.

14.7 Volition and Free Will

It is usually assumed that sufficiently advanced species possess something more
than has yet been discussed in this chapter, namely volition – what is commonly
referred to as free will. The vehicle through which free will is exercised is just as
commonly taken to be consciousness. And the degree of success with which the
animal uses this free will is generally regarded as being dependent upon the
animal's intelligence. It is thus appropriate that we continue by taking a brief
look at these attributes.

What is consciousness? *Are* members of some other species conscious, as the preceding paragraph might be taken to imply? Some people believe that these enigmas will remain forever beyond our intellectual grasp, and that the so-called explanatory gap – why there should even *be* such a thing as consciousness – will never be closed. Those who are less pessimistic – and their number includes the present author – feel that the mystery surrounding the subject will disappear when science has adequately tackled the complex reality of behaviour in advanced species such as our own. The linchpin of behaviour would seem to be the organism's reaction to stimuli, as we have just been discussing in connection with the bee, and such responses are divided into unconscious reflexes and the conscious reactions associated with volition. The distinction is sharp because consciousness of a stimulus frequently produces no visible reaction at all, only thought. However, what *is* thought, and what purpose does consciousness actually serve?

In the epiphenomenal view, it serves no real purpose; it merely emerged during evolution as a sort of free bonus. Although it might seem that there must be situations which demand consciousness, one should bear in mind all the complicated things that can be accomplished without it, such as engaging in conversation while driving a car in unfamiliar territory or while sight-reading a novel piano score. With sufficient practice, as Roger Carpenter has stressed, *any* action becomes a reflex. This is difficult to gainsay, but what about that practice; isn't consciousness needed when we are acquiring expertise?

We will return to that central issue, but let us consider evidence that has been assumed to bolster the importance of the unconscious. Max Velmans has posed the surprising question: *Is human information processing conscious?* And he asks us to articulate the following sentence: THE FOREST RANGER DID NOT PERMIT US TO ENTER THE STATE PARK WITHOUT A PERMIT. There is a six-letter word which occurs twice, and yet the reader will have given it two different pronunciations, and hence meanings, because of the different contexts. Moreover, when the sentence is read at normal speed, there is insufficient time for the individual words to be contemplated; the decision as to how a word will be pronounced is made *un*consciously.

That claim of insufficient time is endorsed by the work of Benjamin Libet and his colleagues. They exploited the fact that the cerebral cortex is exposed during brain surgery, allowing its electrochemical activity to be monitored directly with a small probe, while the patient is awake and able to describe what is felt. (Permission was naturally obtained from the patient beforehand, and there was no risk of discomfort because the brain itself has no pain centres.) The cerebral region receiving signals associated with touch is the somatosensory cortex, and it comprises a distorted map of the body's surface. The primary signals corresponding to all five senses enter the cerebral cortex at the sensory lobes which lie posterior to the motor cortex. The motor-planning function is handled in the frontal lobe which lies anterior to the somatosensory area. After Libet and his colleagues had located the area in that cerebral region corresponding to the hand, they showed that signals reach there about 40 ms after the hand has been touched (see Figure 14.10). They then stimulated this hand area of the cortex, thereby giving the patient the sensation that the actual hand was being touched. But the patient became aware of this direct stimulation only if it was continued for at least 350 ms. These temporal signatures were then employed to establish the time-course of sensation, by comparing the moments of awareness of actual and pseudo touchings. The biggest surprise was that conscious awareness of a real stimulus occurs only after about 500 ms has elapsed. So those decisions of pronunciation when reading the Velmans test sentence seem indeed to have been made unconsciously. However, we will be reconsidering this interpretation later in the chapter.

The most important follow-up studies of this issue were carried out by Roger Carpenter and Matthew Williams, who made observations on eye movements. Although such movements might seem to have only marginal relevance, we should note that vision accounts for almost half of the cerebral cortex in humans; it is our senior sense. Each subject initially fixed his gaze on a centrally located cross, and was told to let his eyes swivel freely in the direction of a suddenly appearing target (that is, to let the eyes make a saccade). The target was a coloured disc of light, and the subject was not told exactly when it would

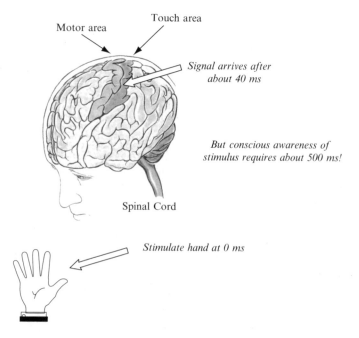

Figure 14.10 It takes about 40 ms for nerve signals to reach that part of the somatosensory (touch) area responsive to stimulation of the hand, and similar delays apply to auditory and visual stimulation, but consciousness of such unprovoked stimuli requires 500 ms

appear, nor whether it would show up on the left or the right of the central cross. The subject was exposed to numerous trials, and the experimenters were able to surreptitiously vary the left–right statistics (for example, 25% left, 75% right).

When Carpenter and Williams analysed the saccadic latencies (that is, the delays before the eyes moved), they found consistency with a purely deterministic decision-making process. Even more intriguingly, they discovered a subset in their data which corresponded to unusually rapid saccades – the subject's gaze shifts would occasionally jump the gun (see below). The biggest surprise was revealed when the statistics showed that those unusually rapid gaze shifts were just as deterministic and random as their normal counterparts; when eye movements jump the gun, there is no implication of conscious 'guilt' on the part of the subject. During subsequent studies with Douglas Hanes, Carpenter discovered that eye-movement decisions are subject to two opposing forces: one ordering the eyes to make a saccade and the other to countermand this. Earlier experiments by Libet, which involved subjects occasionally not following through on an initial intention to bend a finger, and noting when they became aware of this change of heart, had also revealed such a veto mechanism.

These issues are so important that we should do what biophysics demands of us: we should enquire as to the quantitative evidence for these conclusions. Let us take a closer look, therefore, at the manner in which Carpenter and Williams recorded and analysed their data. They interpreted saccadic latency in terms of the time required for the nervous system to decide whether or not the target is actually present, such decisions being clearly unconscious because the observed latencies were well below the 500 ms delay-to-consciousness measured by Libet and his colleagues. Given the uncertain conditions under which the subjects were performing, with implicit knowledge of the left–right distribution developing only as the trials proceeded, the underlying decision processes were necessarily stochastic. Let the actual presence of a target trigger a linear rise of signal strength in a neural decision unit, from its initial level s_0 toward a saccade-provoking threshold s_{thresh}, at rate r. The data showed that that manipulation of the left–right distribution influenced s_0 in a manner that would be expected if the decision signal was essentially a neural estimate of the log likelihood of the (unconscious) hypothesis that the target was present. Let us take a look at that concept of likelihood.

The crux of the issue is that the likelihood L of the hypothesis H that the target is present, relative to a reference hypothesis H_{ref}, is modified by the observation O, and thereby updates the likelihood to L^*. The observation O therefore increases support for H, relative to H_{ref}, by increasing the logarithm of the perceived likelihood by a fixed amount ΔL, where

$$\log L^* = \log L + \Delta L \qquad (14.2)$$

and

$$\Delta L = \log \frac{\text{prob}\langle O|H\rangle}{\text{prob}\langle O|H_{\text{ref}}\rangle} \tag{14.3}$$

where 'prob' naturally indicates probability. If the subject's nervous system is (unconsciously) gathering information at a fixed rate, about the presence of the target, during a particular trial, there will thus be a linear rise in the log likelihood, toward the threshold value.

If the variation of r is gaussian, with mean value and variance μ and σ^2, respectively, the latency will be $(s_{\text{thresh}} - s_0)/r$ on any one trial (see Figure 14.11), and the distribution of latencies will fall on a straight line with median $(s_{\text{thresh}} - s_0)/\mu$. Moreover, the straight line will intercept the $t = \infty$ axis at the point $\mu/(\sigma\sqrt{2})$. We should note in particular that the latter point is independent of s_0 and s_{thresh}, so the straight lines for the various left–right distributions should all pass through the same intercept. As one can see from the Carpenter–Williams results reproduced in Figure 14.12, which shows how the cumulative probability varies with saccadic latency, that is precisely what is observed, and this provides strong evidence that their model is a good one.

As one can see from Figure 14.12, about 5% of the experimental data are produced by saccades with latencies shorter than would be expected. The data

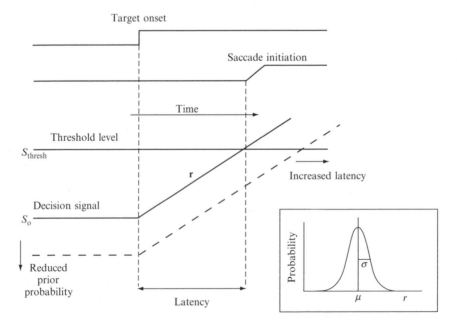

Figure 14.11   Carpenter and Williams' model of the proposed decision process for initiating saccades

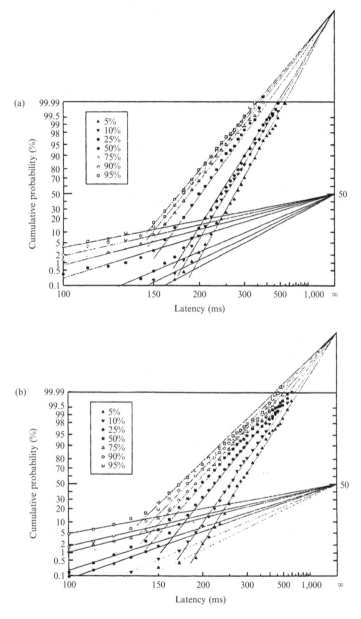

Figure 14.12 Plots of the cumulative probability of saccadic activation after a given delay, for two human subjects and for various left-right distributions of visual stimuli, revealing the existence of express saccades (data published by Roger Carpenter and Matthew Williams)

points for these express saccades also lie on straight lines, though the latter have shallower slopes, and the lines for the different left–right distributions again pass through a common intercept, which in this case is 50%. That value merely reflects the fact that the eyes will have a 50–50 chance of moving in the wrong direction if the saccade is initiated before the triggering target appears, and that in such cases the eyes will ultimately reach the target rather late. What causes such jumping of the gun? It is certainly not a product of consciousness, because the straight-line characteristic is merely a manifestation of an underlying mechanistic process, as we have already seen. Moreover, the express latencies are even further removed from the above-cited 500 ms than are their non-express counterparts. Their occurrence seems to indicate a pronounced increase of the prior probability (see Figure 14.11), at certain moments, and this could loosely be compared with the over-anxiety that makes some sprinters jump the gun in competitive events. The important message from these studies is that such heightened anticipation, and the associated hair-trigger reflexes, are not indicative of a truly free will. As Willard Quine has noted, freedom of will is nothing more than freedom to do as one will. We could add – in the present author's view, at least – that what one will is dictated by a mechanism that is deterministic and unconscious.

14.8 What Purpose Does Consciousness Serve?

If we are apprised too late of our intentions to act – as the observations of Benjamin Libet and his colleagues appear to suggest – and if we are not party to the processes underlying our decisions, what use could consciousness possibly serve? It might seem that one promising strategy would be to focus on the issues of awareness and thought. They are unavoidably implicated in consciousness, after all, and many of our thoughts do not even lead to motor acts. One might argue that thinking before acting is obviously a good thing, and that this provides the evolutionary advantage, but that line of reasoning tells us nothing about the mechanism of thought. Moreover, it sheds no light on the so-called qualia (singular: quale), that is to say the raw sensations of consciousness – the redness of red (as opposed to the mere fact of the colour red), the painfulness of pain (as opposed to the mere fact of pain), and so on. We require a biological theory of consciousness: one which closes the explanatory gap, and tells us what thought and qualia actually are, in terms of the underlying anatomy and physiology. So to reiterarate, if consciousness follows half a second after the neuronal events that really control our decisions, what evolutionary advantage could it *possibly* confer? Couldn't all human endeavour be achieved just as well by an unfeeling machine, devoid of consciousness? This is now known as the zombie argument, the word zombie referring to behaviour as an unthinking automaton, not to something from voodoo.

The present author suspects that the great step forward involved acquisition, during an individual's lifetime, of a new type of reflex: the context-specific reflex. Such acquisition, according to the present author's theory of consciousness, requires, *inter alia*, the ability to react to something that has not yet happened!

The key to understanding consciousness might therefore lie in a proper appreciation of what it means to be able to anticipate. Such anticipation could be related to any of the four basic survival functions – feeding, fighting, fleeing or procreating – though the decisive advantage may have been most closely related to the avoidance of immediate threats to life.

It is important to understand that this anticipation is not a simple reaction to already-apparent danger; even quite lowly creatures have this in their behavioural repertoires. Neither should it be confused with classical conditioning. By repeatedly pairing the sound of a bell, say, with administration of an electric shock to the soles of a rat's feet, one can ultimately get that creature to jump at the mere sound of the bell. However, the innocuous sound of the bell has simply become a surrogate for the danger; there is no anticipation involved. Moreover, as we learned in Chapter 13, even the simple aquatic mollusc *Aplysia* can be classically conditioned, and few would claim that this creature possesses consciousness.

The anticipation in question is more subtle. It involves being able to react evasively because the animal can predict that a currently-unfolding series of events may lead to danger, even though there is not yet any actual sign of a threat. Let us imagine that the anticipation can look 3 s into the future. Then although consciousness lags 0.5 s behind the unconscious mechanism making the prediction, it would nevertheless be 2.5 s *in advance of* the manifest emergence of the danger. That temporal advantage could sometimes make the difference between life and death. This conjecture is not particularly bold; we have already seen that Libet and Carpenter, and their respective colleagues, found evidence of a veto mechanism.

What is the need for consciousness; couldn't our zombie be just as good an anticipator? We have reached the pivotal point, because it is now necessary to postulate that *the context-dependent anticipatory mechanism cannot function without consciousness*. Because this isn't the same as challenging the supremacy of the unconscious in that mechanism, we have the shocking implication that *consciousness is, in effect, the servant of the unconscious*. This goes one step further than Sigmund Freud; he stressed the importance of the unconscious – here we are saying that it is the *real* controller of our actions. These are strong words. If true, they indicate that our concept of what it means to be a conscious being is ripe for revision. However, it would also mean that the explanatory gap has been closed. Let us now try to establish why consciousness should be a prerequisite for viable context-dependent anticipation, and through that for the acquisition of novel context-specific reflexes.

The familiarity of anticipation makes it an easy faculty to underestimate. Newly-born children are incapable of it; it is acquired through experience. We have already noted that movement is any animal's only externally-directed action. In fact, every muscular movement is essentially a question put to the surroundings. For example, every walking step we take asks the question: *is the ground still there*? Such questions are primarily put unconsciously. Only when we encounter something unusual, a hole perhaps or an initially overlooked tree root, does the answer break through into consciousness. However, whether or not an event is unusual must itself be established by experience. The infant, attempting to walk, encounters an obstacle, falls, and feels pain. Its nervous system records the correlation between cause and effect, and knowledge is thereby acquired. One could go further and claim that muscular movement, overt or merely simulated, is our *only* means of gaining knowledge. There is thus a counterpart of Figure 14.1, namely Figure 14.13, in which it is the environment that is the black box. The plausibility of this idea can be verified through the expedient of closing one's eyes and reaching for an object in one's vicinity. Such reaching, and indeed all other motor acts, can be regarded as questions put to the environment. However, it is important to bear in mind that not all such questions are 'asked' consciously.

It is fascinating to note that such black-box reversal also applies to the bacterium, whose case we discussed in Chapter 10, and the general applicability of the concept is hinted at by the investigation of honeybee behaviour, reported in 2000 by Mandyan Srinivasan and his colleagues. Karl von Frisch had assumed that their ability to (unconsciously) measure distance stemmed from a mechanism that keeps track of the amount of energy used during a foraging expedition. Srinivasan and his colleagues demonstrated that this is not the case. They made their bees fly through a tube decorated with black and white stripes, which the insects had to traverse on their way to a source of nectar, just 0.5 m away. By studying the bees' subsequent waggle dance, they could see that the

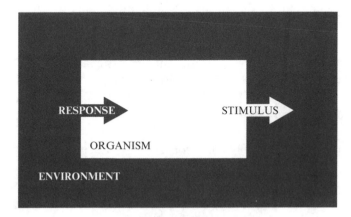

Figure 14.13 A more realistic picture of the first-person perception of consciousness reverses Figure 14.1 and views the environment as the black box. Knowledge of the latter is obtained through the person's self-paced probing, and detection of the resulting environmental feedback. But a distinction has to be made between such feedback and the unprovoked signals also impinging on the senses. This requires the attention that is a hallmark of consciousness

creatures had been duped into grossly overestimating the distance to the food. Because the striped pattern was masquerading as the alternations of light and dark usually encountered by a foraging bee, it became apparent that the insect measures distance by (unconsciously) gauging the amount of environmental feedback resulting from its own motion. This is essentially a sophisticated counterpart of the mechanism seen in the bacterium. Both creatures, *and all others I suggest*, gain information by asking questions of their environment through the agency of their own movement.

There is still something missing in the case of higher animals, however, namely *volition*. If the questions put to the environment are to enter consciousness, they must be accompanied by the necessary desire to attend to, and concentrate on, what is being probed. This means that the object of attention must be brought into working memory. In the case of an item heard, Alan Baddeley suggested that this temporary storage involves what he called the phonological loop. For an item seen, the corresponding neural circuitry constitutes what he called the visuospatial sketchpad.

Where in the brain are these circuits located? Given that we can pay attention to something while remaining muscularly passive, it might seem obvious that they are to be found exclusively in the sensory cortex (see Figure 14.14). This is in conflict with experimental data, however. An investigation carried out by Shigeo Kinomura and his colleagues, using the imaging technique known as positron emission tomography, provided direct proof of the obligatory participation of the premotor cortex during attention. Their study also revealed simultaneous activity in the thalamic intralaminar nuclei, and naturally in the appropriate areas of the sensory cortex as well. The premotor cortex lies immediately anterior (i.e. in front of) the primary motor cortex, while the thalamic intralaminar nuclei occupy a central slab of the thalamus itself. The sensory cortex is the entire region posterior to the central sulcus.

The concurrent activation of these disparate brain regions underlies the involvement of what are known as schemata (singular: schema). The present author's definition has it that a schema is comprised of *concurrently activated motor-planning and anticipated feedback patterns corresponding to a cognitive element*. One advantage of this definition lies in its generality. It can be applied even to the bacterium, despite that lowly creature's lack of a nervous system. In the bacterium, the patterns in question are merely cascading chemical reactions, the outcome of which leads to the mechanistic and unconscious decision as to whether the flagellum will continue to rotate in the same direction, or to reverse and rotate in the other direction. In the bacterium, therefore, we see a behavioural repertoire that has been telescoped down to a single binary choice: clockwise or anti-clockwise. However, that (unconscious) choice could still be used, logically, as the basis of cognition. The reversal of direction of rotation of the bacterium's flagellum can be regarded as a cognitive event. In the bacterium, therefore, there are only two possible cognitive elements.

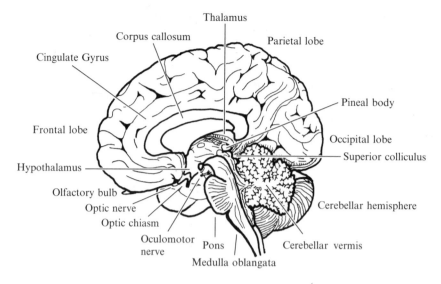

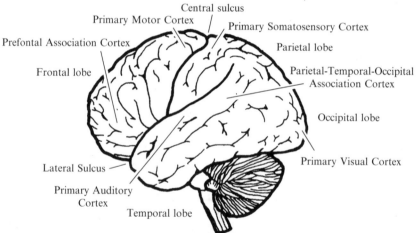

Figure 14.14 Several brain components believed to be indispensable to consciousness can be located with the help of this schematic diagram (the direction of the nose lying to the left). The premotor cortex lies immediately anterior (i.e. in front of) the primary motor cortex, while the thalamic intralaminar nuclei occupy a central slab of the thalamus itself. The sensory cortex is the entire region posterior to the central sulcus

In a creature possessing a nervous system, the above definition of the schema indicates that it is a reproducible co-activation of nerve cells linking a specific pattern of motor-planning activity in the frontal lobe (as exemplified in the studies carried out by Patricia Goldman-Rakic and her colleagues, cited at the end of this chapter) to relevant activity in a sensory region, reproducibility prevailing because schemata are laid down in memory. In the normally endowed adult human being, a good example of an elementary schema is a single phoneme, the co-activation in this case tying together the muscular movements required for its articulation with the corresponding detection, by the subject's own auditory cortex, of the uttered sound. When a baby cries, we hear it probing the environment, capturing correlations between its vocal activations and the auditory feedback it receives through its ears. As the child develops, its success in consolidating individual phonemes into words will depend upon its underlying ability to manipulate schemata. Indeed, this may be the most fundamental definition of intelligence.

14.9 Passive versus Active in Mental Processing

The honeybee and the lamprey, which we discussed earlier, do not possess consciousness because their nervous systems lack certain vital components and connections. When a child attempts its first step, prior attainment of the balanced upright position will have involved failed attempts, with attendant pain. What leads to that discomfort will have been stored as memory of possible sensory feedback resulting from certain self-paced movements. Likewise, the fact that forward motion can be achieved by specific muscular movements will already be part of a repertoire accessible unconsciously. Ultimately, the child hits upon the correct combination and timing of elemental movements and the first successful step is taken. That consolidation into a more complex motor pattern is temporarily deposited in explicit memory (see, for example, the work of Larry Squire cited at the end of this chapter), and subsequently transferred to long-term implicit memory, probably during the frequent periods of sleep which are an infant characteristic (we will return to the question of sleep and dreaming later). Soon, the toddler is able to walk without concentrating on every step, and more complicated foot-related scenarios will enjoy brief sojourns at the centre of the explicit stage.

By way of underpinning the credibility of those conjectures, we will now consider some illustrative sentences, but let us first remind ourselves that a stimulus can be followed by a response after a time delay that varies over a wide range. Indeed, one could say that this range is effectively infinite; a response can be almost instantaneous, whereas at the other extreme it may never actually come, the sensory input merely giving rise to thoughts. Many would regard that latter option as obvious, because there is no obligatory involvement of muscular movement in thought processes. As a case in point, we may note that the reader of these words may be scanning my text while remaining quite motionless. Let us now attempt to demonstrate the weakness of that assumption.

We are going to start by reading two sentences, *aloud*, even though we might be alone as we do so. Let it be stressed that these sentences are to be taken at face value; they contain no deliberate errors or other such artifice. The first sentence is: NO GNUS WERE FOUND IN THE BUSH, BUT SOME ARE SEEN ON

THE VELDT. The second, familiar to all involved in the scientific endeavor; is:
FINISHED FILES ARE THE PRODUCT OF MONTHS OF SCIENTIFIC STUDY
COMBINED WITH THE WISDOM OF YEARS. We are now going to give those
sentences a second reading, but this time we will make things somewhat more
difficult. We will count up the number of times we encounter a specific letter,
while reading each sentence aloud, straight through from start to finish, *without
repeating any sections*. To make things as easy as possible, let us first focus on
the letter N because the first sentence begins with that letter. Here we go: NO
GNUS WERE FOUND IN THE BUSH, BUT SOME ARE SEEN ON THE VELDT.
Having made a mental (or written) note of our tally, we proceed directly to the
second sentence, counting the number of times we encounter an F, because that
is now the initial letter. Here we go: FINISHED FILES ARE THE PRODUCT OF
MONTHS OF SCIEN-TIFIC STUDY COMBINED WITH THE WISDOM OF
YEARS. Let us quickly make a note of the second number, and carry on.

We will now return to that sentence composed by Max Velmans, which we
considered earlier in this chapter. We recall that it was: THE FOREST RANGER
DID NOT PERMIT US TO ENTER THE STATE PARK WITHOUT A PERMIT.
When it is read at normal speed, there is insufficient time for individual words
to enter consciousness before being pronounced. This follows from those
observations reported by Libet and his colleagues, which demonstrated that
it takes about 500 ms for an unprovoked stimulus to make its impression on
consciousness. It takes about 1/10th of that time, roughly 50 ms, to articulate a
single phoneme. Let us enquire, therefore, as to what happens when we read
Velmans's sentence. The six-letter string P-E-R-M-I-T admits of different
pronunciations, per*mit* and *per*mit, though we are apparently unaware of the
fact when reading the sentence.

We use the correct pronunciations because of the differing *contexts*, of
course, and it is interesting to contemplate the actual kinematics of articulation. Noting the above-mentioned 50 ms, and assuming an average of two
phonemes per syllable, we find that we have articulated 12 phonemes, contained in five *unambiguous* words, before we reach the potentially ambiguous
P-E-R-M-I-T, and that about 800 ms have elapsed (allowing a parsimonious
50 ms gap between words). There is thus enough time for the *meaning* of the
first few words to have started entering consciousness, and this enables the
language system to make the correct decision when faced with the task of
correctly intoning P-E-R-M-I-T.

What do we intend to convey by that word *meaning*? Let us assume that
meaning, in the context of language at least, is the province of semantics,
whereas pronunciation of single words usually belongs to the realm of syntax.
Moreover, let us assume that semantics involves the emotions while syntax does
not. This leads to the conclusion that emotion-free syntax is not predicated
on consciousness, given sufficient proficiency with language at least, while
emotion-charged semantics *does* require consciousness. That word proficiency

is important. The present author recalls the first time he encountered the word INEVITABLE, and the mirth of his classmates when he incorrectly placed the stress on the third syllable. Even the teacher found the situation hilarious, and the author's embarrassment certainly involved strong emotions.

So in acquiring reflexes, as when learning to walk or acquiring language, consciousness is a *sine qua non*; but once the necessary familiarity has been gained, individual steps can be taken, and words articulated, unconsciously. We can walk without concentrating on each individual step, once we have become proficient at walking, and we can similarly intone individual words automatically, once we have become proficient users of language.

The reader must be wondering about those tallied Ns and Fs in the other two sentences. Let us return to them, and possibly be astonished. For it turns out that there were six Ns and six Fs. If the reader is well-versed in English, the first six should cause no surprise. However, that second six may come as a shock, because the majority of people count only three Fs; we tend to overlook the F in the word OF, which occurs three times. It might seem that they were overlooked because they appear in words of minor importance, but this explanation is countered by our *not* overlooking the Ns in the three words NO, IN and ON. Our success in scoring that first six is attributable to the fact that pronunciation of all six Ns involves the same phoneme. Those three defaulting Fs, on the other hand, were missed because they were pronounced as if they were Vs. This is a profound point, and not merely an entertaining diversion. It shows that when we search for something, we have to invoke the muscular movements that would be required to characterize it.

Later, we will tie this to the important issue of raw sensation, but meanwhile we may note that these ideas do not harmonize with Ned Block's suggestion that there are two distinct forms of consciousness; phenomenal consciousness and access consciousness. Phenomenal consciousness and access consciousness, according to the view being presented here, must at the deepest level draw upon one and the same mechanism.

If the reader is less than convinced by this argument, let us also note that we can repeat, upon demand, a sentence spoken by another person only if that sentence has been consciously attended to. This too demonstrates that we are concentrating on something only if we set up the muscular patterns that would be required to repeat it. This might seem like a waste of resources, for it suggests that we constantly make such covert preparations, without using them. One would not be able to set up and *retain* the muscular motions required to regurgitate 10 min of another's narrative. The alternative is that the motor patterns, once set up, are retained for a few seconds and then discarded, only to be replaced by words heard more recently. This seems to be the gist of working memory, in the context of listened-to language. In the related context of active language, the theory would suggest that a similar thing happens, emotion-charged semantics being the counterpart of those (consciously) attended-to phrases that we store briefly when concentrating on another person's words.

There is an important connection, here, with the discovery of what are now referred to as mirror neurons, by Vitorio Gallese and his colleagues. They had located neurons in the premotor cortex of a monkey which responded when the monkey picked up a raisin, but which remained inactive when, for example, the monkey picked up a stick. There was nothing particularly unusual in this observation. However, the investigators were amazed to discover that those same neurons also became active when the monkey was observing another monkey (or even a human) picking up a raisin. It was as if the neurons were mirroring the actions of other individuals in their *own* actions. These observations were in harmony with what the present author had already suggested regarding the necessarily active nature of conscious awareness, as illustrated by the above discussion of those missed Fs. In both cases, attention was being mediated by generation of activity in the appropriate neurons in the premotor cortex.

The present author knows of no mirror neuron work extended to human observation of birds taking off to fly, but his assumption would be that this activates the same premotor neurons as those galvanized when the person flaps his arms, such action being the closest we come to imitating a bird. However, it cannot be emphasized too strongly that the mechanism embodied in the present author's theory of consciousness has to take things far beyond mere neural imitation of another's actions. Airplanes also fly, for example, but they do not flap their wings; it is doubtful that the arm-flapping neurons in our premotor cortices spring into activity when we look up at an airplane flying overhead; but there *will* be neural activity in the premotor area when we attend to that airplane. It will be seen in the neurons involved in the raw sensation associated with our observation of it; it will be seen in the neurons activated when we concentrate on the aircraft's form, on its path across the sky, on the sound it makes and, when we have become proficient at language, on our possible intonation of the word *airplane*.

14.10 The Relevant Anatomy and Physiology

From the above definition of a schema, we see why the motor-planning area and the sensory processing area have to be implicated, but why also the thalamic intralaminar nuclei? They enter into suggestive interactions with several other brain components, including the anterior cingulate (see Figure 14.15), intimately

---

Figure 14.15 The black and white lines represent excitatory and inhibitory routes, respectively, not individual axons, and branching of lines indicates duality of routes, not the presence of axon collaterals – the somewhat thicker black lines at the upper right indicate influence in a more general manner; only two representative regions of the cerebellum are shown to avoid unnecessary complexity

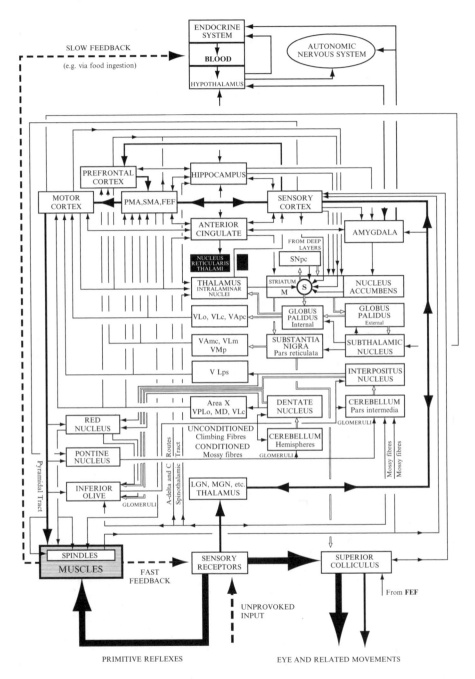

(The abbreviations for thalamic nuclei are: LGN, lateral geniculate nucleus; MGN, medial geniculate nucleus; MD, medialis dorsalis; V Amc, ventralis anterior medialis caudalis; V Apc, vetralis anterior pars principalis; VLc, ventralis lateralis caudalis; VLm, ventralis lateralis pars medialis; VLo, ventralis lateralis pars oralis; VLps, ventralis lateralis pars postrema; VMp, ventralis medialis postralis; VPLo, ventralis posterior lateralis pars oralis.)

involved in the perception of pain, the cerebellum, and the basal ganglia, which themselves are connected to the amygdala, a key structure in the service of emotion. Let us consider them in that order.

When probing our environment, overtly or merely through thought, the system requires a sentinel to watch out for possible danger and redirect the focus of attention accordingly. This task is centred on the anterior cingulate, and on the structure it activates, namely the nucleus reticularis thalami. (The latter brain component was indetified by Francis Crick as the likely mediator of what he called the searchlight of attention.) In order to be one step ahead of such danger, the system needs to be able to draw on the predictive power of the cerebellum (and ultimately also on the prefrontal cortex). Furthermore, there will have to be a component which serves as an adjudicator, evaluating which experiences really were threats to survival, and directing the system to store memories of them for later use. This is the role played by the amygdala.

Then again, the system will require what could loosely be called a clutch mechanism, allowing some planned movements to be executed, while holding back on others because of the anticipated danger. This task appears to be handled collectively by the basal ganglia, which we noted are under the control of the adjudicating amydala. When learning to drive, it is not uncommon for the novice to inch forward in a series of jerks, through imperfectly controlling the clutch. The same result will plague the expert if the clutch itself is malfunctioning. The effect is sadly reminiscent of the jerky movements displayed by victims of Parkinson's disease, which is caused by faulty biochemistry in a component of the basal ganglia known as the substantia nigra (see Figure 14.15).

We can think without acting, act without thinking, act while thinking about that act, and act while thinking about something else. Our acts can be composite, several muscular patterns being activated concurrently, though we appear not to be able to simultaneously maintain two streams of thought. When we think about one thing while doing something else, it is always our thoughts which are the focus of attention. This suggests that there are least two thresholds, the higher associated with overt movement and the lower with thought.

Assuming that the signals underlying competing potential thoughts must race each other to a threshold, as the above-cited work of Carpenter and Williams indicates, it may be highly significant that cortical and thalamic projections form no strong loops, as was noted by Francis Crick and Christof Koch in 1998. The presence of strong loops could make overt movement too automatic. There is also a second possible penalty; thoughts might otherwise establish themselves by default. One should note that overt movement and mere imagery – that is, covert preparations for movement – appear to involve identical brain regions, as has been established by Jean Decety, and independently by Marc Jeannerod. Indeed, Gabriel Kreiman and his colleagues have demonstrated that this equivalaence applies even at the level of *individual neurons*.

Figure 14.15 indicates that the competition is played out in a group of collectively functioning components, these being the sensory, motor and anterior cingulate areas of the cortex, the thalamic intralaminar nuclei (in conjunction with the nucleus reticularis thalami, or NRT for short), the amygdala, and the striatum. In mammals, Ann Graybiel and Clifton Ragsdale Jnr have found that the latter has a heterogeneous structure in which the continuous matrix (denoted by the letter M in Figure 14.15) is interdigitated with the isolated striosomes (S in Figure 14.15 – here represented by a single member). The input to the striosomes appears to be more intimately connected to the components just identified.

Given that its output reaches the primary motor cortex, via the internal segment of the globus palidus, whereas the matrix output does not, it seems that the striosomes may be more essentially related to consciousness. Likewise, the pars intermedia seems to be the more intimately consciousness-related part of the cerebellum, because it has analogous projections. The threshold for overt movement may be exceeded only when both motor-cortex-feeding components are dispatching signals concurrently. The matrix, conversely, appears to serve already-established motor patterns, because its output ultimately reaches the premotor area, PMA, (and supplementary motor area, SMA) and the prefrontal area (PFA). Its cerebellar partner is clearly the hemispherical region. It is worth noting that the cerebellar hemispheres are particularly prominent in the primates, and that they are pre-eminent in humans; they appear to bear much of the responsibility for making us what we are.

The target of the competition for attention appears to be the PMA–SMA, because it receives from all the thalamic nuclei handling basal ganglia (BG) and cerebellar (Cb) output. Attention is influenced by more remote regions of the system, which feed signals to those BG-Cb components. We have already noted the amygdala's control of the BG. The inferior olive seems to play a complementary role for the Cb, sending signals through the climbing fibres when something unexpected occurs and, because a corresponding memory trace will not yet have had time to develop for this novel situation, disturbing the permissive effect of the disinhibition (see the white arrows in Figure 14.15). The periodic shifting of attention, as when we simultaneously converse (or merely think) and drive in a busy thoroughfare, must make considerable demands on the putative (differential) clutch mechanism mentioned earlier, and this could be the dual responsibility of the substantia nigra pars compacta (SNpc) and the sub-thalamic nucleus, which appear to serve as gain controls for the striosome-related and matrix-related routes, respectively. Whence – the present author suggests – our ability to (covertly) think of one thing while overtly doing another.

Thoughts, according to this scheme, are merely simulated interactions with the environment, and their ultimate function is addition of new implicit memories, new standard routes from sensory input to permitted motor output – new optimized context-specific reflexes. The term proxied movement appears to be appropriate here. For a given set of synaptic couplings between PMA–SMA and the primary motor cortex, a specific pattern of output signals from the former will produce a specific sequence of muscular movements. Duplicates of those output signals (the above-mentioned efference copies), dispatched through axon collaterals, will carry the full information sent to the muscles, via the primary motor cortex, but they will not directly produce movement because their target neurons are not immediately concerned with motor output. Those efference-copy signals may be above the threshold for thought, however, and the latter will thus be subtly tied to a pattern of motor output; whence *proxied* movement. This is the mechanism which is believed to underlie that overlooking of those Fs, in the three occurrences of the word OF; the system was searching for Fs by activating the proxied movements embodied in the schema for F, and this is distinct from the schema for V.

The duality of routes, and the fact that these overlap in the PMA–SMA region, could well underlie the interplay between explicit and implicit in brain function. That region might be the common ground where semantics dictates syntax, and where the first step of the infant invokes elemental motor fragments already learned. Given its strategic location, and its connections to other brain parts, the hippocampus is ideally placed to mediate the consolidation needed when complex reflexes are thereby acquired (see Figure 14.15). According to the ideas being expressed here, it ties signals in the motor-planning areas to the activity in the sensory cortex (SC) that results from the corresponding movement, as has been established by Neal Cohen and his colleagues.

When the motor output is straightforwardly related to movement of the body within the environment, neurons in the hippocampus will appear to be place cells, as originally suggested by John O'Keefe and Lynn Nadel, in 1978, but the hippocampus must serve more than mere spatial navigation, as noted by Eleanor Maguire and her colleagues. It captures correlations between all motor patterns and the resulting sensory feedback; and the captured contexts are thereafter available for generalized muscular *navigation*, this naturally including the version employed by the articulatory system in speech.

When is the system updated, so as to incorporate the new schema into the system's existing repertoire of schemata? This could happen continuously throughout the day, of course, but the present author would like to make an alternative suggestion, namely that it occurs during dreaming. Indeed, the present author would like to put forward his own theory of dreaming. He suggests that the updating would most conveniently take place when the body is immobilized, as during dreaming sleep, because it has perforce to activate those existing schemata that are related to the new schema – a candidate for incorporation into the nervous system's repertoire. When the muscles are immobilized, no danger will arise when those existing schemata are activated.

It may be noted, in passing, that this theory automatically explains why the plots of dreams tend to be related to recent events.

As is well known, dreaming is accompanied by the rapid movement of the eyes, under their lids. Indeed, rapid-eye-movement sleep, or REM sleep, is an alternative name for the dreaming state. The eye muscles are thus the only muscles that are *not* immobilized during dreaming. However, this is of little consequence, given that the lids are closed, and that no feedback signals are being received from the environment that would require incorporation into the schemata being activated. So why do the eyes move at all during dreaming? This might be for nothing more sophisticated than mere good house-keeping! The point is that the blood circulatory system requires that we change posture periodically during slumber, to prevent blood clots. (The development of deep-vein thrombosis amongst long-haul airplane passengers has highlighted this problem.) This is why the dreaming period is limited to a few minutes. The periodic frenetic activity of the eye-balls may be desirable because this simply massages the corneas, and ensures adequate blood flow to these vulnerable ocular components.

Returning to the issue of qualia, we note that their emergence must have been underpinned by an evlutionary advantage, which perforce was mediated by muscles if action was involved. Now we saw earlier that the signals which activate muscles are also sent to the spindles, and these compare the desired length of the muscle with its actual length. Only if there is disparity are additional signals then sent to the main muscle fibres, thereby modifying the degree of muscular contraction. Even the muscles of quite lowly creatures display this heterogeneity, but something quite suggestive first emerged with the mammals, namely that copies of those length-disparity signals are sent in the reverse direction, back to the cerebral cortex. The brain is thus kept informed of the current state of each muscle, and of the rate at which that state is currently changing.

With our muscular movements underlying exploration of our environments, and our senses detecting the resultant environmental feedback, while the message-dispatching spindles keep the system informed of the consequences for the motor machinery, all the requirements are fulfilled for viable exploit-ation of that life-preserving veto mechanism. Such dual awareness of our surroundings and their consequence for our bodies is, by definition, conscious-ness.

As we have already noted, the elementary unit of sensation (the quantum of qualia, so to speak) is the quale. Let us link it to that other elementary unit, the schema. We recall that the latter's definition coupled the premotor cortex with the sensory cortex, in sufficiently advanced animals, and that memorized schemata are acquired through our muscular explorations. These do not necessarily involve consciousness, but when things or events catch our atten-tion, the thalamic intralaminar nuclei become activated, and these in turn feed

activation into that basal ganglia clutch, which is also governed by the signifi-
cance-gauging (and emotion-mediating) amygdala. The conclusion must be
that a schema gives rise to a quale when the emotions are participating. That
situation is what we usually associate with the term *intentionality*. Our *un*inten-
tional explorations, on the other hand, involve schemata but not qualia.

It is tempting to write this formally, although it certainly seems premature to
attempt something along the lines of a mathematical equation. Let us suggest,
therefore, that the link between a quale and a shema is given by

$$\text{SCHEMA} \ast \text{ATTENTION} \rightarrow \text{QUALE}$$

this being read as follows: *a schema, when accompanied by attention, leads to a
quale.*

The indications are, then, that an object will not be attended to (and thus not
perceived) unless the signals it provokes have set up *possible* motor acts in the
premotor area, this response putting previously established associations on the
alert. (Such setting up inevitably requires time, the underlying process being an
example of self-organization; this is probably the origin of that 0.5 s delay
observed by Benjamin Libet and his colleagues.) One's perception of the colour
red will thus strike up possible connections to fire-engines, ripe strawberries,
blood and the like; and it will generate activity in those parts of the motor-
planning region that would, if permitted, activate the vocal apparatus required
to articulate the word *red*.

So the body expresses intentionality by preparing for motor acts. Our
spindle-mediated qualia thereby underlie sensations that might not have
seemed to be related to muscles, such as nausea, hunger, thirst and pain. The
sensation of nausea is related to muscular preparations to vomit, for example,
hunger by the implicit rehearsals of bringing food to the mouth and chewing it,
and so on. Revealing evidence also comes from changes accompanying increased stimulus intensity. The sensation we have when listening to music
seems nebulous, until a favourite piece sends shivers down the spine, thereby
exposing the muscular connection. The influence of mental state on the body's
musculature is familiar to the physiotherapist, of course, and the goal of the
masseur is as much a relaxed mind as it is relaxed muscles.

Consciousness, then, arises from the dialogue between our nervous system
and our muscles, and thought is simply internal simulation of the body's
interactions with the environment. This means that there could be no disem-
bodied consciousness. All mammals possess the anatomy that permits the vital
brain–muscle exchanges and they are probably all conscious, when awake. If
this is true, the explanatory gap has been closed, and the hoary mind–body
schism exposed as a sham, because body is intimately involved in mind. It is a
tale that may offend high-flown intellectualism, but the indications are that we
know with our bodies. Thinking is a bodily function.

In dictionaries, consciousness is usually characterized as an awareness of one's surroundings, of one's situation, and of one's thoughts and feelings. However, such definitions shed no light on the question of where this awareness is being registered. Where is one's me-ness? A strong hint comes from considering the nature of raw sensations. Pain, in particular, reveals their utility, through what we have been referring to as their associated qualia. They enable us to pinpoint something requiring attention (part of the body in this case); and phantom limb pain is special because it is felt in a body part that no longer exists. This suggests that a sensation felt in one part of the body must be a collective product of processes taking place in some or all of the others. How and where is the responsibility apportioned amongst those surviving members? The PMA–SMA would seem to be involved because that region has to ensure an absence of conflict between the muscles in any planned motor sequence.

The impressive mirror-aided 'amputation' of a phantom limb, by Vilayanur Ramachandran was accomplished by making the patient accept as real a limb merely seen in reflection. This probably gave the BG and Cb the wherewithal to readjust the PMA–SMA's locally effective neuronal connections. Likewise, the sensation associated with something perceived beyond the body – a red object, say – must be linked to the anticipation to react. Such anticipation requires measurement of *rate* of change. Without this capability, reliable prediction would be impossible. Only one type of component is imbued with that higher-order capacity; the sub-class of muscle spindles comprising nuclear bag fibres. It is interesting that PMA–SMA has direct links with the spinal cord, this route paralleling the pyramidal tract. If these connections run to the spindles (via intermediate synapses, of course), as indicated in Figure 14.15, they could co-operate with the route returning signals to the SC, such return being an exclusively mammalian anatomical feature.

It is still a moot point as to whether thought *has* to involve the spindles. In the circuit shown in Figure 14.15, one can trace routes which activate the attention-directing inner loops linking the PMA(SMA), anterior cingulate, NRT and ILN, and also the Cb, BG and amygdala, without necessarily provoking activity in the spindles. It thus seems possible that thought might be possible even when the system is denied access to the spindles, in the adequately trained individual at least. This could explain why people suffereng from the dreadful locked-in syndrome nevertheless retain access to consciousness.

14.11 Intelligence and Creativity

The premotor and supplementary motor cortices are essentially the higher counterparts of the pattern generators located in the spinal cord, the classic example of which are those found in the lamprey, which we discussed above. The difference is that although a variety of motor patterns can be generated by the latter, the available repertoire cannot be altered. The patterns generated by the PMA–SMA, on the other hand, show far greater sophistication, and they *can* be reprogrammed. It is this latter fact which suggests a simple definition of intelligence. *Intelligence is the ability to consolidate already-learned motor patterns into more complex composites, such consolidation sometimes being merely overt, rather than covert.*

This definition has been discussed in the context of autism, by the present author, and the prediction that autistic children would encounter difficulties in executing and learning motor sequences was subsequently confirmed by Philip Teitelbaum and his colleagues. A normal child, lying on its back and wanting to roll over onto its front, soon learns that this can be readily accomplished if first the head, then the shoulders, and finally the hips are swiveled in the same direction. If the timing of this sequence is correct, the supine-prone transition requires a minimum of effort. Autistic infants appear to experience considerable difficulty in learning this simple motor sequence. Indeed, the sequence does not even occur in their failed attempts. Instead, they awkwardly arch their backs and ultimately fall into the desired position.

In general, when a new motor pattern is being acquired, both the means and the ends will be coded in currently active patterns of neuronal signals. Also, there must be interactions between these patterns because the goal will influence the route through muscular hyperspace by which it is to be achieved. This, after all, is what is implied by the use of the term schema. The signalling pattern related to the goal will have to persist for a longer time than that related to elemental schemata, because it has to provide the necessary continuity. In the now-standard terminology, there must be a working memory, and there is abundant evidence that this implicates the prefrontal cortex (PFC), as discussed above. The PFC probably dictates *sequences* of elementary muscular sequences, but it must be borne in mind that the sophistication of the latter will depend upon what the individual has already learned. A ballet dancer would regard as an elementary motor pattern a muscular sequence which the novice would find quite difficult.

The circuitry depicted in Figure 14.15 indicates how the mechanism might function. The dentate nucleus of the cerebellum (DN) and the SNpr both feed signals not only to the SMA–PMA region, but also to the PFC, and there is a route from the latter to the former. Now the persistence time for signals in the PFC is longer than it is for the SMA–PMA (as established by the work of Joaquin Fuster, for example), so the former area seems ideally suited to the dual job of dictating the sequence of sequences and of providing the necessary temporal continuity. As was also mentioned above, the system's ability to simultaneously handle both means and ends, *in a context-specific manner*, may depend upon that matrix-striosome heterogeneity in the striatum.

With sufficient experience, schemata become permanently laid down in memory, and even sequences of muscular sequences become second nature.

The cliché is the manifestation of such consolidation in the language domain. It is a noteworthy feature of the underlying anatomy, shown in Figure 14.15, that it includes direct connections between the SC and the PFC. And the Cb's contribution to such ingrained behaviour appears similarly to be served by the direct route connecting the SC to the Cb, via the pontine nucleus. In the case of the former route, the hippocampus is ideally situated to record the correlations between the sequencing signals dispatched to the muscles and the resulting temporally patterned feedback from the environment, because of its widespread reciprocal connections with virtually the entire cerebrum except the motor cortex (see, for example, the work of Larry Squire, cited at the end of this chapter). It is in this sense that the adjective *context-specific* is being used here, rather than in the more limited sense of background clues.

However, there might not be a sharp distinction between these connotations because even the information stemming from the background will often not be unvarying, because of the animal's own movements In any event, there is evidence that the hippocampus is implicated in context-dependent memory processing. In view of the recent findings of Robert Stickgold, it seems likely that the consolidation process is one of the main processes accomplished by sleep, both of the rapid eye movement (REM) and non-REM types. An important aspect of this consolidation will be the detection of novelty, because only novel correlations need be candidates for memorizing. This was touched on above, in the present author's theory of dreaming

The views that have been expressed here would lead us to expect that the premotor and supplementary motor areas, the anterior cingulate, and possibly also the PFC would all be involved in novelty detection, and this is indeed what is observed (see the work of Endel Tulving, for example, cited at the end of this chapter).

Mere proficiency with the above-discussed consolidation process does not, of itself, guarantee that an individual will be able to invent novel solutions. Intelligence, of the relevant type, might be a prerequisite for creativity, but it is not sufficient. Roger Carpenter's ingenious race-to-threshold mechanism received strong support from his own experimental results, discussed above; and his equally insightful ideas about the randomization of behaviour, published in 1999, appear to hold what might be a vital clue to the neuronal mechanism underlying creativity. However, that mechanism will produce only familiar winners, in the absence of new runners.

Such fresh contenders might be the product of unexpected correlations being captured by those areas involved in the detection of novelty. The correlations could then be returned by the feedback connections shown in Figure 14.15, the importance of these connections in learning having been stressed by Wolf Singer in 1995. Indeed, the cause of novelty might be particularly well served by the fact that those feedback connections achieve a wider spread than their feed-forward counter-parts, as discovered by Semir Zeki and Stewart Shipp in

1988. If this scenario proves to be reliable, it would indicate that an individual's anatomical endowment, with respect to the actual spread of connections between areas, is a vital ingredient in creativity. It would indicate that a creative mind is something that one is born with, though the fact that synapses continue to be formed during an infant's first 2 years also indicates that richness of experience at that tender age can be a contributory factor.

Mention of the early years of life tempts us to make a brief digression, and suggest a litmus test for the possession of consciousness in a species. Given that self-paced probing of the environment plays such a vital role in the present author's theory of consciousness, one might guess that consciousness must be present in those species whose young indulge in play.

Returning to creativity, the question arises as to the venue for the putative race, and Figure 14.15 strongly suggests that the competing signals impinge upon the premotor and supplementary motor areas, and probably also on the PFC. The race's judge is probably the amygdala, given its control of access to the BG, and the competition must also be influenced by attention, which would involve the anterior cingulate, the NRT and the thalamic ILN.

14.12 A Final Word

To the best of the present author's knowledge, this is the first book on biophysics to include discussions of consciousness and biological intelligence.This is reflected in the relatively large number of citations to the writings of other authors, at the end of this chapter. The relevant literature is probably not so accessible to people studying biophysics, so they may appreciate this additional help with approaching unfamiliar territory.

Some people may indeed find inclusion of these topics rather surprising, not to mention inadvisable. I would vigorously defend the coverage given to these issues, for they provide a natural conclusion to the story that has been unfolding in the preceding 13 chapters. We have progressed from the realm of atoms and molecules, up through successive levels of structure, and even considered the movement of entire organisms, in the service of those aspects of behaviour most directly related to survival. *En route*, we have also considered the acquisition and use of energy. What could be more natural than to round off the tale with a look at how the entire organism is controlled, as it seeks to function optimally in its environment?

The novelty of bringing such issues within the compass of biophysics is attested to by the relative paucity of mathematical treatment. This chapter has included a few equations, but the bulk of the treatment has been perforce more qualitative than I would have liked. The situation is currently changing, however, and those with a taste for the mathematical will hopefully be pleased by my intention to increase gradually the amount of quantitative treatment of behaviour in future editions. The future for biophysics promises to be a bright one.

Exercises

14.1 Write an essay on conscious and unconscious reactions to stimuli.

14.2 Write an essay on the significance of the experiments of Benjamin Libet and his colleagues.

14.3 Write an essay with the title: Do we have free will?

14.4 List the types of behaviour you would want to observe in a computer simulated pseudo-person before concluding that it possessed consciousness.

Further reading

Baddeley, A. D. and Hitch, G. J., (1974). Working memory. In: Bower, G. H. (Ed.), *The Psychology of Learning and Motivation*, vol. 8. Academic Press, New York.

Berthier, N. E. *et al.*, (1993). Distributed representation of limb motor programs in arrays of adjustable pattern generators. *Journal of Cognitive Neuroscience* **5**, 56–78.

Block, N., (1995). On a confusion about a function of consciousness. *Behavioural and Brain Science* **18**, 227–247.

Carpenter, R. H. S., (1996). *Neurophysiology* Arnold, London.

Carpenter, R. H. S., (1999). A neural mechanism that randomizes behaviour. *Journal of Consciousness Studies* **6**, 13–22.

Carpenter, R. H. S. and Williams, M. L. L., (1995). Neural computation of log likelihood in the control of saccadic eye movements. *Nature* **377**, 59–62.

Cohen, N. J. *et al.*, (1999). Hippocampal system and declarative (relational) memory: summarizing the data from functional neuroimaging studies. *Hippocampus* **9**, 83–98.

Cotterill, R. M. J., (1995). On the unity of conscious experience. *Journal of Consciousness Studies* **2**, 290–312.

Cotterill, R. M. J., (1997). Navigation, consciousness and the body: mind 'problem'. *Psyke and Logos* **18**, 337–341.

Cotterill, R. M. J., (1998). *Enchanted Looms Conscious Networks in Brains and Computers*. Cambridge University Press, Cambridge.

Cotterill, R. M. J., (2001). Cooperation of the basal ganglia, cerebellum, sensory cerebrum and hippocampus: possible implications for cognition, consciousness, intelligence and creativity. *Progress in Neurobiology* **64**, 1–33.

Crick, F., (1984). The function of the thalamic reticular complex: the searchlight hypothesis. *Proceedings of the National Academy of Science, USA.* **81**, 4586–4590.

Crick, F. and Koch, C., (1998). Constraints on cortical and thalamic projections: the no-strong-loop hypothesis. *Nature* **391**, 245–250.

Decety, J., (1996). Do imagined and executed actions share the same neural substrate? *Cognitive Brain Research* **3**, 87–93.

Dowling, J. E., (1987). *The Retina: An Approachable Part of the Brain*. Harvard University Press, Cambridge, MA.

Dum, R. P. and Strick, P. L., (1991). The origin of corticospinal projections from the premotor areas in the frontal lobe. *Journal of Neuroscience* **11**, 667–689.

Fuster, J. M., (1985). The prefrontal cortex: mediator of cross-temporal contingencies. *Human Neurobiology* **4**, 169–179.

Gallese, V. *et al.*, (1996). Action recognition in the premotor cortex. *Brain* **119**, 593–609.

Goldman-Rakic, P. S., (1992). Working memory and the mind. *Scientific American* **267** (3), 72–79.

Graybiel, A. M. and Ragsdale, C. W., (1978). Histochemically distinct compartments in the striatum of human, monkeys, and cat demon-strated by acetylcholinesterase staining. *Proceedings of the National Academy of Science, USA*. **75**, 5723–5726.

Grillner, S. *et al.*, (1995). Neural networks that co-ordinate locomotion and body orientation in the lamprey. *Trends in Neurosciences* **18**, 270–279.

Hammer, M., (1997). The neural basis of associative reward learning in honeybees. *Trends in Neurosciences* **20**, 245–252.

Hanes, D. P. and Carpenter, R. H. S., (1999). Countermanding saccades in humans: evidence for a race-to-threshold process. *Vision Research* **39**, 2777–2791.

Hoover, J. E. and Strick, P. L., (1999). The organization of cerebellar and basal ganglia outputs to primary motor cortex as revealed by retrograde transneuronal transport of herpes simplex virus type 1. *Journal of Neuroscience* **19**, 1446–1463.

Houk, J. C. and Wise, S. P., (1995). Distributed modular architectures linking basal ganglia, cerebellum, and cerebral cortex: their role in planning and controlling action. *Cerebral Cortex* **5**, 95–110.

Jeannerod, M., (1999). To act or not to act: perspectives on the representation of actions. *Quarterly Journal of Experimental Psychology* **52A**, 1–29.

Kinomura, S. *et al.*, (1996). Activation by attention of the human reticular formation and thalamic intralaminar nuclei. *Science* **271**, 512–515.

Kreiman, G., Koch, C. and Fried, I., (2000). Imagery neurons in the human brain. *Nature* **408**, 357–361.

Kuffler, S. W., Nicholls, J. G. and Martin, A. R., (1984). *From Neuron to Brain: A Cellular Approach to the Function of the Nervous System*. Sinauer, Sunderland, MA.

Libet, B. *et al.*, (1979). Subjective referral of the timing for a conscious sensory experience: a functional role for the somatosensory specific projection system in man. *Brain* **102**, 193–224.

Maguire, E. A. *et al.*, (1998). Knowing where and getting there a human navigation network. *Science* **280**, 921–924.

Matthews, P. B. C., (1972). *Mammalian Muscle Receptors and their Central Actions*. Edward Arnold, London.

Mel, B. W., (1994). Information processing in dendritic trees. *Neural Computation* **6**, 1031–1085.

Middleton, F. A. and Strick, P. L., (1994). Anatomical evidence for cerebellar and basal ganglia involvement in higher cognitive function. *Science* **266**, 458–461.

Milner, B., Petrides, M. andSmith, M. L., (1985). Frontal Lobes and the temporal organization of memory. *Human Neurobiology* **4**, 137–142.

O'Keefe, J. and Nadel, L., (1978). *The Hippocampus as a Cognitive Map*. Clarendon Press, Oxford.

Oscarsson, O. and Rosén, I., (1963). Projection to cerebral cortex of large muscle-spindle afferents in forelimb nerves of the cat. *Journal of Physiology* **169**, 924–945.

Pardo, J. V. *et al.*, (1990). The anterior cingulate cortex mediates processing selection in the Stroop attentional conflict paradigm. *Proceedings of the National Academy of Science, USA*. **87**, 256–259.

Procyk, E., Tanaka, Y. L. and Joseph, J. P., (2000). Anterior cingulate activity during routine and non-routine sequential behaviors in macaques. *Nature Neuroscience* **3**, 502–508.

Purpura, K. P. and Schiff, N. D., (1997). The thalamic intralaminar nuclei: a role in visual awareness. *The Neuroscientist* **3**, 8–15.

Quine, W. V. O., (1960). *Word and Object*. M.I.T. Press, Cambridge, MA.

Rainville, P. *et al.*, (1997). Pain affect encoded in human anterior cingulate but not somatosensory cortex. *Science* **277**, 968–971.

Ramachandran, V. S. and Hirstein, W., (1998). The perception of phantom limbs. *Brain* **121**, 1603–1630.

Rescorla, R. A. and Wagner, A. R., (1972). A theory of Pavlovian conditioning: variations in the effectiveness of reinforcement and nonreinforcement. In: *Classical Conditioning II: Current Research and Theory* A. H. Black and W. F. Prokasy, eds. Appleton-Century-Crofts, New York.

Sherrington, C. S., (1924). Problems of muscular receptivity. *Nature* **113**, 892–894.

Singer, W., (1995). Development and plasticity of cortical processing architectures. *Science* **270**, 758–764.

Squire, L. R., (1992). Memory and the hippocampus: a synthesis from the findings with rats, monkeys and humans. *Psychological Review* **99**, 195–231.

Srinivasan, M. V. *et al.*, (2000). Honeybee navigation: nature and calibration of the odometer. *Science* **287**, 851–853.

Stickgold, R., (1998). Sleep: off-line memory processing. *Trends in Cognitive Sciences* **2**, 484–492.

Teitelbaum, P. *et al.*, (1998). Movement analysis in infancy may be useful for early diagnosis of autism. *Proceedings of the National Academy of Science, USA.* **95**, 13982–13987.

Tulving, E. *et al.*, (1994). Novelty encoding networks in the human brain: positron emission tomography data. *Neuroreport* **5**, 2525–2528.

Velmans, M., (1991). Is human information processing conscious? *Behavioural and Brain Sciences* **14**, 651–669.

Von Frisch, K., (1993). *The Dance Language and Orientation of Bees*. Harvard University Press, Cambridge, MA.

Zeki, S., (1993). *A Vision of the Brain*. Blackwell, Oxford.

Zeki, S. M. and Shipp, S., (1988). The functional logic of cortical connections. *Nature* **335**, 311–317.

Appendix A: Elements of Quantum Mechanics

A.1 Quantization of Energy

Toward the end of the 19th century, it appeared that most physical phenomena had received reasonably satisfactory explanations, but there were three issues which continued to defy the theoreticians. The first of these stemmed from the failure by Albert Michelson and Edward Morley to detect what was known as the ether drift effect, and it indicated that light is an electromagnetic wave that requires no medium for its propagation. Albert Einstein concluded that the speed of light is a constant, and this ultimately led to his theory of relativity. That theory later indicated the need for corrections to certain of the parameters we will be encountering in this appendix, and a particularly spectacular consequence was Paul Dirac's prediction of the existence of anti-particles. The subsequent discovery of the positron confirmed that prediction, and the mutual annihilation of a positron and an electron, to produce two gamma particles, is now exploited in the technique of positron emission tomography, which permits imaging of activity in the brain. Einstein's ideas could thus be said to have had an important impact on biophysics.

The remaining two of the above difficulties concerned what was known as the ultraviolet catastrophe and the sharp spectral lines associated with the elements. Let us consider those problems in turn. When an object such as a piece of metal is heated, it emits radiation whose spectral composition depends upon its temperature. Heated tungsten, for example, glows red when its temperature T reaches about $1000\,\mathrm{K}$ (about $700\,°\mathrm{C}$), and it gives off primarily shorter wavelengths (that is to say, higher frequencies) when it is further heated to incandescence, as in a switched-on light bulb. The intensity $I(\lambda, T)$ is small at sufficiently long wavelengths and it also tends to zero as the wavelength λ approaches zero. It is the maximum of the intensity spectrum, at some intermediate wavelength, which changes its position with temperature.

Theories of that time were not in agreement with these spectral characteristics. They could reproduce the behaviour at long wavelengths, and they correctly predicted that $I(\lambda, T)$ would increase as λ decreases. However, they predicted no falling away of $I(\lambda, T)$ as λ tends toward zero, that is to say they predicted no spectral maximum. This was a catastrophe for the theories, but not for us and other organisms, because according to these analyses any hot object would have been a lethal weapon, giving off large amounts of ultraviolet and even shorter wavelength radiation.

A case in point was the theory produced by Lord Rayleigh (John William Strutt) and James Jeans, according to which thermal radiation emanates from oscillating (and thus periodically accelerated) electrical charges, as is the case in a transmitting antenna. Their derived relationship was

$$I(\lambda, T) = \frac{2\pi c k_B T}{\lambda^4} \qquad (A.1)$$

where c is the speed of light and k_B is Boltzmann's constant (see Appendix C), and it predicts that $I(\lambda, T)$ would become infinite for zero wavelength.

Max Planck realized that the discrepancy was arising from the assumption that the energy emitted by the electromagnetic oscillators could vary continuously. He suggested, instead, that the energy is constrained to adopt discrete values, which became known as quanta (singular: quantum). Because there will be a statistical distribution of these quantized oscillations, the above expression has to be modified, to become

$$I(\lambda, T) = \frac{2\pi h c^2}{\lambda^5 (e^{hc/\lambda k_B T} - 1)} \qquad (A.2)$$

in which the constant h could be adjusted to fit the data. The latter is referred to as Planck's constant, and it has the value $h = 6.626 \times 10^{-34} \, \text{J s}$.

Albert Einstein soon realized that Planck's hypothesis could be used to explain a key aspect of the photoelectric effect. When the surface of a material is irradiated with light, electrons will be emitted, but only if the wavelength of the incident light is less than a certain threshold value (which is equivalent to saying that the frequency must exceed a certain threshold, because $c = \nu\lambda$, where ν is the frequency). The threshold varies from material to material. The emission of an electron occurs if the energy with which it is bound to the material can be overcome, that energy being known as the work function Φ. Einstein saw that this condition would be fulfilled only if the energy of the incident light photons is greater than or equal to Φ, and he put forward the following expression for the kinetic energy of the ejected electrons

$$\frac{1}{2}\mu_e v^2 = h\nu - \Phi \qquad (A.3)$$

in which μ_e is the mass of the electron and v is its speed. The term $h\nu$ thus represents the photon energy, and Equation (A.3) also explains why the kinetic energy of the emitted electrons is linearly proportional to the frequency of the incident light, why that kinetic energy is independent of the intensity of the light, and why electrons are emitted *immediately* when the light falls upon the surface. At a single stroke, Einstein had established the corpuscular nature of light and given decisive endorsement to Planck's quantum theory.

A.2   Atomic Structure

Niels Bohr invoked the quantum ideas of Planck and Einstein in his strikingly simple model for the structure of atoms, which explained the above-cited puzzle of the sharp spectral lines. Ernest Rutherford had discovered that almost all the mass of an atom is concentrated in a small region at its centre, the positively-charged nucleus, and Joseph (J. J.) Thomson had discovered the very much smaller, and negatively-charged, electron. However, this raised the further difficulty of explaining how the electrons avoided being drawn into the nucleus because of the attraction between the positive and negative charges.

Bohr assumed that the electrons lie in orbits around the nucleus, and he argued that there must be a lowest-energy state of an atom, now also called a ground state, which corresponded to the orbit that lies closest to the nucleus. He also realized that a consequence of quantization would be that the various orbits would be associated with only certain discrete energies. An electron in the ground state, according to this model, would be unable to mutually annihilate with a positive charge on the nucleus because the associated energy change would be only a fraction of a quantum, and such a fraction could not exist.

A transition from one orbit to another would lead to the emission or absorption of a photon, depending on which of the orbits was associated with the larger energy, and Bohr's expression for the photon energy was

$$\mathscr{E}_i - \mathscr{E}_f = h\nu \tag{A.4}$$

where $\mathscr{E}_i$ and $\mathscr{E}_f$ are the energies of the initial and final states, respectively. If $\mathscr{E}_f > \mathscr{E}_i$, the situation corresponds to absorption. Because the various energies are constrained by the quantum hypothesis to have only certain discrete values, the transitional energies will be similarly discrete, and this is what causes the observed spectral lines to be sharp. It is important to understand that in Bohr's model of the atom, it is meaningless to ask what is happening when an electron is in an intermediate state, between two orbits (as one could for an artificial satellite, for example), because only quantum states can be defined.

As exciting as these developments were, other puzzles soon appeared on the scene. In 1927, Clinton Davisson and Lester Germer showed that electrons have a wavelength, which they determined, and George Thomson demonstrated that electrons can give rise to diffraction patterns, just as Thomas Young had done for light. How was the wave nature of the electron to be reconciled with Bohr's model of the atom?

In fact, these experiments confirmed a bold hypothesis that had been put forward in 1923 by Louis de Broglie, which saw *all* objects as having both particle and wave properties. He had suggested that a particle's momentum p, which of course equals μv, is related to its wavelength by the expression

$$\lambda = \frac{h}{p} = \frac{h}{\mu v} \tag{A.5}$$

and we should not be surprised to find Planck's constant appearing in the numerators. In practice, it is only for pieces of matter with atomic and sub-atomic sizes that de Broglie's wave-particle duality really manifests itself, but strictly speaking, a thrown baseball, for example, does have an associated wavelength. (It is instructive to work out what this wavelength actually is.)

To complete this brief review of the foundations of quantization, we should consider the principle introduced by Werner Heisenberg in 1927 and an inter-pretation introduced by Max Born one year later. Heisenberg argued that if the position of a particle is determined to within a precision of Δr and its momentum is simultaneously determined to within a precision of Δp, the product of those two uncertainties can never be less than a number of the order of $h/2\pi$. Heisenberg's uncertainty principle is thus captured in the expression

$$\Delta r \bullet \Delta p \geq \frac{h}{2\pi} \tag{A.6}$$

and this shows that one's efforts to measure one parameter more precisely will be thwarted by the need to accept less precision in the simultaneous determin-ation of the other parameter. The uncertainty principle makes itself felt only for particles of atomic and sub-atomic dimensions.

The fundamental uncertainty which rules at the microscopic level led Born to suggest an interpretation of the wave aspect of a particle that was decreed by de Broglie's hypothesis of wave-particle duality. If that wave is represented math-ematically by the magnitude of its spatial and temporal variation, Born argued, the *probability* of finding the particle at any position, at a given time, will be proportional to the product of that wave function and the wave function's complex conjugate. In general, we may describe the wave function by the expression

$$\Psi = \psi e^{-2\pi i v t} \tag{A.7}$$

where ψ, the amplitude, depends only on position, and ν is the frequency. As is standard when dealing with complex numbers, $i = \sqrt{(-1)}$, and in general we would expect Ψ to represent a damped oscillatory wave.

The complex conjugate of a function is obtained by replacing i by $-i$ wherever it appears, and it is designated by an asterisk, thus Ψ^*. Using expression (A.7), it is easy to show that

$$\Psi \Psi^* = \psi^2 \tag{A.8}$$

and this gives the probability of finding the particle, according to Born's hypothesis. The particle must be found somewhere or other, of course, and this gives us the boundary condition that the wave function must die away to zero as infinity is approached. In one dimension, for example, we have

$$\int_{-\infty}^{\infty} \Psi \Psi^* \, dx = 1 \tag{A.9}$$

and this is equivalent to saying that there must be zero-amplitude nodes at plus and minus infinity. If instead we had considered the situation in three dimensions, the equivalent result would have led us to conclude that there has to be a spherical node at infinity.

A.3   The Wave Equation

In classical mechanics, as was demonstrated by Isaac Newton, the motion of a particle can be described through an equation linking its mass and acceleration to the force acting on the particle. In 1927, Erwin Schrödinger had been speculating upon the changes to Newton's equation that would be required in view of wave-particle duality, and he realized that a new equation was needed that would describe the dynamics of a particle's wave characteristic. His wave equation, now known as the Schrödinger equation, links the continuous characteristics of a wave with the discrete characteristics of a particle.

Before proceeding to what could be called the wave–particle equation, let us consider the simple case of a wave of amplitude A and angular frequency ω travelling with a speed v in the x direction. The displacement y of this wave, for any pair of values x and t, is given by

$$y = A \cos \omega \left(t - \frac{x}{v} \right) \tag{A.10}$$

and by deriving the two second derivatives, $\partial^2 y/\partial x^2$ and $\partial^2 y/\partial t^2$, we can easily establish the validity of the general wave equation in one dimension, namely

$$\frac{\partial^2 y}{\partial x^2} = \frac{1}{v^2}\frac{\partial^2 y}{\partial t^2} \tag{A.11}$$

By analogy, we write the quantum-mechanical wave function by replacing y by $\Psi(x, t)$, though we have to accept that in this case, as discussed above, the counterpart of the displacement of the function is *not* a directly measurable quantity. We have

$$\Psi(x, t) = Ae^{-i\omega(t-x/v)} \tag{A.12}$$

and by replacing ω by $2\pi\nu$ and v by $\lambda\nu$, we get

$$\Psi(x, t) = Ae^{-2\pi i(\nu t - x/\lambda)} \tag{A.13}$$

Moreover, recalling two earlier results, $\mathscr{E}_{tot} = h\nu$ and $\lambda = h/p$, where $\mathscr{E}_{tot}$ is the total energy of the particle, we can rewrite this in the form

$$\Psi(x, t) = Ae^{-(2\pi i/h)(\mathscr{E}_{tot}t - px)} \tag{A.14}$$

Now if we differentiate Equation (A.14) twice with respect to x, we get

$$\frac{\partial^2 \Psi(x, t)}{\partial x^2} = -\frac{4\pi^2 p^2}{h^2}\Psi(x, t) \tag{A.15}$$

while differentiating once with respect to t produces

$$\frac{\partial \Psi(x, t)}{\partial t} = -\frac{2\pi i \mathscr{E}_{tot}}{h}\Psi(x, t) \tag{A.16}$$

Ignoring relativistic effects (see above), the total energy $\mathscr{E}_{tot}$ of a particle is the sum of its kinetic and potential ($\mathscr{E}_{pot}$) energies, and noting that the kinetic energy $\mu v^2/2$ can be rewritten as $p^2/2\mu$, using expressions we encountered earlier, we arrive at

$$\mathscr{E}_{tot} = \frac{p^2}{2\mu} + \mathscr{E}_{pot} \tag{A.17}$$

As a final preliminary, let us multiply both sides of Equation (A.17) by $\Psi(x, t)$. This produces

$$\mathscr{E}_{tot}\Psi(x, t) = \frac{p^2}{2\mu}\Psi(x, t) + \mathscr{E}_{pot}\Psi(x, t) \tag{A.18}$$

We can use Equation (A.16) to note that

$$\mathscr{E}_{tot}\Psi(x, t) = -\frac{h}{2\pi i}\frac{\partial\Psi(x, t)}{\partial t} \tag{A.19}$$

and Equation (A.15) to note that

$$p^2\Psi(x, t) = -\frac{h^2}{4\pi^2}\frac{\partial^2\Psi(x, t)}{\partial x^2} \tag{A.20}$$

Substituting Equations (A.19) and (A.20) into Equation (A.18), we obtain what is known as the time-dependent Schrödinger's equation, namely

$$\frac{h}{2\pi i}\frac{\partial\Psi(x, t)}{\partial t} = \frac{h^2}{8\pi^2\mu}\frac{\partial^2\Psi(x, t)}{\partial x^2} - \mathscr{E}_{pot}\Psi(x, t) \tag{A.21}$$

the generalization to three dimensions being obvious.

In many problems of interest, the potential energy does not vary with time, and simplification of Equation (A.21) is thus possible. We note that Equation (A.14) can be rearranged to produce

$$\Psi(x, t) = \psi e^{-(2\pi i\mathscr{E}_{tot}/h)t} \tag{A.22}$$

where

$$\psi = Ae^{(2\pi ip/h)x} \tag{A.23}$$

and ψ is only a function of x. Substituting (A.22) into (A.21) gives

$$-\mathscr{E}_{tot}\psi e^{-(2\pi i\mathscr{E}_{tot}/h)t} = \frac{h^2}{8\pi^2\mu}e^{-(2\pi i\mathscr{E}_{tot}/h)t}\frac{\partial^2\psi}{\partial x^2} - \mathscr{E}_{pot}\psi e^{-(2\pi i\mathscr{E}_{tot}/h)t} \tag{A.24}$$

Dividing throughout by the common exponential factor produces what is known as the time-independent Schrödinger's equation, namely

$$\frac{\partial^2\psi}{\partial x^2} + \frac{8\pi^2\mu}{h^2}\left(\mathscr{E}_{tot} - \mathscr{E}_{pot}\right)\psi = 0 \tag{A.25}$$

and here too the generalization to three dimensions is straightforward.

The result of applying Equation (A.25) to a specific problem will naturally depend on how $\mathscr{E}_{pot}$ varies with position, and on any boundary conditions that apply. (This should not surprise us; the solution of any differential equation requires specification of the boundary conditions.) In practice, the equation will produce allowable solutions only for a number of discrete values of the parameters and the corresponding forms of ψ, these being referred to as eigenvalues and eigenfunctions, respectively. (The German word 'eigen' means 'own' or 'belonging to'.) The great importance of Equation (A.25) is that it permits one to calculate *all* the measurable features of a system. As we shall see in Appendix B, it was applied to the hydrogen atom with spectacular success. This development provided the vital bridge between physics and chemistry, a bridge which later was to have an enormous impact on the study of biology at the molecular level. In that respect, Equation (A.25) may be regarded as one of the corner stones of modern biophysics.

A.4 Quantum Mechanical Tunnelling

In classical mechanics, an object travelling with insufficient energy to surmount a barrier will merely be reflected by it. In quantum mechanics, the amplitude of a particle-wave is non-zero everywhere except at a node, and this implies the possibility of simultaneous transmission and reflection. This gives rise to the phenomenon of quantum mechanical tunnelling, a good example of the counter-intuitive nature of the quantum domain. Let us analyse the situation on the basis of the principles established in the preceding sections of this appendix.

We assume a barrier of height $\mathscr{E}_0$ and width a, the wall at the incident side coinciding with the plane $x = 0$. In other words, we have $\mathscr{E}_{pot}(x) = 0$ for $x < 0$ and $x > a$, while $\mathscr{E}_{pot}(x) = \mathscr{E}_0$ for $0 < x < a$. In the regions where $\mathscr{E}_{pot}(x) = 0$, the solutions of Equation (A.25) will have the form

$$\psi(x) = Ae^{i\alpha x} + Be^{-i\alpha x} \tag{A.26}$$

for $x \leq 0$, and

$$\psi(x) = Ce^{i\alpha x} \tag{A.27}$$

for $x \geq a$, where

$$\alpha = \left(8\pi^2 \mu \mathscr{E}_{tot}/h^2\right)^{1/2} \tag{A.28}$$

In the region $0 < x < a$, the solution of Equation (A.25) becomes

$$\psi(x) = Fe^{-\beta x} + Ge^{\beta x} \qquad (A.29)$$

where

$$\beta = \left(8\pi^2 \mu \{\mathscr{E}_{pot} - \mathscr{E}_{tot}\}/h^2\right)^{1/2} \qquad (A.30)$$

Of the five coefficients, A, B, C, F and G, it is clear that G must be zero, because the wave function would otherwise increase exponentially within the barrier. The other four are all non-zero. This means that there are waves travelling in both the positive and negative directions in the region where $x \le 0$. We are particularly interested in the transmission coefficient, which, by Born's principle discussed above, involves the *squares* of the relevant amplitudes. That coefficient is given by

$$\left|\frac{C}{A}\right|^2 = \left[1 + \frac{\mathscr{E}_{pot}^2 \sinh^2 \beta a}{4\mathscr{E}_{tot}(\mathscr{E}_{pot} - \mathscr{E}_{tot})}\right]^{-1} \qquad (A.31)$$

When $\mathscr{E}_{tot} > \mathscr{E}_{pot}$, even a classical particle would be able to penetrate the barrier, but the quantum case is nevertheless fascinating because the wave function within the barrier is oscillatory, and the transmission is perfect whenever an integral number of wavelengths can be fitted into the barrier width. When $\mathscr{E}_{tot} < \mathscr{E}_{pot}$, the wave amplitude decreases exponentially within the barrier, whereas it is oscillatory on either side, as shown in Figure A.1.

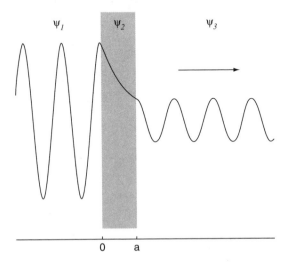

Figure A.1   The amplitude of the wave function on either side of the energy barrier, and also within that barrier, for the case in which the total energy is less than that of the barrier

Quantum mechanical tunnelling is of importance in those biological molecules in which electrons move from one site to another. There is much evidence of such movement in the processes involved in energy conversion, the process of photosynthesis being perhaps the most familiar example. Tunnelling is employed on the experimental side in what is known as the scanning tunnelling microscope, which is described elsewhere in this book.

Exercises

A.1   Demonstrate that Planck's expression for the intensity of radiation emitted from a heated body predicts zero intensity at zero wavelength, and a spectral maximum which decreases in wavelength as the temperature increases.

A.2   In a quantum mechanical tunnelling experiment, the thickness of the barrier was doubled, while all the other variables were maintained at their initial values. Calculate the change in the transmission coefficient.

Further reading

Beiser, A., (1963). *Concepts of Modern Physics*. McGraw-Hill, New York.
Flint, H. T., (1931). *Wave Mechanics*. Methuen, London.
Heitler, W., (1950). *Elementary Wave Mechanics*. Clarendon Press, Oxford.
Mott, N. F., (1962). *Elements of Wave Mechanics*. Cambridge University Press, Cambridge.
Pauling, L. and Wilson, E. B., (1935). *Introduction to Quantum Mechanics*. McGraw-Hill, New York.
Schiff, L. I., (1955). *Quantum Mechanics*. McGraw-Hill, New York.

Appendix B: The Hydrogen Atom

B.1 The Hamiltonian

A useful shorthand manner of writing the Schrödinger wave equation employs an operator known as the Hamiltonian, $\mathcal{H}$ (named after William Hamilton), an operator being a prescription for a set of mathematical operations. In Appendix A, we encountered several expressions in which the wave function was subjected to such operations. Using this notation, we may replace Equation (A.25) by

$$\mathcal{H}\Psi = \mathcal{E}_{tot}\Psi \tag{B.1}$$

where

$$\mathcal{H} = \frac{p^2}{2m} + \mathcal{E}_{pot} = -\frac{h^2}{8\pi^2\mu}\left(\frac{\partial^2}{\partial x^2} + \frac{\partial^2}{\partial y^2} + \frac{\partial^2}{\partial z^2}\right) + \mathcal{E}_{pot} \tag{B.2}$$

p being the momentum and μ the mass of the particle whose wave function is to be determined. The prescriptions for taking the second derivatives, written within the parentheses, are collectively known as the Laplacian operator (after Pierre de Laplace), which is often simply denoted by the symbol ∇^2. The potential energy, $\mathcal{E}_{pot}$, will in general be a function of the spatial coordinates. When a specific potential function is inserted into Equation (B.1), the differential equation that is produced will have solutions only for certain values of the underlying parameters. As we noted in Appendix A, the wave functions corresponding to those values are referred to as eigenfunctions, and the corresponding energies are known as eigenvalues. We now proceed to apply this theoretical machinery to the case of the simplest atomic system, namely the hydrogen atom.

B.2 The Hydrogen Atom

The question of the nature of the allowed quantum states of the hydrogen atom is the most important problem in the field of atomic and molecular structure, not only because the theoretical treatment of this atom is simpler than that of all other atoms and molecules but also because it forms the basis for the discussion of more complex systems.

As with all problems in which the eigenvalues and eigenfunctions are to be derived, we start with the time-independent Schrödinger equation

$$\nabla^2 \Psi + \frac{8\pi^2 \mu_e}{h^2} \left(\mathscr{E}_{tot} - \mathscr{E}_{pot} \right) \Psi = 0 \tag{B.3}$$

where μ_e is now the mass of the electron. As was discussed in Appendix A, $\psi\psi^*$ will give us the probability density for the electron distribution. In the case of the hydrogen atom, the potential energy $\mathscr{E}_{pot}$ has a rather simple form. There is a charge $+q_e$ on the nucleus, while the orbiting electron has a charge of $-q_e$. Because the interaction is of the Coulomb type, we will thus have $\mathscr{E}_{pot} = -q_e^2/r$, where r is the distance between the nucleus and the electron. Because the potential energy depends only upon the distance, and from the symmetry of the situation, it is convenient to work in spherical polar co-ordinates. Figure B.1 shows the relationship of these co-ordinates to the more familiar Cartesian co-ordinates (after René Descartes, of course).

In that coordinate system, the Laplacian operator becomes

$$\nabla^2 = \frac{1}{r^2} \frac{\partial}{\partial r} \left(r^2 \frac{\partial}{\partial r} \right) + \frac{1}{r^2 \sin^2 \theta} \frac{\partial^2}{\partial \phi^2} + \frac{1}{r^2 \sin \theta} \frac{\partial}{\partial \theta} \left(\sin \theta \frac{\partial}{\partial \theta} \right) \tag{B.4}$$

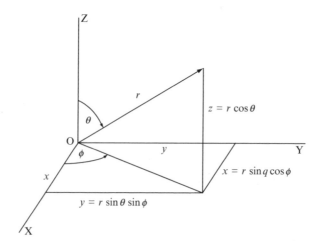

Figure B.1 The relationship between the Cartesian and spherical polar coordinate systems

So our Schrödinger equation is now

$$\frac{1}{r^2}\frac{\partial}{\partial r}\left(r^2\frac{\partial\Psi}{\partial r}\right) + \frac{1}{r^2\sin^2\theta}\frac{\partial^2\Psi}{\partial\phi^2} + \frac{1}{r^2\sin\theta}\frac{\partial}{\partial\theta}\left(\sin\theta\frac{\partial\Psi}{\partial\theta}\right)$$
$$+ \frac{8\pi^2\mu_e}{h^2}\left(\mathscr{E}_{tot} + \frac{q_e^2}{r}\right)\Psi = 0 \tag{B.5}$$

The variables can be separated by noting that we may write

$$\Psi(r,\theta,\phi) = R(r)\Theta(\theta)\Phi(\phi) \tag{B.6}$$

Substituting Equation (B.6) in (B.5), and dividing throughout by $R\Theta\Phi$, we have

$$\frac{1}{Rr^2}\frac{\partial}{\partial r}\left(r^2\frac{\partial R}{\partial r}\right) + \frac{1}{\Phi r^2\sin^2\theta}\frac{\partial^2\Phi}{\partial\phi^2} + \frac{1}{\Theta r^2\sin\theta}\frac{\partial}{\partial\theta}\left(\sin\theta\frac{\partial\Theta}{\partial\theta}\right)$$
$$+ \frac{8\pi^2\mu_e}{h^2}\left(\mathscr{E}_{tot} + \frac{q_e^2}{r}\right) = 0 \tag{B.7}$$

We note that if this equation were multiplied throughout by $r^2\sin^2\theta$ the second term would contain only the variable φ, while no other term would be a function of this variable. This would have to mean that the second term would be equal to a constant. We call this constant $-m^2$, and thus have already established that

$$\frac{d^2\Phi}{d\phi^2} = -m^2\Phi \tag{B.8}$$

This is the first of three separated wave equations, the solutions of which, when taken together and inserted in Equation (B.6) will give us the wave functions that we seek. Our Schrödinger equation has now become

$$\frac{1}{R}\frac{\partial}{\partial r}\left(r^2\frac{\partial R}{\partial r}\right) - \frac{m^2}{\sin^2\theta} + \frac{1}{\Theta\sin\theta}\frac{\partial}{\partial\theta}\left(\sin\theta\frac{\partial\Theta}{\partial\theta}\right) + \frac{8\pi^2\mu_e r^2}{h^2}\left(\mathscr{E}_{tot} + \frac{q_e^2}{r}\right) = 0 \tag{B.9}$$

The second and third terms of this equation are independent of r, while the first and fourth terms do not involve θ. Hence we can equate each of these two parts of the equation to a constant; the θ terms to $-\beta$, say, and therefore the r

terms to $+\beta$. After multiplying the resulting two separated equations by Θ and R/r^2, respectively, we obtain

$$\frac{1}{\sin\theta}\frac{d}{d\theta}\left(\sin\theta\frac{d\Theta}{d\theta}\right) - \frac{m^2\Theta}{\sin^2\theta} + \beta\Theta = 0 \qquad (\text{B.10})$$

and

$$\frac{1}{r^2}\frac{d}{dr}\left(r^2\frac{dR}{dr}\right) - \frac{\beta R}{r^2} + \frac{8\pi^2\mu_e}{h^2}\left(\mathscr{E}_{\text{tot}} + \frac{q_e^2}{r}\right)R = 0 \qquad (\text{B.11})$$

We now have our three separated wave equations, (B.8), (B.10) and (B.11), which, when solved, will give us the desired wave functions and allowed energy levels. The strategy will now be as follows. We will first find that Equation (B.8) possesses acceptable solutions only for certain values of the parameter m. Using these values of m in Equation (B.10), we will then find that (B.10) gives acceptable solutions only for certain values of β. Finally, introducing these values of β in Equation (B.11), we will discover that this equation yields acceptable solutions only for certain values of $\mathscr{E}_{\text{tot}}$. *En route*, we will find the forms of the functions Φ, Θ and R, each of which will be normalized to unity, in view of Equation (B.6) and the requirement that

$$\iiint \Psi\Psi^* dV = 1 \qquad (\text{B.12})$$

which is the same as saying that the electron must be located somewhere or other in space (dV being the element of volume). In principle, the integration stretches to infinity, in all directions, but in practice the probability of finding the electron more than a few Ångström units (i.e. units of 0.1 nm) away from the nucleus is essentially zero (that unit being named after Anders Ångström, of course).

B.3   Solution of the Φ Equation

Equation (B.8) has acceptable solutions

$$\Phi_m(\phi) = \frac{1}{\sqrt{2\pi}}\exp\left(im\phi\right) \qquad (\text{B.13})$$

if, and only if, the m are positive or negative integers, or zero. The wave function is then single-valued at $\phi = 0$ (which is of course identical with $\phi = 2\pi$). The denominator is a normalization constant. The constant integer m is known as the magnetic quantum number.

B.4 Solution of the Θ Equation

Equation (B.10) can be modified into a recognizable form by using the transformation

$$z = \cos\theta \tag{B.14}$$

and thus

$$P(z) = \Theta(\theta) \tag{B.15}$$

Because we now have $\sin^2\theta = 1 - z^2$ and $d\Theta/d\theta = (dP/dz)(dz/d\theta)$ $= -(dP/dz)\sin\theta$, this changes Equation (B.10) to

$$\frac{d}{dz}\left[(1-z^2)\frac{dP}{dz}\right] + \left[\beta - \frac{m^2}{1-z^2}\right]P = 0 \tag{B.16}$$

Equation (B.16) happens to be the differential equation satisfied by the associated Legendre functions, $P_l^{|m|}(z)$, of degree l and order $|m|$, which themselves are defined by

$$P_l^{|m|}(z) = (1-z)^{|m|/2}\frac{d^{|m|}}{dz^{|m|}}P_l(z) \tag{B.17}$$

where the $P_l(z)$ are the Legendre polynomials, as generated by the equation

$$\sum_{l=0}^{\infty}P_l(z)t^l = (1 - 2tz - t^2)^{-1/2} \tag{B.18}$$

(It was Adrien-Marie Legendre who first explored these functions.) The condition that Equation (B.16) has this type of solution is that

$$\beta = l(l+1) \tag{B.19}$$

where

$$l = |m|, |m| + 1, |m| + 2, \dots \tag{B.20}$$

and where l is another integer known as the azimuthal quantum number. The azimuthal contribution to the wave function follows from the known solutions of the differential Equation (B.16), and we have, finally,

$$\Theta_{lm}(\theta) = \left\{ \frac{(2l+1)(l-|m|)!}{2(l+|m|)!} \right\}^{1/2} P_l^{|m|}(\cos\theta) \tag{B.21}$$

in which the symbol ! indicates a factorial, as usual.

B.5 Solution of the R Equation

Using Equations (B.19) and (B.20) in Equation (B.11), our R equation becomes

$$\frac{1}{r^2}\frac{d}{dr}\left(r^2\frac{dR}{dr}\right) + \left\{ -\frac{l(l+1)}{r^2} + \frac{8\pi^2\mu_e}{h^2}\left(\mathscr{E}_{tot} + \frac{q_e^2}{r}\right) \right\}R = 0 \tag{B.22}$$

Bound states, in which the electron has insufficient energy to escape out of its orbit around the nucleus, occur for negative values of $\mathscr{E}_{tot}$. Introducing the symbols

$$\alpha^2 = -\frac{8\pi^2\mu_e\mathscr{E}_{tot}}{h^2} \tag{B.23}$$

and

$$\lambda = \frac{4\pi^2\mu_e q_e^2}{h^2\alpha} \tag{B.24}$$

and changing the independent variable by the following substitutions

$$\rho = 2\alpha r \tag{B.25}$$

and

$$S(\rho) = R(r) \tag{B.26}$$

makes our equation become

$$\frac{1}{\rho^2}\frac{d}{d\rho}\left(\rho^2\frac{dS}{d\rho}\right) + \left\{ -\frac{1}{4} - \frac{l(l+1)}{\rho^2} + \frac{\lambda}{\rho} \right\}S = 0 \tag{B.27}$$

By the further substitution

$$S(\rho) = \rho^2\left\{ \exp\left(-\frac{\rho}{2}\right) \right\}L(\rho) \tag{B.28}$$

we finally obtain the equation

$$\rho\frac{d^2L}{d\rho^2} + [2(l+1) - \rho]\frac{dL}{d\rho} + (\lambda - l - 1)L = 0 \tag{B.29}$$

This differential equation is satisfied by the associated Laguerre polynomials of degree $(n - l - 1)$ and order $(2l + 1)$,

$$L_{n+1}^{2l+1}(\rho)$$

which are related to the normal Laguerre polynomials, $L_r(\rho)$, by

$$L_r^s(\rho) = \frac{d^s}{d\rho^s} L_r(\rho) \tag{B.30}$$

the left-hand side being of degree $(r{-}s)$ and order s. The Laguerre polynomials themselves (first discussed by Edmond Laguerre) are generated by

$$\sum_{r=0}^{\infty} \frac{L_r(\rho)}{r!} u^r = \{\exp[-\rho u/(1 - u)]\}/(1 - u) \tag{B.31}$$

The condition that Equation (B.29) can be identified with the differential equation for the associated Laguerre polynomials is that

$$\lambda = n \tag{B.32}$$

where

$$n = l + 1, l + 2, l + 3, \ldots \tag{B.33}$$

The n are thus integers and they are known by the name total quantum number. Working back through the various substitutions, we find the final form of the radial contribution to the wave function. It is

$$R_{nl}(r) = -\left\{ \left(\frac{2}{na_0}\right)^3 \frac{(n - l - 1)!}{2n[(n + 1)!]^3} \right\}^{1/2} e^{-\rho/2} \rho^l L_{n+1}^{2l+1}(\rho) \tag{B.34}$$

in which

$$\rho = \frac{2r}{na_0} \tag{B.35}$$

and

$$a_0 = \frac{h^2}{4\pi^2 \mu_e q_e^2}$$ (B.36)

B.6 Quantum Numbers and Energy Levels

Rewriting Equations (B.33), (B.20), and the earlier stated condition on the values of m, we have:

Total quantum number $= n = 1, 2, 3, 4, \ldots$

Azimuthal quantum number $= l = 0, 1, 2, 3, \ldots, n-1$

Magnetic quantum number $= m = -l, -l+1, \ldots, +l-1, +l$

Using Equations (B.23), (B.24) and (B.32), we have the expressions for the energy levels

$$\mathcal{E}_{tot} = -\frac{2\pi^2 \mu_e q_e^4}{h^2 n^2}$$ (B.37)

and we note that this involves only the total quantum number. There are $2l + 1$ independent wave functions with given values of n and l, and there are n^2 independent wave functions with a given value of n, that is, with the same energy. The $2l + 1$ wave functions with the same n and l are said to form a completed subgroup, and the n^2 wave functions with the same n constitute a completed group. There is the additional useful rule that the sum of the probability distribution functions for a given value of l and all values of m is constant.

B.7 Wave Functions

The universally accepted nomenclature uses the numerals 1, 2, 3, and so on to designate the value of the total quantum number, and the letters s, p, d and f for l values of 0, 1, 2, and 3, respectively. These letters are of historical origin, and they refer to the various sets of observed spectral lines: **s**harp, **p**rincipal, **d**iffuse and **f**ine. The collective symbol $1s^2 2s^2 2p^3$ indicates that an atom has seven electrons, two of which have $n = 1$, $l = 0$, two have $n = 2$, $l = 0$, and three have $n = 2$, $l = 1$. The electron states associated with the various total quantum number are usually designated by the capital letters K, L, M, etc., this so-called shell nomenclature being that used in connection with excitation by external radiation. The following is a list of the wave functions for the first two shells of the hydrogen atom.

K shell $n = 1, l = 0, m = 0$

$$\Psi_{1s} = \frac{1}{\sqrt{\pi}} \left(\frac{1}{a_0}\right)^{3/2} e^{-r/a_0}$$

L shell $n = 2, l = 0, m = 0$

$$\Psi_{2s} = \frac{1}{4\sqrt{2\pi}} \left(\frac{1}{a_0}\right)^{3/2} \left(2 - \frac{r}{a_0}\right) e^{-r/2a_0}$$

$n = 2, l = 1, m = 0$

$$\Psi_{2pZ} = \frac{1}{4\sqrt{2\pi}} \left(\frac{1}{a_0}\right)^{3/2} \left(\frac{r}{a_0}\right) e^{-r/2a_0} \cos\theta$$

$n = 2, l = 1, m = \pm 1$

$$\Psi_{2pX} = \frac{1}{4\sqrt{2\pi}} \left(\frac{1}{a_0}\right)^{3/2} \left(\frac{r}{a_0}\right) e^{-r/2a_0} \sin\theta \cos\varphi$$

$$\Psi_{2pY} = \frac{1}{4\sqrt{2\pi}} \left(\frac{1}{a_0}\right)^{3/2} \left(\frac{r}{a_0}\right) e^{-r/2a_0} \sin\theta \sin\varphi$$

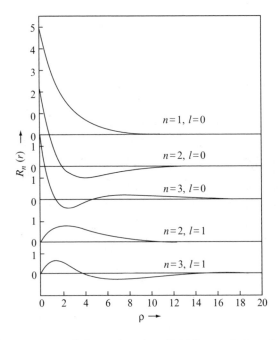

Figure B.2   Hydrogen-atom radial wave functions $R_n(r)$ for $n = 1, 2,$ and 3 and $l = 0$ and 1

We close this appendix by contemplating the spatial variations of the radial (see Figures B.2 and B.3) and angular (see Figures B.4 and B.5) contributions to the wave functions, for some of the states with low quantum numbers.

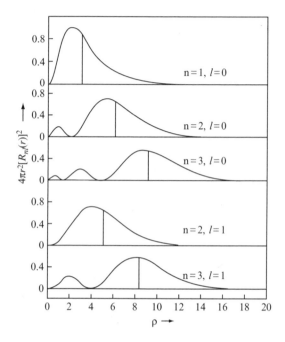

Figure B.3 Electron probability distribution functions $4\pi r^2[(R_{nl}(r)]^2$ for the hydrogen atom

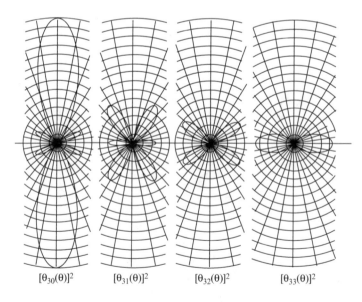

Figure B.4 Polar graphs of the function $[\theta_{lm}(\theta)]^2$ for $l = 3$ and $m = 0, +1, +2,$ and $+3$

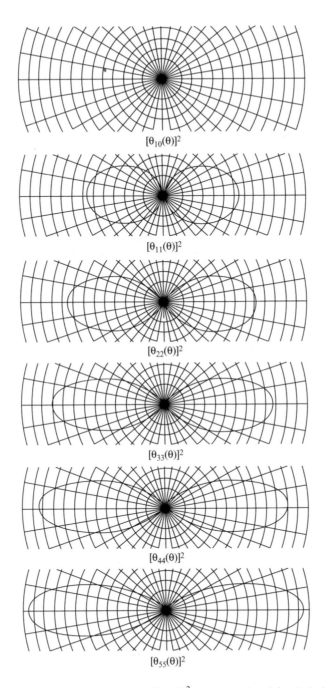

$[\theta_{10}(\theta)]^2$

$[\theta_{11}(\theta)]^2$

$[\theta_{22}(\theta)]^2$

$[\theta_{33}(\theta)]^2$

$[\theta_{44}(\theta)]^2$

$[\theta_{55}(\theta)]^2$

Figure B.5   Polar graphs of the function $[\theta_{lm}(\theta)]^2$ for $m = \pm l$ and $l = 0, 1, 2, 3, 4,$ and $5,$ showing the concentration of the function about the xy-plane with increasing l

Exercises

B.1 Write an essay on the importance to chemistry of the quantum mechanical treatment of the hydrogen atom.

B.2 Write the Hamiltonian for the one-dimensional harmonic oscillator, in which the potential energy varies as the square of the displacement from the zero-energy equilibrium position. Solve the resulting Schrödinger equation, and find the permitted energy levels of the system.

Further reading

Herzberg, G., (1944). *Atomic Spectra and Atomic Structure*. Dover, New York.

Moore, W. J., (1972). *Physical Chemistry*. Longman, Englewood Cliffs, NJ.

Pauling, L., (1970). *General Chemistry*. Freeman, San Francisco.

Pauling, L. and Wilson, E. B., (1935). *Introduction to Quantum Mechanics*. McGraw-Hill, New York.

Richtmyer, F. K. and Kennard, E. H., (1950). *Introduction to Modern Physics*. McGraw-Hill, New York.

Appendix C: Thermal Motion

C.1 Ideal Gases

The classical treatment of an ideal gas seeks to relate macroscopically observable properties, such as the volume, temperature and pressure to the microscopic parameters that characterize the individual atomic or molecular motions. Empirical relationships between macroscopic properties had been established, examples being Boyle's Law (named for Robert Boyle), which states that pressure P varies inversely as the volume V, when the temperature T is held constant, and Charles's Law (named for Jacques Charles), which states that the volume is directly proportional to the temperature, when the pressure is held constant. Collectively, these laws yield the equation of state for an ideal gas, namely

$$PV = nRT \tag{C.1}$$

where n is the number of moles of the gas present and R is the universal gas constant ($= 8.31\,\text{J K}^{-1}\,\text{mol}^{-1}$). An important fact that ultimately emerged from that early work was that the temperature in question was the absolute temperature, or Kelvin temperature (named for Lord Kelvin, of course), the zero point of which corresponds to the situation in which all molecular speeds have dropped to zero.

Turning to the microscopic level, we note that the pressure acting on the wall of a container will arise from the collisions made with it by the moving molecules. The momentum of a molecule will be changed during such a collision, and this results in a force on the wall, force being proportional to the rate of change of momentum, as first shown by Isaac Newton. The contribution that this microscopic event makes to the pressure follows from the fact that pressure is simply force per unit area.

A given collision will change a molecule's velocity v_x in the x-direction by $2v_x$, as it rebounds from the containment wall, so the wall will experience a minute momentum increment $2mv_x$ in the x-direction, where m is the molecule's mass. (In contrast to Appendices A and B, we do not need to use the symbol μ for mass, because there will be no need to reserve the symbol m for a quantum number.) The number of molecules with velocity v_x striking unit area of the wall per second is clearly $n_v v_x$, where n_v is the number density of the molecules (that is to say the number of molecules per unit volume, rather than their total mass in that volume). The total momentum in the x- direction imparted to unit area of the wall in 1 s is thus $mn_v \sum v_x^2$, the factor of 2 having disappeared because, on average, half the molecules will be travelling away from the wall at any instant, rather than toward it. The x- component of the pressure can then be found directly from this sum, and the overall pressure follows by summing over the x, y and z components, namely

$$P = \frac{1}{3} n_v m \bar{v}^2 \tag{C.2}$$

because

$$\sum v^2 = \sum (v_X^2 + v_Y^2 + v_Z^2) \tag{C.3}$$

$\bar{v}^2$ being the mean-square velocity.

However, $m\bar{v}^2/2$ is the mean kinetic energy of a molecule, $\mathscr{E}_{kin}$, so Equation (C.2) can be modified to read

$$P = \frac{2}{3} n_v \mathscr{E}_{kin} \tag{C.4}$$

Combining Equations (C.1) and (C.4), and thus making the desired link between the macroscopic and microscopic domains, we arrive at

$$\mathscr{E}_{kin} = \frac{3\,nRT}{2\,n_v V} \tag{C.5}$$

Now $n_v V$ is simply the number of molecules in volume V, and if $n = 1$, that is to say if there is one mole of the gas present, $n_v V$ will be the number of molecules in one mole. That number is known as Avogadro's number, N_0 ($= 6.024 \times 10^{23}$ mol^{-1}). (This designation honours Amedeo Avogadro, of course.) Our Equation (C.5) has thus become

$$\mathscr{E}_{kin} = \frac{3\,RT}{2\,N_0} \tag{C.6}$$

Finally, we can invoke the relationship

$$R = N_0 k_B \tag{C.7}$$

which defines Boltzmann's constant k_B ($= 1.38 \times 10^{-23}$ JK^{-1}), to obtain

$$\mathscr{E}_{kin} = \frac{3}{2} k_B T \tag{C.8}$$

Equation (C.8) is of paramount importance in thermodynamics and statistical mechanics. Because the molecules are moving in three-dimensional space, we see that the kinetic energy associated with each dimension, that is to say with each degree of freedom, is $k_B T/2$. One does well to commit that simple fact to memory.

It is frequently of interest to know the relative numbers of molecules travelling with various *speeds* (as opposed to velocities), at any instant. The distribution of speeds is found experimentally to be constant, for a given temperature, despite the fact that the speed of an individual molecule perpetually fluctuates, as it gains and loses kinetic energy through collisions with other molecules. That the distribution is nevertheless constant demonstrates that the losses are precisely offset by the gains, when one considers the system as a whole. As the popular saying has it: what is lost on the roundabouts is gained on the swings.

We are seeking knowledge of the fraction of molecules having speeds in the range v to $v + dv$, in the usual manner of the calculus, so it is appropriate to choose the letter f and denote this fraction by $f(v)dv$. The rate of collisions between molecules with speed v_1 and molecules with speed v_2 will be $\alpha_{12} f(v_1) f(v_2)$, where α_{12} is a coefficient whose form we will need to establish. Similarly the rate of collisions which change the speeds of two molecules from v_3 and v_4 to v_1 and v_2 will be $\alpha_{34} f(v_3) f(v_4)$, where α_{34} could in principle be a coefficient that is not identical to α_{12}. Nevertheless, because the total kinetic energy will remain the same, we will find that

$$v_1^2 + v_2^2 = v_3^2 + v_4^2 \tag{C.9}$$

If equilibrium prevails, the two collision rates must be equal, so we find that $\alpha_{12} = \alpha_{34}$, and thus that

$$f(v_1)f(v_2) = f(v_3)f(v_4) \tag{C.10}$$

which implies

$$\log_e[f(v_1)] + \log_e[f(v_2)] = \log_e[f(v_3)] + \log_e[f(v_4)] \tag{C.11}$$

It is clear that Equations (C.9) and (C.11) can be made compatible if $\log_e f(v)$ is proportional to v^2, that is to say

$$f(v) = A \exp\left[-\beta v^2\right] \tag{C.12}$$

the minus sign being required because the fraction ultimately must decrease with increasing speed, in the limit of sufficiently high speed. The constants A and β remain to be determined. Because we are interested in speed, irrespective of direction, the molecules entering the interval v to $v + dv$ can be imagined as having done so by dint of travelling in all possible directions within the solid angle 4π. They will have travelled from many different starting points, of course, since their most recent collision (which left them having a speed lying in the above interval). However, we can turn things around by translating all those different starting positions so as to locate them at one and the same origin. The molecules can then be thought of as all travelling outward from that origin, into the entire solid angle 4π. One unit of time later, they will be located within a shell of thickness dv on the surface of a sphere having radius v, because by definition none of them will have suffered a further collision. We thus see that the number $N(v)$ of these molecules, in the above-identified interval will be

$$N(v) = 4\pi v^2 f(v) = 4\pi v^2 A \exp(-\beta v^2) \qquad (C.13)$$

We can determine the unknown constants through consideration of the total number N of molecules and the total kinetic energy $\mathscr{E}_{kin,\ total}$. We have, respectively,

$$N = \int_0^\infty N(v)dv \qquad (C.14)$$

and

$$\mathscr{E}_{kin,\ total} = N\frac{3}{2}k_BT = \int_0^\infty \frac{1}{2}mv^2 N(v)dv \qquad (C.15)$$

When we substitute Equation (C.13) in Equations (C.14) and (C.15), we will obtain integrals with arguments containing the second and the fourth power of the speed, respectively. These integrals are standard forms, and their values are known to be

$$\int_0^\infty x^2 \exp(-ax^2)dx = \frac{1}{4a}\left(\frac{\pi}{a}\right)^{1/2} \qquad (C.16)$$

and

$$\int_{0}^{\infty} x^4 \exp(-ax^2)dx = \frac{3}{8a^2}\left(\frac{\pi}{a}\right)^{1/2} \tag{C.17}$$

So we find, finally, that

$$N(v) = 4\pi v^2 f(v) = 4\pi v^2 N\left(\frac{m}{2\pi k_B T}\right)^{3/2} \exp\left(-\frac{mv^2}{2k_B T}\right) \tag{C.18}$$

This formula reveals the values of those unknown constants, and it fulfils our original aim of determining the fraction of molecules that possess a given instantaneous speed. It is known as the Maxwell–Boltzmann distribution, in honour of James Clerk Maxwell and Ludwig Boltzmann, who first derived it.

One sees from Equation (C.18) that $N(v)$ is zero when $v = 0$, that it rises to a maximum that increases with increasing temperature, and that it ultimately drops away again toward zero. The maximum for the gaseous form of hydrogen at 0°C lies around 10^3 m s^{-1}, and that value is increased about three-fold if the temperature is raised to 1000°C.

C.2   Liquids

It would be natural to ask how the above analyses must be modified when the gas being considered cannot be regarded as perfect. This will be the case, for example, when the molecules can be no longer be regarded as individual entities. That is the situation when the gas becomes so dense that it is approaching the liquid state. Johannes van der Waals found that he could account for the fact that the liquid–gas transition disappears above the critical temperature, by noting that each molecule occupies a certain amount of volume (which is thus not available to the other molecules) and that the molecules interact with each other over a certain range of distance, rather than merely when they are in direct contact.

This many-body problem becomes particularly severe when one tries to analyse the liquid state itself. Let us start to approach the issue by first considering one of Albert Einstein's rare lapses. When trying to explain the specific heat of solids, almost 100 years ago, he made the assumption that the individual atoms are free to vibrate independently about their mean positions. Peter Debye, on the other hand, reasoned that this could not be the case because there must be forces that keep the atoms in their particular positions. In other words, Einstein was ignoring what provided this book with its third chapter, namely interatomic forces. Debye more realistically analysed the situation in a solid, and he realized that the atomic displacements would take the form of waves, now known as phonons, to which the atoms contribute co-operatively.

We will find it very useful to think in terms of configuration hyperspace, which is admittedly a difficult concept when first encountered. The positions of the N individual atoms in an assembly of atoms will be described by the instantaneous values of their Cartesian co-ordinates: x_1, y_1, z_1, x_2, y_2, z_2, and so on up to x_N, y_N and z_N. When all the atoms are at rest, they can be imagined as each being located at the origin of their own particular coordinate system. However, in order to describe the instantaneous displacements of all the individual atoms, there is no reason why we should not move all those individual origins so as to coincide at a single collective origin, simultaneously serving all $3N$ co-ordinates. This overall origin will be the reference point for a hyperspace having $3N$ dimensions, and the totality of the instantaneous displacements will be describable by a *single point* in that hyperspace.

In the Einstein model referred to above, the motion of an individual atom would be seen to influence just three of the $3N$ dimensions, whereas the Debye model would appear more complicated because the above-mentioned single point would move in configuration hyperspace in such a way that many of the $3N$ dimensions were involved. It is a remarkable (and indeed beautiful) thing that a wave can be described by a single point, so long as the point is moving in high-dimensional space.

How does the energy of the system vary as a function of the individual displacements? Because the lowest-energy state of a crystal clearly corresponds to the situation in which each atom is at its equilibrium position, the energy hypersurface will have the form of a bowl in $3N$ dimensions, that is to say a hyper-bowl (not to be confused with a hyperbole, of course). At the other extreme, in the low-density gas, the energy hypersurface will be flat and constant at all places except those where two atoms are in collision, that latter condition corresponding to a local peak.

The big question is: what does the energy hypersurface look like in the liquid state? If we had reliable knowledge of this, we would be much closer to having what, in the year 2002, we do not have, namely a theory of the liquid state itself. One approach guesses that the liquid energy hypersurface is like a series of bowls, inter-connected by relatively narrow bits of surface resembling viaducts. However, a rival theory put forward in 1986 by Rodney Cotterill and Jens Madsen sees the departure from the crystal-like hyper-bowl as being more global. They proposed that the liquid energy hypersurface resembles the $3N$ dimensional counterpart of a Mexican hat, a hyper-Mexican hat in fact.

Now the significant thing about such a hyper-Mexican hat lies in it permitting the single descriptive point referred to above to *move around without the overall energy actually rising*. This suggests that in a liquid the atoms move not by jumping over energy barriers, as they certainly do in a crystal for example, but rather by collectively moving so that the single point in hyperspace runs around the curved valleys inside the brim of the hyper-Mexican hat. There will naturally be an entropic factor hidden in the required co-operativity, and this serves as a timely reminder of the importance of thinking in terms of the free energy, rather than just the internal energy. If this picture turns out to be correct, it would have repercussions for protein structures.

Exercises

C.1   Derive an expression for the most probable speed of a molecule in a gas, assuming that this obeys the Maxwell–Boltzmann distribution law.

C.2   At what temperature would the average velocity of helium atoms equal the escape velocity from (a) the earth [1.12×10^4m s^{-1}], and (b) the moon [2.73×10^3m s^{-1}]?

Further reading

Brillouin, L., (1953). *Wave Propagation in Periodic Structures*. Dover, New York.
Cotterill, R. M. J. and Madsen, J. U., (1986). Localized fluidity modes and the topology of the constant-potential-energy hypersurfaces of Lennard-Jones matter. *Physical Review B* **33**, 262–268.
Cottrell, A. H., (1964). *The Mechanical Properties of Matter*. Wiley, London.
Frenkel, J., (1955). *Kinetic Theory of Liquids*. Dover, New York.
Pryde, J. A., (1966). *The Liquid State*. Hutchinson University Library, London.
Roberts, J. K. and Miller, A. R., (1951). *Heat and Thermodynamics*. Blackie, London.
Serber, R. A., (1990). *Physics for Scientists and Engineers*. Saunders, Philadelphia.

Appendix D: Probability Distributions

D.1 Bernoulli Trials and the Binomial Distribution

Repeated independent trials (or elements) in which there are only two possible outcomes are known as Bernoulli trials (after Jacob Bernoulli) if the probabilities of the outcomes remain constant. Let those probabilities be p for success and q for failure, which implies that $p + q = 1$. In a succession (or probability aggregate) of n such trials, and assuming that we are interested merely in the total number of successes, not in when they occur during the sequence, the probability $b(k;n,p)$ of just k successes is found to be given by

$$b(k;n,p) = \binom{n}{k} p^k q^{n-k} \tag{D.1}$$

This is known as the binomial distribution because the right-hand side is the kth term in the binomial expansion of $(q + p)^n$, first explored by Isaac Newton. It follows that

$$b(0;n,p) + b(1;n,p) + b(2;n,p) + \ldots + b(n;n,p) = (q + p)^n = 1 \tag{D.2}$$

Tabulations of the binomial distribution are available.

D.2 The Poisson Approximation

If the probability p is small, but the number of trials n is large, such that their product $\lambda = np$ is nevertheless not negligible, it proves convenient to use an approximation for $b(k;n,p)$ first described by Siméon Poisson. In order to derive his expression, we start by noting that for $k = 0$ we have

$$b(0;n,p) = (1 - p)^n = \left(1 - \frac{\lambda}{n}\right)^n \tag{D.3}$$

Taking logarithms, and using the Taylor expansion (named for Brook Taylor), we have

$$\log_e \{b(0;n,p)\} = n \log_e \left\{ 1 - \frac{\lambda}{n} \right\} = -\lambda - \frac{\lambda^2}{2n} - \dots \tag{D.4}$$

which, for large n, gives

$$b(0;n,p) \approx \exp(-\lambda) \tag{D.5}$$

It is readily shown from Equation (D.1) that

$$\frac{b(k;n,p)}{b(k-1;n,p)} \doteq \frac{(n-k+1)p}{kq} = 1 + \frac{(n+1)p-k}{kq} \tag{D.6}$$

For a given value of k and a sufficiently large value of n, and under the conditions we are considering, this leads to

$$\frac{b(k;n,p)}{b(k-1;n,p)} = \frac{\lambda - (k-1)p}{kq} \approx \frac{\lambda}{k} \tag{D.7}$$

We may therefore conclude that there is a succession of relationships which run

$$b(1;n,p) \approx \lambda b(0;n,p) \approx \lambda \exp(-\lambda) \tag{D.8}$$

$$b(2;n,p) \approx \frac{1}{2} \lambda b(1;n,p) \approx \frac{1}{2} \lambda^2 \exp(-\lambda) \tag{D.9}$$

and in general

$$b(k;n,p) \approx \frac{\lambda^k}{k!} \exp(-\lambda) \tag{D.10}$$

This is the Poisson approximation to the binomial distribution. When n is sufficiently large, we may regard $b(k;n,\lambda/n)$ as being essentially $p(k;\lambda)$, in which case Equation (D.10) may be written

$$p(k;\lambda) = \frac{\lambda^k}{k!} \exp(-\lambda) \tag{D.11}$$

The right-hand side of Equation (D.11) is known as the Poisson distribution.

D.3   The Normal, or Gaussian, Distribution

Let us now generalize to the situation in which each element (or trial) in a probability aggregate can have any of s distinguishable properties (as in the throwing of dice, for which $s = 6$), rather than just two (as in the tossing of a coin). In the measurement of a property that is continuous (such as macroscopic length, position or mass), rather than discrete (as in the coin and dice examples), s may be effectively infinite. If n_i is the number of times the ith property occurs, in a total of n elements, n_i/n is known as the relative frequency of that property. The probability of the ith property is defined as the following limit

$$\lim_{n \to \infty} \frac{n_i}{n} = w_i \tag{D.12}$$

It was implicit in our earlier use of the term distribution that this is simply the totality of w_i in a probability aggregate, and it is clear that

$$\sum_i w_i = 1 \tag{D.13}$$

A distribution of discrete elements is sometimes said to be arithmetical, while one comprising continuous elements is sometimes referred to as being geometrical. We will analyse the continuous situation, which is particularly relevant to biophysics, and let the continuous property be x. As defined by Equation (D.12), $w(x)$ is obviously zero, whereas the probability that x lies in the range x to $x + \Delta x$ will be finite (and proportional to Δx if that interval is sufficiently small). That probability is thus $w(x)\Delta x$, the function $w(x)$ being known as the probability density. The continuous counterpart of Equation (D.13) is clearly

$$\int w(x)dx = 1 \tag{D.14}$$

this being the case if the integral extends over the entire range of x. If a function $f(x)$ is defined for all x having non-vanishing probability, the mean of $f(x)$ with respect to the distribution $w(x)$ is given by

$$\bar{f} = \int f(x)w(x)dx \tag{D.15}$$

while the dispersion of that function is

$$D(f) = \int \left[f(x) - \bar{f} \right]^2 w(x)dx \tag{D.16}$$

In the case where the function $f(x)$ is simply the variable x itself, the equivalent versions of Equations (D.15) and (D.16) would be

$$\bar{x} = \int xw(x)dx \tag{D.17}$$

and

$$D(x) = \sigma^2 = \int [x - \bar{x}]^2 w(x)dx \tag{D.18}$$

The dispersion in this case is σ^2, σ being known as the standard deviation, the latter clearly being a measure of the spread of $w(x)$ about its mean value. The rth moment of the distribution is defined by

$$\overline{x^r} = \int x^r w(x)dx \tag{D.19}$$

and higher moments do not always exist for distributions having an infinite range. The dispersion of $w(x)$ can be expressed in terms of the first and second moments because from Equation (D.18) we have

$$\sigma^2 = \overline{x^2} - 2x^2 + x^2 = \overline{x^2} - x^2 \tag{D.20}$$

In some cases, a geometrical distribution can be expanded in terms of its moments, if these exist. And it frequently simplifies things if the expansion is made about $\bar{x}$. In other words, we take $\bar{x} = 0$, $\overline{x^2} = \sigma^2$, and so on. It then transpires that

$$w(x) = \frac{1}{\sigma\sqrt{2\pi}}\exp\left(-\frac{x^2}{2\sigma^2}\right) \tag{D.21}$$

Equation (D.21) is the very important normal distribution, also known as the Gaussian distribution (named in honour of Karl Gauss, though it had been used earlier in probability theory by Abraham DeMoivre and Pierre-Simon Laplace), and it occurs frequently in all manner of experiments and theories. As with the binomial distribution, tabulations of the normal distribution are readily available.

Further reading

Feller, W., (1968). *An Introduction to Probability Theory and its Applications.* Wiley, New York.
Margenau, H. and Murphy, G. M., (1956). *The Mathematics of Physics and Chemistry.* Van Nostrand, Princeton, NJ.
Zernike, F., (1928). *Handbuch der Physik, III.* Springer, Berlin.

Appendix E: Differential Equations

A differential equation is one in which differential coefficients of various orders occur, together with functions of both the independent and dependent variables, but no arbitrary constants. The order of a differential equation is the highest order of derivative that is present, while the degree of a differential equation is that of the highest power of the highest order of derivative that is present.

Two important examples of second-order, first-degree differential equations that occur in Chapters 5 and 12, respectively, will provide us with examples of how such equations can be solved. The spatial and temporal variations of concentration C were found in Chapter 5 to be described by Fick's second law of diffusion, namely

$$D\frac{\partial^2 C}{\partial x^2} = \frac{\partial C}{\partial t} \tag{E.1}$$

which we will find it convenient to write in the form

$$a^2\frac{\partial^2 C}{\partial x^2} = \frac{\partial C}{\partial t} \tag{E.2}$$

We can separate the variables by writing $C = S(x) \cdot T(t)$, that is to say by making the concentration the product of independent spatial and temporal functions. It is then found that

$$\frac{a^2}{S}\frac{\partial^2 S}{\partial x^2} = \frac{\dot{T}}{T} \tag{E.3}$$

By equating the right-hand side to $-a^2 k^2$, where k is an arbitrary constant, we find that

$$T_k = A_1 \exp\left(-a^2 k^2 t\right) \tag{E.4}$$

A_1 being a constant, while the S function must satisfy

$$\frac{\partial^2 S}{\partial x^2} + k^2 S = 0 \tag{E.5}$$

The solution of Equation (E.5) is the standard form

$$S_k = A_2 \exp(ikx) + A_3 \exp(-ikx) \tag{E.6}$$

where A_2 and A_3 are further constants. The product of Equations (E.4) and (E.6) is the desired general solution, and the values of the various constants can be found by applying the appropriate boundary conditions. It is a reasonably straight-forward matter to show that the solution has the form stated in Chapter 5, which comprises the error function. The latter is clearly related to the normal distribution function that was discussed in Appendix D.

In Chapter 12, we encountered an equation having the form

$$\frac{R_m}{R_i}\frac{d^2 V_{x,t}}{dx^2} = V_{x,t} + R_m C_m \frac{dV_{x,t}}{dt} \tag{E.7}$$

The essence of this equation is captured in the simpler form

$$\frac{\partial^2 V}{\partial x^2} = \frac{\partial V}{\partial t} + V \tag{E.8}$$

which has the form of Equation (E.1), but with an additional linear term. It proves useful to introduce the transformation

$$V = W \exp(-t) \tag{E.9}$$

which leads to the relationships

$$\frac{\partial V}{\partial t} = -W \exp(-t) + \frac{\partial W}{\partial t}\exp(-t) \tag{E.10}$$

and

$$\frac{\partial V}{\partial x} = \frac{\partial W}{\partial x}\exp(-t) \tag{E.11}$$

and

$$\frac{\partial^2 V}{\partial x^2} = \frac{\partial^2 W}{\partial x^2} \exp(-t) \tag{E.12}$$

Inserting Equations (E.10), (E.11) and (E.12) in (E.8), we obtain

$$\frac{\partial^2 W}{\partial x^2} \exp(-t) = -W \exp(-t) + \frac{\partial W}{\partial t} \exp(-t) + V = \frac{\partial W}{\partial t} \exp(-t) \tag{E.13}$$

This reduces directly to

$$\frac{\partial^2 W}{\partial x^2} = \frac{\partial W}{\partial t} \tag{E.14}$$

Equation (E.14) is essentially the diffusion equation, (E.1), which we have already discussed. It is not surprising, therefore, that the solution of Equation (E.7) also involves the error function (though in this case, its complement, *erfc*), as noted in Chapter 12.

Further reading

Blakey, J., (1953). *University Mathematics*. Blackie, London.

Margenau, H. and Murphy, G. M., (1956). *The Mathematics of Physics and Chemistry*. Van Nostrand, Princeton, NJ.

Sten-Knudsen, O., (2002). *Biological Membranes: Theory of Mass Transport, Membrane Potential and Electrical Impulses*. Cambridge University Press, Cambridge.

Name Index

Subject Index